ELEMENTS OF LOGIC
AND FOUNDATIONS OF MATHEMATICS
IN PROBLEMS

WIKTOR MAREK and JANUSZ ONYSZKIEWICZ

Warsaw University

ELEMENTS OF LOGIC AND FOUNDATIONS OF MATHEMATICS IN PROBLEMS

D. REIDEL PUBLISHING COMPANY

DORDRECHT:HOLLAND/BOSTON:U.S.A./LONDON:ENGLAND

PWN–POLISH SCIENTIFIC PUBLISHERS

WARSZAWA

Library of Congress Cataloging in Publication Data

Marek, Wiktor.
Elements of logic and foundations of mathematics in problems.
Translation of: Elementy logiki i teorii mnogości w zadaniach/W. Marek and J. Onyszkiewicz.
Includes bibliography.
1. Logic, Symbolic and mathematical. 2. Set theory. I. Onyszkiewicz, Janusz. II. Title.
QA9.M31913 511.3 81-17836
ISBN 90-277-1084-8 AACR2

Translated by *Elżbieta Tarantowicz-Marek*

This translation has been made from
ELEMENTY LOGIKI I TEORII MNOGOŚCI W ZADANIACH
published by Państwowe Wydawnictwo Naukowe, Warszawa 1972

Distributors for Albania, Bulgaria, Chinese People's Republic, Cuba, Czechoslovakia, German Democratic Republic, Hungary, Korean People's Democratic Republic, Mongolia, Poland, Romania, the U.S.S.R., Vietnam and Yugoslavia

ARS POLONA
Krakowskie Przedmieście 7, 00-068 Warszawa 1, Poland

Distributors for the U.S.A. and Canada
KLUWER BOSTON INC.,
190 Old Derby Street, Hingham, MA 02043, U.S.A.

Distributors for all remaining countries
KLUWER ACADEMIC PUBLISHERS GROUP,
P.O. Box 322, 3300 AH Dordrecht, Holland.

D. Reidel Publishing Company is a member of the Kluwer Group.

PRINTED IN POLAND

TABLE OF CONTENTS

PREFACE

Set theory emerged in the nineteenth century as a result of investigations of the foundations of mathematical analysis. Fairly soon, due to their generality, the set theoretical concepts and methods penetrated deeply into every mathematical discipline and now, set theory is considered to be the most fundamental mathematical theory. According to the great French mathematician Denjoy, set theory was a breakthrough on a scale not known since Leibnitz and Newton.

The career of mathematical logic was also quite spectacular – it is enough to mention its applications to psychology, linguistic and computing sciences.

The concepts of logic and set theory have turned out to be both basic and fairly intuitive. As a result, the whole programme of mathematics education has been recently changed to incorporate them and make extensive use of them.

These developments have not been sufficiently matched by the available literature which could be used by people without extensive mathematical training. Most of the existing textbooks on set theory or mathematical logic put the main emphasis on theory and do not give enough examples and simple problems which could familiarize the student with the new concepts.

The idea of our book is to provide a variety of exercises covering all the basic set theoretical and logical concepts. We tried to include most of the typical problems with their solutions. Hints are given to help solve the more difficult ones. In each chapter the exercises are arranged according to their difficulty. Every group of typical problems begins with very easy ones. Most of them could be answered by a student at secondary school. We decided to include so many easy problems for the benefit of those who have not encountered before the concepts in question. Sometimes there are several problems of a similar character. The idea behind this is that, by solving all of them, the student can get more confidence in handling new concepts, which is especially important for those who study individually, without a tutor. We also believe that easy problems will make our book more useful in schools, too.

Chapter 9 is of a special character. It introduces problems requiring knowledge of several previous ones and for this reason the exercises contained in it could be used to check the student's progress.

The book does not give a full exposition of set theory and logic. It is meant as supplementary reading. Nevertheless, for the convenience of the reader, all the basic definitions and facts are given either at the beginning of the chapter or in the text of the appropriate exercises. With regard to advanced topics, however, we decided to rely more heavily on some additional facts, which can be found in textbooks like Kuratowski and Mostowski "Set Theory".

At the end of book there are exercises on Boolean algebras and mathematical induction. We decided to include these topics to make our book as comprehensive as possible.

Some advanced problems are marked with asterisk.

PROBLEMS

CHAPTER 1

PROPOSITIONAL CALCULUS

A *valuation* is a function mapping the set of propositional variables into the set $\{0, 1\}$ where 0 is the logical value of *falsity* and 1 is the logical value of *truth*. Such a function can be extended in natural way to the set of all propositional formulae.

A *tautology* is a formula which takes value 1 for every valuation.

An *n-argument propositional functor* (*propositional connective*) is a function which maps the set of all n-tuples $\langle x_1, \ldots, x_n \rangle$ (where x_i is falsity or truth) into the set $\{0, 1\}$.

Thus we have four one-argument functors:

	0	1
A_0	0	0
A_1	0	1
A_2	1	0
A_3	1	1

and 16 two-argument functors:

	00	01	10	11
B_0	0	0	0	0
B_1	0	0	0	1
B_2	0	0	1	0
B_3	0	0	1	1
B_4	0	1	0	0
B_5	0	1	0	1
B_6	0	1	1	0
B_7	0	1	1	1
B_8	1	0	0	0
B_9	1	0	0	1
B_{10}	1	0	1	0

B_{11}	1	0	1	1
B_{12}	1	1	0	0
B_{13}	1	1	0	1
B_{14}	1	1	1	0
B_{15}	1	1	1	1

Among the functors A_i's and B_j's $(0 \leqslant i \leqslant 3, 0 \leqslant j \leqslant 15)$ some have traditional names. The functor A_2 is called *negation* and is denoted by $\sim$ and the functors B_1, B_7, B_{13}, B_9, B_{14}, B_8 are called respectively *conjunction* ($\wedge$), *alternative* ($\vee$), *implication* ($\Rightarrow$), *equivalence* ($\Leftrightarrow$), *Sheffer's stroke* ($/$) and *simultaneous refutation functor.*

1.1. Define the conjunction functor by means of alternative and negation.

1.2. Define the alternative functor by means of conjunction and negation.

1.3. Define the alternative functor by means of implication and negation.

1.4. Prove that neither implication nor disjunction is definable by means of the alternative and the conjunction.

1.5. Prove that neither alternative nor conjunction is definable by means of the equivalence and the negation.

1.6. Prove that all one- and two-argument propositional functors are definable by means of implication and A_0.

1.7. Prove that each of the functors B_8 and B_{14} (Sheffer's stroke and simultaneous refutation) is sufficient to define all the other propositional functors (both one- and two-argument ones).

1.8. Prove that the functors B_8 and B_{14} are the only binary functors with the property stated in 1.7. Prove that every n-argument functor (n being arbitrary) is definable by means of B_8. Prove the same property of B_{14}.

1.9. Let an expression (formula) f contain n propositional variables. How many valuations has one to consider in order to check if f is a tautology?

Prove that the following expressions are tautologies (1.10–1.32)

1.10. $p \Rightarrow p$.

1.11. $p \Rightarrow (q \Rightarrow p)$.

1.12. $p \Rightarrow (q \Rightarrow p \wedge q)$.

1.13. $[p\Rightarrow(q\Rightarrow r)]\Rightarrow[(p\Rightarrow q)\Rightarrow(p\Rightarrow r)]$.

1.14. $p\Leftrightarrow\sim\sim p$ *(law of double negation).*

1.15. $p\vee\sim p$ *(law of the excluded middle).*

1.16. $\sim(p\wedge\sim p)$ *(law of contradiction).*

1.17. $\sim(p\wedge q)\Leftrightarrow(\sim p\vee\sim q)$ *(de Morgan's law).*

1.18. $\sim(p\vee q)\Leftrightarrow(\sim p\wedge\sim q)$ *(de Morgan's law).*

1.19. $(p\Rightarrow q)\Leftrightarrow(\sim q\Rightarrow\sim p)$ *(transposition rule).*

1.20. $[(p\Rightarrow q)\Rightarrow p]\Rightarrow p$ *(Pierce's law).*

1.21. $(p\Rightarrow q)\Leftrightarrow(\sim p\vee q)$.

1.22. $(\sim p\Rightarrow p)\Rightarrow p$ *(Clavius' law).*

1.23. $\sim p\Rightarrow(p\Rightarrow q)$ *(Duns–Scotus' law).*

1.24. $(p\wedge q)\Rightarrow p$.

1.25. $p\Rightarrow(p\vee q)$.

1.26. $(p\wedge q\Rightarrow r)\Leftrightarrow[p\Rightarrow(q\Rightarrow r)]$.

1.27. $[p\vee(q\vee r)]\Leftrightarrow[(p\vee q)\vee r]$.

1.28. $[p\wedge(q\wedge r)]\Leftrightarrow[(p\wedge q)\wedge r]$.

1.29. $[p\wedge(q\vee r)]\Leftrightarrow[(p\wedge q)\vee(p\wedge r)]$.

1.30. $[p\vee(q\wedge r)]\Leftrightarrow[(p\vee q)\wedge(p\vee r)]$.

1.31. $[(p\Rightarrow q)\wedge\sim q]\Rightarrow\sim p$.

1.32. $[(p\Rightarrow q)\wedge p]\Rightarrow q$.

Check if the following expressions are tautologies (1.33–1.61)

1.33. $[(p\vee q)\wedge\sim p]\Rightarrow q$.

1.34. $(p\Rightarrow q)\Rightarrow[(p\wedge r)\Rightarrow q]$.

1.35. $(p\Rightarrow q)\Rightarrow[p\Rightarrow(q\vee r)]$.

1.36. $p\Rightarrow[(\sim p)\vee q]$.

1.37. $[(p\vee q)\wedge(p\Rightarrow q)]\Rightarrow(q\Rightarrow p)$.

1.38. $p\vee[(\sim p\wedge q)\vee(\sim p\wedge\sim q)]$.

1.39. $\sim[p\wedge(\sim p\wedge q)]$.

1.40. $p\Rightarrow[(\sim q\wedge q)\Rightarrow r]$.

1.41. $[(p\Rightarrow q)\wedge(q\Rightarrow p)]\Rightarrow(p\vee q)$.

1.42. $[(p\vee q)\Rightarrow(p\vee\sim q)]\Rightarrow(\sim p\vee q)$.

1.43. $[(p\wedge q)\vee(p\Rightarrow q)]\Rightarrow(p\Rightarrow q)$.

1.44. $[(p\Rightarrow q)\wedge(q\Rightarrow r)]\Rightarrow(p\Rightarrow r)$.

1.45. $[(p\Rightarrow q)\wedge(r\Rightarrow s)]\Rightarrow(p\vee r\Rightarrow q\vee s)$.

1.46. $(p\wedge q\Rightarrow r)\Rightarrow[(p\Rightarrow r)\wedge(q\Rightarrow r)]$.

1.47. $[(p\Rightarrow q)\wedge(r\Rightarrow s)]\Rightarrow[(p\wedge r)\Rightarrow(q\wedge s)]$.

1.48. $\{[(p\wedge q)\Rightarrow r]\wedge(p\vee q\Rightarrow\sim r)\}\Rightarrow(p\wedge q\wedge r)$.

1.49. $[p \Rightarrow (q \Rightarrow r)] \Leftrightarrow [q \Rightarrow (p \Rightarrow r)]$.
1.50. $(p \vee q \vee r) \Rightarrow \{\sim p \Rightarrow [(q \vee r) \wedge \sim p]\}$.
1.51. $[\sim (p \Rightarrow q) \wedge (q \Rightarrow p)] \Rightarrow (p \wedge \sim q)$.
1.52. $[(p \Rightarrow q) \wedge (r \Rightarrow s)] \Rightarrow [(p \wedge s) \Rightarrow (q \vee r)]$.
1.53. $[(p \Rightarrow q) \Rightarrow (q \Rightarrow r)] \Rightarrow [(r \Rightarrow p) \Rightarrow (q \Rightarrow p)]$.
1.54. $(p \Rightarrow q) \Leftrightarrow [(p \wedge q) \Leftrightarrow p]$.
1.55. $[(p \Rightarrow q) \vee (p \Rightarrow r) \vee (p \Rightarrow s)] \Rightarrow [p \Rightarrow (q \vee r \vee s)]$.
1.56. $[(p \Rightarrow q) \wedge (r \Rightarrow q) \wedge (s \Rightarrow q)] \Rightarrow [(p \wedge r \wedge \sim s) \Rightarrow q]$.
1.57. $\{[(p \wedge q) \Rightarrow r] \wedge [(p \wedge q) \Rightarrow \sim r]\} \Rightarrow (\sim p \wedge \sim q \wedge \sim r)$.
1.58. $[(\sim p \wedge q) \vee (p \vee \sim q)] \Rightarrow \{[p \Rightarrow (q \vee r)] \Rightarrow (p \Rightarrow r)\}$.
1.59. $[(p \vee q) \wedge (r \vee s)] \Rightarrow \{[(p \Rightarrow q) \vee (p \Rightarrow r)] \wedge [(q \Rightarrow p) \vee (q \Rightarrow p)]\}$.
1.60. $[(p \Rightarrow q) \wedge (r \Rightarrow s) \wedge (t \Rightarrow u)] \Rightarrow [(p \wedge r \wedge t) \Rightarrow (q \wedge s \wedge u)]$.
1.61. $[(p \vee q) \Rightarrow r] \Rightarrow [(p \Rightarrow r) \vee (q \Rightarrow r)]$.

1.62. Is the following proposition true? if a natural number a is prime, then if a is composite then a is equal to 4.

1.63. Is the following proposition true? if a natural number a is divisible by 3, then if a is not divisible by 3 then a is divisible by 5.

1.64. Is the following proposition true? if from the fact that a triangle ABC has all sides equal it follows that its angles are all equal and the triangle ABC has unequal angles, then its sides are not equal.

1.65. Is the following proposition true? if A is a quadrangle and the angles of A are all equal, then from the fact that A is a quadrangle it follows that all its sides are equal.

1.66. Is the following proposition true? if a natural number a is divisible both by 3 and by 5, then if a is not divisible by 3 then a is not divisible by 5.

1.67. Is the following proposition true? if a number a is divisible both by 2 and 7 it follows that if a is not divisible by 7 then a is divisible by 3.

1.68. Is the following proposition true? if it is not true that either line L is parallel to line M or line P is not parallel to line M, then either line L is not parallel to M or P is parallel to M.

1.69. Is the following proposition true? if John does not know logic, then if John knows logic then John was born in the fourth century B.C.

1.70. Is the following proposition true? John knows logic if and only if it is not true that it is not true that John knows logic.

1.71. Is the following proposition true? if from the fact that a function f is differentiable at the point x_0 it follows that the function f is continuous at the point x_0, then from the fact that the function f is continuous at the point x_0 it follows that the function f is differentiable at the point x_0.

1.72. Prove that if proposition p is false, then for every proposition q:

(a) $p \vee q$ is equivalent to q.

(b) $p \wedge q$ is equivalent to p.

1.73. Prove that if proposition p is true then for every proposition q:

(a) $p \vee q$ is equivalent to p.

(b) $p \wedge q$ is equivalent to q.

1.74. Let us identify the logical values 0 and 1 with the natural numbers 0 and 1, respectively. The functors may then be expressed as follows:

$$\sim p = 1 - p,$$

$$p \wedge q = \min(p, q) = pq,$$

$$p \vee q = \max(p, q) = p + q - pq,$$

$$p \Rightarrow q = 1 - p + pq.$$

Prove that under such an interpretation an expression is a tautology if and only if under every valuation it takes the value 1.

* **1.75.** Let us identify the logical values 0 and 1 with the natural numbers 1 and 0 respectively. Find expressions analogous to that given in 1.74. Prove the corresponding theorem (notice that instead of 1 one should put 0 in the assertion of the theorem).

1.76. Prove that if the expression Ψ is a tautology then the expression

$$\Phi_1 \Rightarrow (\Phi_2 \Rightarrow \ldots \Rightarrow (\Phi_n \Rightarrow \Psi) \ldots)$$

is also a tautology.

1.77. Prove that if the expression Ψ is a tautology then the expression

$$\sim \Psi \Rightarrow (\Phi_1 \Rightarrow (\Phi_2 \Rightarrow \ldots \Rightarrow (\Phi_{n-1} \Rightarrow \Phi_n) \ldots)$$

is also a tautology.

1.78. Consider an expression of the form

$$\underbrace{(\ldots((p\Rightarrow p)\Rightarrow p)\Rightarrow p)\ldots)\Rightarrow p}_{n}.$$

For what n is this formula a tautology?

1.79. Define $p^0 = p$, $p^1 = \sim p$. Consider an expression of the form

$$(*)\qquad (\ldots(p^{i_0}\Rightarrow p^{i_1})\Rightarrow\ldots)\Rightarrow p^{i_{n-1}}.$$

For what sequences $\langle i_0, \ldots, i_{n-1}\rangle$ is formula (*) a tautology?

1.80. A formula Θ is in the disjunctive normal form if and only if it is of the form

$$\Theta\colon\ \Phi_1 \vee \ldots \vee \Phi_n,$$

where every formula Φ_j is of the form

$$\Phi_j\colon\ \Psi_{j_1} \ldots\ldots \wedge \Psi_{j_s}$$

and each Ψ_{j_l} is a propositional variable or its negation.

Prove that for every formula Θ there is a formula Θ_1 in the disjunctive normal form such that Θ is equivalent to Θ_1, i.e. $\Theta\Leftrightarrow\Theta_1$ is a tautology.

1.81. A formula Θ is in the conjunctive normal form if and only if it is of the form

$$\Theta\colon\ \Phi_1 \ldots\ldots \wedge \Phi_n,$$

each of the formulas Φ_i is in the form

$$\Phi_i\colon\ \Psi_{j_1} \vee \ldots \vee \Psi_{j_l},$$

and each of the formulas Ψ_{j_r} is a propositional variable or its negation.

Prove that every formula is equivalent to a formula in the conjunctive normal form.

* **1.82.** Prove that a formula built of the propositional variables only by means of the connective $\Leftrightarrow$ is a tautology if and only if every variable occurs in it an even number of times.

1.83. Find a formula Ψ in the disjunctive normal form which is equivalent to a given formula Φ where Φ is:

(a) $\quad p \wedge [q \vee (\sim p \wedge r)],$

(b) $\quad p \Leftrightarrow [q \Rightarrow (q \Rightarrow p)],$

(c) $p \Rightarrow [q \wedge (\sim p \Leftrightarrow q)]$,

(d) $p \vee [(q \wedge p) \Rightarrow (q \Leftrightarrow \sim p)]$.

1.84. For each of the formulas (a)–(d) of 1.83, find an equivalent formula in conjunctive normal form.

1.85. Show that if an expression does not contain connectives other than $\Leftrightarrow$, then changing in it the order of brackets in an arbitrary way leads to a formula which is equivalent to the original expression.

1.86. Assume that an expression Φ is built of propositional variables by means of the connectives $\wedge$ and $\vee$ only. Let Φ_d denote the expression arising from Φ by changing each symbol $\wedge$ to $\vee$ and conversely.

Let Φ^* be a formula in which every propositional variable is changed to its negation.

Prove that $\sim \Phi_d \Leftrightarrow \Phi^*$ is a tautology.

1.87. Prove that if $\Phi \Leftrightarrow \Psi$ is a tautology then $\Phi_d \Leftrightarrow \Psi_d$ is also a tautology.

Using the following rules of the real numbers arithmetic: $a/b < 0 \Leftrightarrow ab < 0$, $a/b > 0 \Leftrightarrow ab > 0$, $a/b \leqslant 0 \Leftrightarrow ab \leqslant 0 \wedge b \neq 0$, $a/b \geqslant 0 \Leftrightarrow ab \geqslant 0 \wedge b \neq 0$ and the tautologies of the propositional calculus, find the solutions of inequalities 88–97.

1.88. $\dfrac{x^2-4}{x^2-9} \geqslant 0$.

1.89. $\dfrac{x^3+x-2}{x^2-x+12} \geqslant 0$.

1.90. $x^2 \geqslant 0$.

1.91. $\dfrac{x^2-6x+5}{x^2-4} \geqslant 0$.

1.92. $\dfrac{x^3-2}{x^2-1} \geqslant 0$.

1.93. $\dfrac{x^2-7}{x^3+8} \leqslant 0$.

1.94. $\dfrac{x^2-9}{x^2-3x-4} < 0$.

1.95. $\dfrac{x^2-16}{x^2-x+2} \leqslant 0$.

1.96. $\dfrac{x^2-2x+1}{x^2-2x+8} \geqslant 0$.

1.97. $\dfrac{x^2-5x+6}{x^2-2x+1} > 0$.

* **1.98.** Prove the following theorem: if the implications $p_1 \Rightarrow q_1, \ldots, p_n \Rightarrow q_n$ are true and, moreover, the propositions $(p_1 \vee \ldots \vee p_n)$ and $\sim(q_i \wedge q_j)$ are true for $i \neq j$, then the implications $q_1 \Rightarrow p_1, \ldots, q_n \Rightarrow p_n$ are all true.

In this case we say that the statements $p_1, \ldots, p_n, q_1, \ldots, q_n$ form a *closed family of statements.*

1.99. Prove that if in the triangles ABC and $A'B'C'$ $AB = A'B'$ and $AC = A'C'$ then $\sphericalangle A > \sphericalangle A' \Leftrightarrow BC > B'C'$.

1.100. Show that the statements $p_1, p_2, p_3, q_1, q_2, q_3$ about the triangles ABC, $A'B'C'$ for which

$$AC = A'C', \quad AB = A'B',$$

$$p_1 : \sphericalangle A > \sphericalangle A', \quad q_1 : BC > B'C',$$

$$p_2 : \sphericalangle A = \sphericalangle A', \quad q_2 : BC = B'C',$$

$$p_3 : \sphericalangle A < \sphericalangle A', \quad q_3 : BC < B'C'.$$

form a closed family of statements.

The *length of a formula* is defined inductively with respect to its complication as follows: the length of the propositional variable is equal to 1 and if $\mathrm{lh}\,\Phi = a$, $\mathrm{lh}\,\Psi = b$ then $\mathrm{lh}\,(\Phi \wedge \Psi) = \mathrm{lh}\,(\Phi \vee \Psi) = \mathrm{lh}\,(\Phi \Rightarrow \Psi) = a + b + 1$, $\mathrm{lh}\,(\sim \Phi) = \mathrm{lh}\,\Phi + 1$.

1.101. Find the shortest formula Φ equivalent to a given Φ.

(a) $(p \wedge q \wedge s) \vee (p \wedge \sim q \wedge \sim r) \vee (p \wedge q \wedge \sim s) \vee \sim(p \wedge r \Rightarrow q)$,

(b) $\sim p \Rightarrow \sim\sim q$,

(c) $(p \wedge q) \vee \sim(\sim p \Rightarrow q)$,

(d) $(q \wedge r \wedge s \wedge \sim q) \vee (p \wedge \sim q \wedge \sim p) \vee (r \wedge s)$.

The above notation required the use of brackets. It is also common (especially in computer applications) to use the so-called *bracketless Łukasiewicz notation* (very often called the *Polish notation*),

We transform formulas written in the bracket formalism into bracketless formulas by means of the following algorithm.

Propositional variables are not changed. Assume that we are given formulas Φ_1 and Φ_2 (already in bracketless notation) and we want to form a new formula by means of a binary connective S (where S can be equal to C, D, E, I, denoting, respectively, conjunction, alternative, equivalence, and implication functors). We simply write $\mathrm{S}\Phi_1\Phi_2$, and this is the bracketless version of the formula $(\Phi_1)\mathrm{S}(\Phi_2)$. In the case of negation we simply write $\mathrm{N}\Phi$ instead of $\sim(\Phi)$ (Φ is supposed to be already in bracketless notation).

If the formula Φ_1 is written in the Łukasiewicz notation, then its translation into the previously introduced bracket formalism is performed as follows:

(a) If after the connective N there is a propositional variable, then instead of Np we write $(\sim p)$.

(b) If after one of the connectives C, D, E, I occur the propositional variables p and q, then we write, respectively, $(p \wedge q)$, $(p \vee q)$, $(p \Leftrightarrow q)$, $(p \Rightarrow q)$.

(c) Every expression already given in bracket notation we treat as if it were a propositional variable, using a) or b). For instance $\mathrm{CD}p\mathrm{N}qr = \mathrm{CD}p(\sim q)r = \mathrm{C}(p \vee \sim q)r = (p \vee \sim q) \wedge r$.

1.102. Transform into the Łukasiewicz notation the following formulas:

(a) $p \wedge [q \vee (\sim p \Rightarrow r)]$,

(b) $p \Rightarrow [q \Rightarrow (r \Rightarrow p)]$,

(c) $p \Leftrightarrow [p \wedge \sim(\sim q \Leftrightarrow q)]$,

(d) $(p \Rightarrow q) \Rightarrow [(q \Rightarrow r) \Rightarrow (\sim r \Rightarrow \sim p)]$,

(e) $[(p \vee q) \wedge \sim p] \Rightarrow q$,

(f) $(p \vee q) \wedge (\sim p \Rightarrow q)$.

1.103. Show the bracket notation form of the following bracketless expressions.

(a) DDC*pq*C*p*N*q*DCN*pq*CN*p*N*q*,

(b) EC*p*D*qr*DC*pq*C*pr*,

(c) II*pq*IN*q*N*p*

(d) DC*pq*NND*q*N*r*,

(e) INC*pq*CN*pq*.

* **1.104.** Show that a formula in bracketless notation which is not a propositional variable must start with the symbol of a connective. When is the connective N?

The *rules of proof* (*derivation rules*) are operations which produce true statements from true premises. Thus the rules of proof allow us to get new statements from those already accepted as true. The most important of the rules of proof are: the *detachment rule* (*modus ponens*), which leads from the statements Φ and $\Phi \Rightarrow \Psi$ to the statement Ψ, and the *substitution rule*, which allows us to derive $\Phi(a)$ (for every a from the range of the variable x) from the statement "for all x, $\Phi(x)$".

1.105. From the appropriate instances of the tautologies of 1.11 and 1.13, derive by means of the detachment rule the tautology

$$p \Rightarrow p.$$

1.106. From the same tautologies derive

$$(q \Rightarrow r) \Rightarrow [(p \Rightarrow q) \Rightarrow (p \Rightarrow r)].$$

1.107. Show the validity of the following rule of proof:

"From $\Phi(p)$ and $p \Leftrightarrow \Psi$ derive $\Phi(\Psi)$" where: p is a propositional variable and $\Phi(\Psi)$ arises from Φ by substituting the expression Ψ in place of p in some of the occurrences of p in Φ (not necessarily all).

CHAPTER 2

ALGEBRA OF SETS

If a is an element of A, then this is written as "$a \in A$" (read: a belongs to A). Instead of $\sim a \in A$ we often write $a \notin A$; $a, b, \ldots, d \in A$ is an abbreviation of the formula $a \in A \wedge b \in A \wedge \ldots \wedge d \in A$.

Sets are defined by specifying their elements. Therefore sets with the same elements are identical. This is the so-called *extensionality principle*, which may be stated as follows:

$$A = B \Leftrightarrow \text{for every } x \; (x \in A \Leftrightarrow x \in B).$$

Elements of a set may be defined by listing them. In particular, $\{a, b, c, d\}$ is the set consisting exactly of objects a, b, c, and d.

A set may also be defined by means of a condition which its elements must satisfy. Thus, having a condition (property) Φ, we may define a set A as follows:

$$x \in A \Leftrightarrow \Phi(x).$$

The set A defined in this manner consists of exactly those objects which have property Φ. (Note that at this point we do not discuss the problem what properties determine the sets.) In such a situation we often write $A = \{x : \Phi(x)\}$.

The set of elements belonging to the set B and satisfying the condition Φ is denoted by:

$$\{x \in B : \Phi(x)\}.$$

Sets can in turn belong to other sets, sometimes called families of sets. We assume, however, that there is no family of sets $\{A_n : n \in \mathcal{N}\}$, such that, for all n, $A_{n+1} \in A_n$. In particular it follows that for every A, $\sim(A \in A)$.

The set which does not contain any elements is called the *empty set*. The extensionality principle implies that there is exactly one such set. We denote that set by O. Also the symbol Ø can be used.

We admit the existence of objects which are not sets. Thus, if a is not a set, then there is no x such that $x \in a$. O has a similar property but $O \neq a$ because a is not a set.

We write $A \subset B$ (which we read A is included in B) if every element of A is an element of B. Thus

$$A \subset B \Leftrightarrow \text{for every } x \ (x \in A \Rightarrow x \in B).$$

In this case we say – equivalently – that A is a *subset* of B.

The relationship $A \subset B$ is called *inclusion*; among its properties let us state the following:

$$A \subset A,$$

$$A \subset B \wedge B \subset C \Rightarrow A \subset C,$$

$$A \subset B \wedge B \subset A \Rightarrow A = B.$$

List the elements of the following sets (Problems 2.1–2.28):

2.1. $\{a, b, c\}$.
2.2. $\{a, b, a\}$.
2.3. $\{a\}$.
2.4. O.
2.5. $\{\{a\}\}$.
2.6. $\{O\}$.
2.7. $\{\{a, b\}, \{a\}\}$.
2.8. $\{\{\{a\}\}, \{a\}, a\}$.
2.9. $\{\{a, b, c\}, c\}$.
2.10. $\{\{a, b\}, \{\{a, b\}\}, O\}$.
2.11. $\{x \in \mathcal{N} : x \leqslant 2\}$.
2.12. $\{x \in \mathcal{N} : x^2 \leqslant 7\}$.
2.13. $\{x \in \mathcal{N} : x^2 < 7\}$.
2.14. $\{x \in \mathcal{N} : x = 2\}$.
2.15. $\{x \in \mathcal{N} : x = 2 \vee x = 3\}$.
2.16. $\{x \in \mathcal{N} : x^2 < 0\}$.
2.17. $\{x \in \mathcal{N} : x = -1\}$.
2.18. $\{x \in \mathcal{N} : x \geqslant 0\}$.
2.19. $\{x \in \mathcal{N} : x^2 = 4\}$.

2.20. $\{x \in \mathscr{Q} : x^2 = 4\}$.
2.21. $\{x \in \mathscr{Q} : x^2 = 2\}$.
2.22. $\{x \in \mathscr{N} : x^2 - 8x + 1 < 0\}$.
2.23. $\{x \in \mathscr{N} : |3 - x| < 3\}$.
2.24. $\{x \in \mathscr{R} : x^2 + 4x + 4 \leqslant 0\}$.
2.25. $\{x \in \mathscr{R} : x^2 = 2\}$.
2.26. $\{x \in \mathscr{Q} : (x+1)^2 \leqslant 0\}$.
2.27. $\{x \in \mathscr{R} : x^2 < 3\}$.
2.28. $\{x \in \mathscr{R} : x^2 + 1 \geqslant 0\}$.

Assuming that different letters are used to denote different objects, check what inclusions hold among the following sets A and B (Problems 2.29–2.50).

2.29. $A = \{a, b, c, d\}$, $B = \{a, c, d\}$.
2.30. $A = \{a, b\}$, $B = \{a, c, d\}$.
2.31. $A = O$, $B = \{a, b, c\}$.
2.32. $A = O$, $B = \{a\}$.
2.33. $A = \{\{a\}, a, O\}$, $B = \{a\}$.
2.34. $A = \{\{a, b\}, \{c, d\}, c, d\}$, $B = \{\{a, b\}, c\}$.
2.35. $A = \{\{a, b\}, \{a\}, b, O\}$, $B = \{\{a\}, b, \{O\}\}$.
2.36. $A = \{x \in \mathscr{N} : x > 2\}$, $B = \{y \in \mathscr{N} : y > 2\}$.
2.37. $A = \{x \in \mathscr{R} : x > 0\}$, $B = \{y \in \mathscr{N} : y > 0\}$.
2.38. $A = \{x \in \mathscr{N} : x^2 > 4\}$, $B = \{x \in \mathscr{N} : x > 2\}$.
2.39. $A = \{ax + b : a, b \in \mathscr{R}\}$, $B = \{x + y : y \in \mathscr{R}\}$.
2.40. $A = \{ax^2 + bx + c : a, b, c \in \mathscr{R}\}$, $B = \{ax + b : a, b \in \mathscr{R}\}$.
2.41. $A = \{ax + b + 2 : a, b \in \mathscr{R}\}$, $B = \{bx + a : a, b \in \mathscr{R}\}$.
2.42. $A = \{(a-1)x^2 + (b+1)x + c : a, b, c \in \mathscr{R}\}$,
$B = \{zx^2 + tx + v : z, t, v \in \mathscr{Q}\}$.

Remark: The sets A and B in problems 2.39–2.42 are sets of functions.

2.43. $A = \{x \in \mathscr{Q} : x^2 \leqslant 1\}$, $B = \{x \in \mathscr{R} : 2x^3 - 5x^2 + 4x = 1\}$.
2.44. $A = \{x \in \mathscr{Q} : x^2 < 0\}$, $B = \{x \in \mathscr{Q} : x^6 + 7x^5 - 3x = 0\}$.
2.45. $A = \{x \in \mathscr{Q} : x^2 + x - 2 \leqslant 0\}$, $B = \{y \in \mathscr{R} : y^2 + y - 2 = 0\}$.
2.46. $A = \{x \in \mathscr{R} : x^3 - 5x + 3 = 0\}$, $B = \{x \in \mathscr{Q} : x^3 - 5x + 4 \leqslant 0\}$.
2.47. A – the set of isosceles triangles, B – the set of equilateral triangles.
2.48. A – the set of squares in the plane, B – the set of rectangles.
2.49. A – the set of rhombuses with at least one right angle, B – the set of rectangles.

2.50. A – the set of polygons with the circumference equal to 4, B – the set of squares with the area equal to 1.

Now let a, b, c, and d be different from the empty set. What relationship must hold between them for the following equalities to be true (Problems 2.51–2.56)?

2.51. $\{b,c\}=\{b,c,d\}$.

2.52. $\{a,b,a\}=\{a,b\}$.

2.53. $\{\{a\},\{a,b\}\}=\{\{c\},\{c,d\}\}$.

2.54. $\{\{a,b\},c\}=\{\{a\},c\}$.

2.55. $\{\{a,b\},\{d\}\}=\{\{a\}\}$.

2.56. $\{\{a,O\},b\}=\{\{O\}\}$.

From given sets A and B we can form new sets $A \cup B$, $A \cap B$, and $A-B$, called, respectively, their *union*, *intersection* and *difference*. Those sets are defined as follows:

$$x \in A \cup B \Leftrightarrow x \in A \vee x \in B,$$
$$x \in A \cap B \Leftrightarrow x \in A \wedge x \in B,$$
$$x \in A - B \Leftrightarrow x \in A \wedge x \notin B.$$

The sets A and B are *disjoint* if $A \cap B=O$. If all the sets under consideration are subsets of a certain fixed set then this set is called a *space* and usually denoted by X. Then, instead of $X-A$ we write $-A$. Sometimes the set $-A$ is denoted by A'. It is usually called the *complement of set A*.

Assuming that different small letters are used to denote different objects which are not sets calculated $A \cup B$, $A \cap B$, $A-B$, and $B-A$ for the following sets A and B (Problems 2.57–2.62):

2.57. $A=\{a,b,c\}$, $B=\{c,d\}$.

2.58. $A=\{\{a,b\},c\}$, $B=\{c,d\}$.

2.59. $A=\{x,y,\{z\}\}$, $B=\{a,x,y\}$.

2.60. $A=\{\{a,\{a\}\},a\}$, $B=\{a,\{a\}\}$.

2.61. $A=\{a,\{a\},\{b\}\}$, $B=\{\{a\},\{b\}\}$.

2.62. $A=\{\{a,\{b\}\},c,\{c\},\{a,b\}\}$, $B=\{\{a,b\},c,\{b\}\}$.

Find – as before – $A \cup B$, $A \cap B$, $A-B$, and $B-A$ for each of the following A and B (Problems 2.63–2.66):

2.63. $A=\{x \in \mathcal{N} : x<3\}$, $B=\{x \in \mathcal{N} : x \geqslant 3\}$.

2.64. $A=\{x \in \mathcal{N} : x<0\}$, $B=\{x \in \mathcal{N} : x=2\}$.

2.65. $A=\{x \in \mathcal{R} : x<1\}$, $B=\{x \in \mathcal{N} : x<1\}$.

2.66. $A=\{x \in \mathcal{R} : x<1\}$, $B=\{x \in \mathcal{R} : x<2\}$.

2.67. Let the space X be the set of all polygons in the plane and let A be the set of isosceles triangles, B the set of equilateral triangles and C the set of rectangular triangles.

Find the sets:

(a) $(A \cap B) \cap C$,

(b) $(A \cap -B) \cap C$,

(c) $(-A) \cap (B \cap C)$

(d) $(-A) \cap (C \cap -B)$,

(e) $(A \cap B) \cap (-C)$

Prove that, given the space X, for all sets A, B, and C the following equalities hold (Problems 2.68–2.79):

2.68. $A \cup B = B \cup A$.

2.69. $A \cap B = B \cap A$.

2.70. $A \cap (B \cup C) = (A \cap B) \cup (A \cap C)$.

2.71. $A \cup (B \cap C) = (A \cup B) \cap (A \cup C)$.

2.72. $A \cup (B \cup C) = (A \cup B) \cup C$.

2.73. $A \cap (B \cap C) = (A \cap B) \cap C$.

2.74. $O \cap A = O$.

2.75. $X \cap A = A$.

2.76. $O \cup A = A$.

2.77. $X \cup A = X$.

2.78. $A \cup -A = X$.

2.79. $A \cap -A = O$.

2.80. Prove that

$$A \subset B \Leftrightarrow A \cup B = B \Leftrightarrow A \cap B = A \Leftrightarrow A - B = O.$$

Show that for all sets A, B, C, and D the following equalities hold (Problems 2.81–2.87):

2.81. $(A \cup B) - C = (A - C) \cup (B - C)$.

2.82. $A - (B - C) = (A - B) \cup (A \cap C)$.

2.83. $(A - B) \cup C = [(A \cup C) - B] \cup (B \cap C)$.

2.84. $A \cup (B - C) = [(A \cup B) - C] \cup (A \cap C)$.

2.85. $A - (B \cup C) = (A - B) - C$.

2.86. $(A - B) \cap (C - D) = (A \cap C) - (B \cup D)$.

2.87. $A - [B - (C - D)] = (A - B) \cup [(A \cap C) - D]$.

Show the following identities of the algebra of sets. They are called *de Morgan Laws* (Problems 2.88–2.89):

2.88. $-(A \cup B)=(-A) \cap (-B)$.

2.89. $-(A \cap B)=(-A) \cup (-B)$.

Check if the following equalities are identities of the algebra of sets. If they are not, provide appropriate counterexamples (Problems 2.90–2.93):

2.90. $A-(B \cup C)=(A-B) \cap (A-C)$.

2.91. $A \cup (A \cap B)=A$.

2.92. $A \cap (A \cup B)=B$.

2.93. $(A \cup B \cup C)-(A \cup B)=C$.

Prove that for all the sets A, B and C the following implications or equivalences hold (Problems 2.94–2.100):

2.94. $(A \subset B) \wedge (C \subset D) \Rightarrow (A \cap C \subset B \cap D)$.

2.95. $(A \subset B) \wedge (C \subset D) \Rightarrow (A \cup C \subset B \cup D)$.

2.96. $(A \subset B) \wedge (C \subset D) \Rightarrow (A-D \subset B-C)$.

2.97. $(A \subset B) \Rightarrow (C-B \subset C-A)$.

2.98. $(A-B=B-A) \Rightarrow A=B$.

2.99. $(A \subset B) \Leftrightarrow [B=A \cup (B-A)]$.

2.100. $(A \subset B) \Leftrightarrow \{(B \subset C) \Rightarrow [(C-A) \cap (C-B)=C-B]\}$.

Find what inclusions must hold between the sets A, B, and C if the following equality is true (Problems 2.101–2.106):

2.101. $(A \cap B) \cup (C \cap B)=B$.

2.102. $(A \cup B) \cap (C \cup B)=B$.

2.103. $(A-C) \cup B=A \cup B$.

2.104. $(A \cup B)-C=(A-C) \cup B$.

2.105. $(A \cup B)-(B \cap C)=A \cap C$.

2.106. $[(A \cap B) \cup C]-A=(A \cap B)-C$.

2.107. Prove that $A \cap B$ is the largest set included both in A and in B, i.e. that every set X included both in A and in B is included in $A \cap B$.

2.108. Prove that $A \cup B$ is the least set containing simultaneously both A and B, i.e. that every set X including both A and B includes also $A \cup B$.

2.109. Prove that $A-B$ is the largest set contained in A and disjoint from B, i.e. that each set X contained in A and disjoint from B is included in $A-B$.

***2.110.** Estimate the cardinality of the least family of sets $\mathscr{A}$ such that $A_i \in \mathscr{A}$ $(i \leqslant n)$ and satisfying the following condition:

Whenever $A, B \in \mathscr{A}$ then also $A \cup B \in \mathscr{A}$.

2.111. Estimate the cardinality of the least family of sets $\mathscr{A}$ such that $A_i \in \mathscr{A}$ $(i \leqslant n)$ and satisfying the following condition:

Whenever $A, B \in \mathscr{A}$ then also $A \cup B \in \mathscr{A}$, $A \backslash B \in \mathscr{A}$.

2.112. Let $A_1, \ldots, A_n$ be sets and let $k \leqslant n$. Prove that the union of all intersections of k-element subfamilies of $\{A_1, \ldots, A_n\}$ is equal to the intersection of all unions of subfamilies of cardinality $n-k+1$ of the family $\{A_1, \ldots, A_n\}$.

2.113. Prove that:

(a) $$A_1 \cup A_2 \cup \ldots \cup A_n = (A_1 - A_2) \cup (A_2 - A_3) \cup \ldots$$
$$\ldots \cup (A_{n-1} - A_n) \cup (A_n - A_1) \cup (A_1 \cap A_2 \cap \ldots \cap A_n),$$

(b) $$(A_1 \cup \ldots \cup A_n) \cap (B_1 \cup \ldots \cup B_m)$$
$$= (A_1 \cap B_1) \cup \ldots \cup (A_1 \cap B_m) \cup \ldots$$
$$\ldots \cup (A_2 \cap B_1) \cup \ldots \cup (A_2 \cap B_m) \cup \ldots$$
$$\ldots \cup (A_n \cap B_1) \cup \ldots \cup (A_n \cap B_m).$$

The *symmetric difference* of sets is the set:

$$A \dot{-} B = (A - B) \cup (B - A).$$

Prove that (Problems 2.114–2.118):

2.114. $A \dot{-} B = B \dot{-} A$.

2.115. $A \dot{-} (B \dot{-} C) = (A \dot{-} B) \dot{-} C$.

2.116. $A \dot{-} O = A$.

2.117. $A \dot{-} A = O$.

2.118. $A \cap (B \dot{-} C) = (A \cap B) \dot{-} (A \cap C)$.

Check if the following equalities are true for arbitrary sets A, B, C (Problems 2.119–2.121):

2.119. $A \dot{-} B = (A \cup B) - (A \cap B)$.

2.120. $A \cup B = (A \dot{-} B) \dot{-} (A \cap B)$.

2.121. $A \cup (B \dot{-} C) = (A \cup B) \dot{-} (A \cup C)$.

Let $A_1, \ldots, A_n$ be arbitrary subsets of the space X. Let A_i^0, A_i^1 denote the sets A_i and $-A_i$, respectively. Every intersection of the form,

$$A_1^{i_1} \cap \ldots \cap A_n^{i_n},$$

where $i_l = 0$ or $i_l = 1$, is called a *constituent.*

2.122. Prove that there are at most 2^n different constituents.

2.123. Prove that different constituents are disjoint.

2.124. Find the union of all constituents.

2.125. Prove that the set A_k is the union of all constituents with $i_k = 0$.

2.126. Prove that every element of the least field of subsets of the space X containing the family $\{A_1, \ldots, A_n\}$ can be represented as the union of a certain number of constituents.

An *ordered pair* $\langle a, b\rangle$ is a set $\{\{a\}, \{a, b\}\}$.

Ordered pairs have the following important property (cf. Problem 2.53):

$$\langle a, b\rangle = \langle c, d\rangle \Leftrightarrow (a = c) \wedge (b = d).$$

An *ordered triple* $\langle a, b, c\rangle$ is the set $\langle\langle a, b\rangle, c\rangle$. In general, the ordered *n-triple* $\langle a_1, \ldots, a_n\rangle$ is the set

$$\langle\langle \ldots \langle\langle a_1, a_2\rangle, a_3\rangle \ldots\rangle, a_n\rangle.$$

The *Cartesian product* $A \times B$ is defined as follows:

$$A \times B = \{\langle a, b\rangle : a \in A \wedge b \in B\}.$$

In general,

$$A_1 \times \ldots \times A_n = \{\langle a_1 \ldots a_n\rangle : a_1 \in A_1 \wedge \ldots \wedge a_n \in A_n\}.$$

2.127. Prove that

$$\langle a_1, \ldots, a_n\rangle = \langle b_1, \ldots, b_n\rangle \Leftrightarrow (a_1 = b_1) \wedge \ldots \wedge (a_n = b_n).$$

Find the Cartesian products $A \times B$ and $B \times A$ for the following sets A and B (Problems 2.128–2.131):

2.128. $A = \{0, 1\}, B = \{1, 2\}$.

2.129. $A = \{0, 1, 2\}, B = \{0, 2, 3\}$.

2.130. $A = \{1\}, B = \{1, 2, 3, 4, 5\}$.

2.131. $A = O, B = \{1, 2, 3\}$.

2.132. Prove that if the sets A and B possess n and m elements, respectively then $A \times B$ and $B \times A$ have nm elements.

2.133. Prove that if $A \times B = B \times A$ then one of the following holds: $A = O$, $A = B$ or $B = O$.

2.134. Find $A \times (B \times C)$, $(A \times B) \times C$, $A \times B \times C$ where $A = \{0, 1\}$, $B = \{1\}$, $C = \{2, 3\}$.

Assuming that points in the plane are ordered pairs $\langle a, b\rangle$ of real numbers where a is the abscissa and b the ordinate of the point, find $A \times B$ and $B \times A$ for the following sets A and B (Problems 2.135–2.139):

2.135. $A = \{x \in \mathscr{R} : 1 < x < 2\}$, $B = \{x \in \mathscr{R} : 0 < x < 1\}$.

2.136. $A = \{x \in \mathscr{R} : 0 < x\}$, $B = \{y \in \mathscr{R} : 0 < y\}$.

2.137. $A = \{y \in \mathscr{R} : -1 < y < 1\}$, $B = \{x \in \mathscr{R} : 0 < x \leqslant 1\}$.

2.138. $A = \{y \in \mathscr{R} : x < 1 \vee 1 < x\}$, $B = \{y \in \mathscr{R} : y^2 > 0\}$.

2.139. $A = \{x \in \mathscr{R} : 0 < x < 1 \vee 2 < x \leqslant 3\}$,
$B = \{x \in \mathscr{R} : 1 < x \leqslant 2 \vee 3 < x \leqslant 4\}$.

Assuming that the points of the three-dimensional cartesian space are ordered triples $\langle a, b, c\rangle$ such that $\langle a, b\rangle$ is the projection of the point $\langle a, b, c\rangle$ on the plane $z = 0$ and c is the projection of the point $\langle a, b, c\rangle$ on the axis z, find $A \times B$ for the following sets A and B (Problems 2.140–2.142):

2.140. $A = \{\langle x, y\rangle : 0 < x < 1 \wedge 0 < y < 1\}$, $B = \{x : 0 < x < 1\}$.

2.141. $A = \{\langle x, y\rangle : x^2 + y^2 < 1\}$, $B = \{x : -1 < x < 1\}$.

2.142. $A = \{\langle x, y\rangle : 1 \leqslant x^2 + y^2 < 4\}$, $B = \{x : 0 < x\}$.

Check if the following equalities are true (Problems 2.143–2.146):

2.143. $A \times (B \cup C) = (A \times B) \cup (A \times C)$.

2.144. $A \times (B \cap C) = (A \times B) \cap (A \times C)$.

2.145. $A - (B \times C) = (A - B) \times (A - C)$.

2.146. $A \cap (B \times C) = (A \cap B) \times (A \cap C)$.

Let X be a fixed space. Assume that for every set $A \subset X$ we are given a set $\bar{A} \subset X$ called the *closure* of the set A and such that the following are true:

$$\overline{A \cup B} = \bar{A} \cup \bar{B}, \quad \bar{\bar{A}} = \bar{A}, \quad A \subset \bar{A}, \quad \bar{O} = O.$$

The *interior* of the set A is the set $\operatorname{Int}(A) = X - \overline{X - A}$.

The *boundary* of the set A is the set $\operatorname{Fr}(A) = \bar{A} - \operatorname{Int}(A)$.

For a given set A we define the *power set* of A, $\mathscr{P}(A)$ as follows:

$$X \in \mathscr{P}(A) \Leftrightarrow X \subset A.$$

The set $\mathscr{P}(A)$ is often denoted by 2^A.

Show the following equalities (Problems 2.147–2.156):

2.147. $\overline{X}=X$.

2.148. $\overline{A}-\overline{B}=\overline{A-B}-\overline{B}$.

2.149. $A\subset B \Rightarrow \overline{A}\subset\overline{B}$.

2.150. $(\overline{A}=A \cap \overline{B}=B) \Rightarrow \overline{A\cap B}=A\cap B$.

2.151. $\mathrm{Int}(\mathrm{Int}\,A)=\mathrm{Int}(A)$.

2.152. $\mathrm{Int}(A\cap B)=\mathrm{Int}(A)\cap\mathrm{Int}(B)$.

2.153. $A\subset B \Rightarrow \mathrm{Int}(A)\subset\mathrm{Int}(B)$.

2.154. $\mathrm{Fr}(A)=\overline{A}\cap(\overline{X-A})=(A\cap\overline{X-A})\cup(\overline{A}-A)$.

2.155. $\mathrm{Fr}(A)\cup A=\overline{A}$

2.156. $\mathrm{Fr}[\mathrm{Int}(A)]\subset\mathrm{Fr}(A)$.

Find $\mathscr{P}(A)$ for the following sets A (Problems 2.157–2.160):

2.157. $A=\{a, b, c\}$,

2.158. $A=O$.

2.159. $A=\{O\}$.

2.160. $A=\{\{\{a\}\}, \{a\}, a\}$.

Prove that (Problems 2.161–2.163):

2.161. $A\subset B \Rightarrow \mathscr{P}(A)\subset\mathscr{P}(B)$.

2.162. $A\neq B \Rightarrow \mathscr{P}(A)\neq\mathscr{P}(B)$.

2.163. $\sim[A=\mathscr{P}(A)]$.

2.164. Is it true that if $X\subset\mathscr{P}(X)$ then X must be empty?

2.165. Prove that if A has exactly n elements then $\mathscr{P}(A)$ has exactly 2^n elements.

CHAPTER 3

PROPOSITIONAL FUNCTIONS. QUANTIFIERS

A propositional function is an expression containing variables which becomes a sentence (true or false) if we replace those variables by the names of objects.

Thus a propositional function of one variable may be interpreted as a property of some objects whereas a propositonal function of several variables may be interpreted as a condition on n-tuples of objects. In particular, a sentence should be treated as a propositional function with no free variables.

Every propositional function $\Phi(x_1, \dots, x_n)$ usually determines a family of sets $X_1, \dots, X_n$ such that the names of the elements of X_i may be substituted for x_i in Φ. Those sets are called the *domains* of corresponding variables.

If $\Phi(x)$ is a propositional function of one variable, then clearly

$$a \in \{x : \Phi(x)\} \Leftrightarrow \Phi(a),$$

and if X is the domain of the variable x, then $\Phi(a) \Rightarrow a \in X$.

Sometimes, instead of saying that the domain of the variable x is X, we say that the function $\Phi(x)$ is defined on the set X.

More generally, if $\Phi(x_1, \dots, x_n)$ is a propositional function of n variables, then

$$\langle a_1, \dots, a_n \rangle \in \{\langle x_1, \dots, x_n \rangle : \Phi(x_1, \dots, x_n)\} \Leftrightarrow \Phi(a_1, \dots, a_n).$$

The sets

$$\{x : \Phi(x)\} \quad \text{and} \quad \{\langle x_1, \dots, x_n \rangle : \Phi(x_1, \dots, x_n)\}$$

are called the *graphs* of the function Φ.

Let us notice that if the sets being the domains of the variables of Φ exist, then so does the graph of Φ. Unfortunately, not every propositional function possesses a graph. This is due to some restrictions on the existence of sets, related to the appearance of so called *antinomies*.

Propositional functions and propositional connectives may be used to build new propositional functions, providing that identical variables in the subformulas have the same domain. Thus $\Phi(x, y) \wedge \Psi(x, z)$ is a propositional function provided the domain of the variable x in Φ is identical with the domain of x in Ψ.

Quantifiers are operations on propositional functions.

Mathematicians generally use quantifiers which formally correspond to the expressions "for every x..." and "there exists an x such that...". We denote them respectively by:

$$\bigwedge_x \quad \text{and} \quad \bigvee_x$$

and call them the *universal quantifier* and the *existential quantifier*.

Thus, if $\Phi(x, y)$ is given, we read $\bigwedge_x \Phi(x, y)$ as: for all x in the domain of Φ, $\Phi(x, y)$ holds. Similarly $\bigvee_x \Phi(x, y)$ is read: there is an x in the domain of Φ such that $\Phi(x, y)$.

In both cases the application of the quantifier transforms a propositional function of two variables x and y into a propositional function of one variable y. Though the formula contains the letter x, the logical value of that function depends only on the substitution for y. The variable x is thus *bounded*, whereas y is *free*. A formula without free variables is a *sentence*.

One introduces the notion of *restricted quantifiers* $\bigwedge_{\Phi(x)} \ldots$, $\bigvee_{\Phi(x)} \ldots$, which denote, respectively, "For all x such that $\Phi(x)$..." and "There exists an x with the property $\Phi(x)$ such that...". Formally, the expression $\bigwedge_{\Phi(x)} \Psi$ is an abbreviation of the expression $\bigwedge_x (\Phi(x) \Rightarrow \Psi)$, whereas $\bigvee_{\Phi(x)} \Psi$ is an abbreviation of the expression $\bigvee_x (\Phi(x) \wedge \Psi)$.

The symbols Φ, Ψ, Θ will be used to denote *predicate variables* which range over propositional functions.

An expression (formula) built of predicate variables, propositional functors and quantifiers is called a *tautology of the predicate calculus* (sometimes also called the *functional calculus*) if that expression is true regardless of the propositional functions substituted for the predicative variables.

Sometimes a formula Φ is called a sentence even though it is written in such a way that it has free variables $x_1, \ldots, x_n$. In this case we consider that formula as an abbreviation of the formula

$$\bigwedge_{x_1} \cdots \bigwedge_{x_n} \Phi(x_1, \ldots, x_n).$$

This notation is often used in algebra, where – for instance – the commutativity of addition is stated as $x+y=y+x$, whereas the precise expression is

$$\bigwedge_{x} \bigwedge_{y} x+y=y+x.$$

Find the graphs of the following propositional functions (Problems 3.1–3.10):

3.1. $x^2-1 \geqslant 0, X=\mathscr{R}$.

3.2. $x=x, \ X=\mathscr{N}$.

3.3. $x \neq x, \ X=\mathscr{Z}$.

3.4. $x+1=2x, \ X=\mathscr{R}^+$.

3.5. $x+1=2x, \ X=\mathscr{R}-\mathscr{R}^+$.

3.6. $|x| \geqslant 1, \ X=\mathscr{C}$.

3.7. $|x|=|x+1|, \ X=\mathscr{R}$.

3.8. $x^2 \geqslant x, \ X=\mathscr{N}$.

3.9. $(x-1)(x+1)=0, \ X=\mathscr{Z}$.

3.10. $|x+1|+|x+2|=1, \ X=\mathscr{R}$.

Assume that the variables of propositional functions Φ_1 and Φ_2 have the same domains. Let Z_1, Z_2 be the graphs of Φ_1 and Φ_2 respectively. Find the graph of the propositional function (Problems 3.11–3.16):

3.11. $\Phi_1 \wedge \Phi_2$.

3.12. $\Phi_1 \vee \Phi_2$.

3.13. $\sim \Phi$.

3.14. $\Phi_1 \Rightarrow \Phi_2$.

3.15. $\Phi_1 \Leftrightarrow \Phi_2$.

3.16. $B_8(\Phi_1, \Phi_2)$.

Find the propositional function with the graph (Problems 3.17–3.20):

3.17. $Z_1 \cap Z_2$.

3.18. $Z_1 \cup Z_2$.

3.19. $Z_1 - Z_2$

3.20. $Z_1 \times Z_2$ (assuming that Φ_1 has no common free variable with Z_2).

A propositional function $\Phi(x_1, \ldots, x_n)$ is *true* in $X_1, \ldots, X_n$ iff for all tuples $\langle a_1, \ldots, a_n \rangle$ $(a_i \in X_i,\ 1 \leqslant i \leqslant n)$, $\Phi(a_1, \ldots, a_n)$ is true. Similarly, we say that $\Phi(x_1, \ldots, x_n)$ is *false* if, for all tuples $\langle a_1, \ldots, a_n \rangle$ $(a_i \in X_i,\ 1 \leqslant i \leqslant n)$, $\Phi(a_1, \ldots, a_n)$ is false.

Let $\Phi(x)$ be a propositional function with the domain X and let Z be the graph of Φ. Find a necessary and sufficient condition for (Problems 3.21 and 3.22):

3.21. $Z = X$.

3.22. $Z = O$.

3.23. Assume that Z_1, the diagram of Φ_1, is equal to Z_2, the diagram of Φ_2. What is the diagram of:

(a) $\Phi_1(x) \Leftrightarrow \Phi_2(x)$,

(b) $\Phi_1(x) \wedge \Phi_2(x)$,

(c) $\Phi_1(x) \vee \sim\Phi_2(x)$.

3.24. Let $\Phi_1(x)$ be a propositional function true in X and Φ_2 any other propositional function. Find the graph of the functions:

(a) $\Phi_1(x) \wedge \Phi_2(x)$,

(b) $\Phi_1(x) \vee \Phi_2(x)$,

(c) $\sim\Phi_1(x) \wedge \Phi_2(x)$.

(d) $\sim\Phi_1(x) \vee \Phi_2(x)$.

Find the graph of the propositional function $\Phi(x, y)$ where the domain of both x any y is the set $\mathscr{R}$ (Problems 3.25–3.55):

3.25. $x = y$.

3.26. $x < y$.

3.27. $x \leqslant y$.

3.28. $x \geqslant y$.

3.29. $x^2 + y^2 \leqslant 1$.

3.30. $x^2 + y^2 = 1$

3.31. $ax + by + c = 0$ $(a, b, c \in \mathscr{R})$

3.32. $x \neq y$.
3.33. $x^2 + y^2 = 0$.
3.34. $x \cdot y = 1$.
3.35. $x \cdot y = 0$.
3.36. $x \cdot y < 1$.
3.37. $ax^2 + bx + c + y = 0 \quad (a, b, c \in \mathscr{R})$.
3.38. $x < |y|$.
3.39. $|x| > y$.
3.40. $(ax)^2 + (by)^2 \leqslant 1 \quad (a, b \in \mathscr{R})$.
3.41. $x^2 + y^2 \geqslant 0$.
3.42. $|x \cdot y| < 0$.
3.43. $x \geqslant y \vee x^2 + y^2 \leqslant 1$.
3.44. $x + y = 0 \vee x \geqslant y$.
3.45. $x + y = 1 \vee x \neq y$.
3.46. $x \geqslant y \wedge x^2 + y^2 = 0$.
3.47. $x \cdot y < 1 \Rightarrow x \cdot y = 1$.
3.48. $\sim(x \cdot y < 1)$.
3.49. $|x \cdot y| < 0 \Rightarrow x^2 + y^2 = 0$.
3.50. $x \geqslant 0 \vee y \geqslant 0$.
3.51. $x < |y| \Rightarrow x^2 + y^2 \geqslant 0$.
3.52. $x \leqslant 0 \vee y = y$.
3.53. $x^2 + y^2 = 0 \vee x = x$.
3.54. $y^2 + 2y - 3 \leqslant 0$.
3.55. $x^3 + 1 > 0$.

Find the graphs of the propositional functions of the variables x, y, z ranging over $\mathscr{R}$ (Problems 3.56–3.62):

3.56. $x^2 + y^2 + z^2 \leqslant 1$.
3.57. $x + y = z$.
3.58. $x + y = 1$.
3.59. $x^2 + y^2 \leqslant 1$.
3.60. $x^2 + y^2 < 4 \wedge |z| < 1$.
3.61. $|x| < 1 \wedge |y| < 1 \wedge |z| < 1$.
3.62. $x^2 + y^2 = z$.

We say that the n-tuple $\langle a_1, \ldots, a_n \rangle$ *satisfies* the propositional function $\Phi(x_1, \ldots, x_n)$ if $\Phi(a_1, \ldots, a_n)$ is true. If there exist $a_1, \ldots, a_n$ such that $\langle a_1, \ldots, a_n \rangle$ satisfies $\Phi(x_1, \ldots, x_n)$, then we say that $\Phi(x_1, \ldots, \ldots, x_n)$ is *satisfiable*.

3.63. Prove that the graph of a propositional function Φ is empty iff Φ is not satisfiable.

3.64. Prove that Φ is not satisfiable if and only if Φ is false.

3.65. Prove that if $X_1, \ldots, X_n$ are domains of $x_1, \ldots, x_n$, respectively, in $\Phi(x_1, \ldots, x_n)$ and $X_i \neq 0$ for $1 \leqslant i \leqslant n$, then if Φ is true then it is satisfiable.

3.66. Prove that the function $\Theta(x, y) = \Phi(x) \vee \Psi(y)$ is satisfiable iff either $\Phi(x)$ is satisfiable or $\Psi(y)$ is satisfiable (or both are satisfiable).

3.67. Prove that the function $\Theta(x, y) = \Phi(x) \wedge \Psi(y)$ is satisfiable iff the functions $\Phi(x)$ and $\Psi(y)$ are both satisfiable. Is the same proposition true for the functions $\Phi(x)$ and $\Psi(x)$?

3.68. Prove that the function $\Theta(x, y) = \Phi(x) \Rightarrow \Psi(y)$ is not satisfiable iff $\Phi(x)$ is true and $\Psi(y)$ false. Is the same proposition true for the functions $\Phi(x)$ and $\Psi(x)$?

3.69. When is the function $\sim\Phi(x)$ satisfiable?

3.70. Prove that if X is the domain of the variable x in $\Phi(x)$ then the graph of the function $\sim\Phi(x)$ is different from X if and only if $\Phi(x)$ is satisfiable.

3.71. Show that if $\Phi(x_1, \ldots, x_n)$ and $\Psi(x_1, \ldots, x_n)$ are propositional functions such that, for every $a_1, \ldots, a_n$ from the apropriate domains of the variables,

$$\Phi(a_1, \ldots, a_n) \Leftrightarrow \Psi(a_1, \ldots, a_n),$$

then the graphs of Φ and Ψ are identical.

3.72. Prove the theorem converse to 3.71.

Assuming that $\Phi(x, y, z)$, $\Psi(x, y, z)$, $\Theta(x, y, z)$ are propositional functions with variables x, y, z, find which variables are free and which are bounded in the following formulae (Problems 3.73–3.85):

3.73. $\bigwedge_x \Phi(x, y, z)$.

3.74. $\bigwedge_x \bigwedge_y \Phi(x, y, z)$.

3.75. $\bigvee_z \Phi(x, y, z)$.

3.76. $\bigvee_x \Phi(x, y, z)$.

3.77. $[\bigvee_x \bigwedge_y \Phi(x, y, z)] \Rightarrow \Psi(x, y, z)$.

3.78. $\bigwedge_x \bigwedge_y \Phi(x, y, z) \wedge \bigvee_z \Psi(x, y, z)$.

3.79. $\bigwedge_{x} \Phi(x, y, z) \Rightarrow \{\bigvee_{z} (\bigvee_{y} \Psi(x, y, z) \wedge \bigwedge_{z} \Theta(x, y, z))\}$.

3.80. $\bigvee_{x} [\Phi(x, y, z) \Rightarrow \Psi(x, x, y)] \Rightarrow$

$\Rightarrow \{\bigvee_{x} \bigvee_{z} [\Phi(x, x, y) \wedge \Theta(x, y, y)]\}$.

3.81. $\bigvee_{x} (x<y \vee x<z)$.

3.82. $\bigvee_{x} \bigwedge_{y} [(x<y) \Rightarrow (x<z) \wedge (z<y)]$.

3.83. $\bigwedge_{x} (x|y \wedge x|z \Rightarrow x|z)$.

3.84. $(\bigwedge_{x} \bigvee_{y} x<y) \vee (x<z)$.

3.85. $\bigvee_{x} (x<x \vee x<z)$.

3.86. Let $\Phi(x)$ be a propositional function with the domain of x equal to $A=\{a_1, \ldots, a_n\}$. Prove that:

a) $\bigwedge_{x} \Phi(x) \Leftrightarrow (\Phi(a_1) \wedge \Phi(a_2) \wedge \ldots \wedge \Phi(a_n))$,

b) $\bigvee_{x} \Phi(x) \Leftrightarrow (\Phi(a_1) \vee \Phi(a_2) \vee \ldots \vee \Phi(a_n))$.

3.87. Let $\Phi(x)$ and $\Psi(x)$ be propositional functions defined on the set $A=\{a_1, \ldots, a_n\}$. Show that:

a) $\bigwedge_{\Psi(x)} \Phi(x) \Leftrightarrow \{[\Psi(a_1) \Rightarrow \Phi(a_1)] \wedge [\Psi(a_2) \Rightarrow \Phi(a_2)] \wedge \ldots$

$\ldots \wedge [\Psi(a_n) \Rightarrow \Phi(a_n)]\}$,

b) $\bigvee_{\Psi(x)} \Phi(x) \Leftrightarrow \{[\Psi(a_1) \wedge \Phi(a_1)] \vee [\Psi(a_2) \wedge \Phi(a_2)] \vee \ldots$

$\ldots \vee [\Psi(a_n) \wedge \Phi(a_n)]\}$.

Assuming that x, y and z range over the set $\mathscr{R}$, find the graphs of the following propositional functions (Problems 3.88–3.111):

3.88. $\bigvee_{x} x^2+y^2=1$.

3.89. $\bigvee_{x} x \cdot y=1$.

3.90. $\bigwedge_{x} x^2+y^2=1$.

3.91. $\bigwedge_{x} x \cdot y=1$.

3.92. $\bigvee_{x} x \cdot y \neq 1$.

3.93. $\bigwedge_{x} x \cdot y<1$.

3.94. $\bigvee_{x} x \cdot y<1$.

3.95. $\bigwedge_{x} x^2+1<y$.

3.96. $\bigvee_{x} x^2+y^2=z^2$.

3.97. $\bigwedge_{x} x^2+y^2 \neq z^2$.

3.98. $\bigwedge_x \bigvee_y x^2+y^2=z^2$.

3.99. $\bigvee_x \bigwedge_y x\cdot y=z$.

3.100. $\bigwedge_x \bigvee_y (x<z)\wedge(z<y)$.

3.101. $\bigwedge_x \bigwedge_y x^2+y^2\geqslant z$.

3.102. $\bigvee_x (x^2+y^2=1)\vee(x<x)$.

3.103. $\bigvee_x (x\cdot y=1)\wedge(x=x)$.

3.104. $\bigvee_x \sqrt{1-x^2}=y$.

3.105. $\bigwedge_x \sqrt{1-x^2}=y$.

3.106. $\bigvee_x (x^2+2ax+b=y)\wedge(y>0)$.

3.107. $\bigwedge_x \sin y<x\wedge x<2+\sin y$.

3.108. $\bigvee_x \sin y<x\wedge x<2+\sin y$.

3.109. $\bigvee_x \operatorname{tg} x>y\wedge -\frac{1}{2}\pi<x<\frac{1}{2}\pi$.

3.110. $\bigwedge_x \operatorname{tg} x>y\wedge -\frac{1}{2}\pi<x\wedge x<\frac{1}{2}\pi$.

3.111. $\bigvee_z x=z\cdot\sin z\wedge y=z\cdot\cos z$.

3.112. Let $\Phi(x, y)$ be a propositional function defined on real numbers. What is the geometrical meaning of the operations leading from the graph of the function $\Phi(x, y)$ to the graph of the following function:

(a) $\bigwedge_x \Phi(x, y)$,

(b) $\bigvee_x \Phi(x, y)$,

(c) $\bigwedge_y \Phi(x, y)$,

(d) $\bigvee_y \Phi(x, y)$

3.113. Prove that, given a propositional function $\Phi(x, y)$, if any of graphs of the functions: $\bigvee_x \Phi(x, y)$, $\bigvee_y \Phi(x, y)$ is nonempty, then $\Phi(x, y)$ is satisfiable.

Let $x=y$, $x<y$, $x\leqslant y$ be propositional functions on the set of natural numbers denoting respectively: equality, strong inequality and weak inequality. Using those functions, and also arithmetical operations, symbols for numbers, propositional connectives and quantifiers, write in symbols the following propositional functions (Problems 3.114–3.128):

3.114. x is an even number.

3.115. x is a sum of two squares.

3.116. x is a prime number.

3.117. x is not prime.

3.118. x is the least common multiple of y and z.

3.119. x is the greatest common divisor of y and z.

3.120. x, when divided by 4, gives as the remainder 1 or 2.

3.121. Every number, when divided by 2, gives as the remainder 0 or 1.

3.122. There is a prime number between n and $2n$ (Tschebyscheff's theorem).

3.123. The numbers x and y have the same divisors.

3.124. Every odd number greater than 3 is a sum of two primes (Goldbach's conjecture).

3.125. Every three numbers have the least common multiple.

3.126. Every three numbers have the greatest common divisor.

3.127. There is no largest natural number.

3.128. There is no largest prime number.

Assume now that the variables in the propositional functions $x=y$, $x<y$ and $x\leqslant y$ range over real numbers. Using them and the symbols for arithmetic operations ($x+y$, x^y, $x\cdot y$, $|x|$ etc.), write in symbols the following formulas (Problems 3.129–3.135):

3.129. No square is smaller than 0.

3.130. The function $f(x)$ has exactly one root.

3.131. Between any two real numbers there is another one.

3.132. There is no largest real number.

3.133. x is not a square of a real number.

3.134. x is not a root of at most third degree of some number.

3.135. $f(x)$ is a decreasing function.

Using the same symbols as before and, in addition, the propositional function $n\in\mathcal{N}$ (meaning: n is a natural number) write in symbols the following formulas (Problems 3.136–3.150):

3.136. The sequence $\{a_n\}$ is increasing.

3.137. The sequence $\{a_n\}$ takes only positive values.

3.138. The sequence $\{a_n\}$ converges.

3.139. The sequence $\{a_n\}$ is bounded.

3.140. The sequence $\{a_n\}$ is eventually constant.

3.141. If the sequence $\{a_n\}$ is eventually constant, then it is convergent.

3.142. If the sequence $\{a_n\}$ is bounded, then it possesses a convergent subsequence.

3.143. The function $f(x)$ is continuous at x_0.

3.144. If the function $f(x)$ is continuous in a closed segment $\langle a, b\rangle$, then $f(x)$ is bounded in it.

3.145. The function $f(x)$ is uniformly continuous in the closed segment $\langle a, b\rangle$.

3.146. a is the supremum of numbers in $\mathcal{R}$.

3.147. a is the infimum of numbers in $\mathcal{R}$.

3.148. If $f(x)$ is continuous in the segment $\langle a, b\rangle$, then it reaches both the supremum and the infimum in this segment.

3.149. If $f(x)$ and $g(x)$ are continuous, then $f(x)\cdot g(x)$ is also continuous.

3.150. If $f(x)$ and $g(x)$ are uniformly continuous, then $f(x)+$ $+g(x)$ is also uniformly continuous.

Prove that the following expressions are tautologies of the predicate calculus (Problems 3.151–3.159):

3.151. $\sim \bigvee_x \Phi(x) \Leftrightarrow \bigwedge_x \sim\Phi(x)$.

3.152. $\sim \bigwedge_x \Phi(x) \Leftrightarrow \bigvee_x \sim\Phi(x)$.

3.153. $\bigvee_x \bigwedge_y \Phi(x, y) \Rightarrow \bigwedge_y \bigvee_x \Phi(x, y)$.

3.154. $\bigvee_x [\Phi(x) \vee \Psi(x)] \Leftrightarrow \bigvee_x \Phi(x) \vee \bigvee_x \Psi(x)$.

3.155. $\bigvee_x [\Phi(x) \wedge \Psi(x)] \Rightarrow \bigvee_x \Phi(x) \wedge \bigvee_x \Psi(x)$.

3.156. $\bigwedge_x [\Phi(x) \wedge \Psi(x)] \Leftrightarrow \bigwedge_x \Phi(x) \wedge \bigwedge_x \Psi(x)$.

3.157. $\bigwedge_x \Phi(x) \vee \bigwedge_x \Psi(x) \Rightarrow \bigwedge_x [\Phi(x) \vee \Psi(x)]$.

3.158. $\bigwedge_x [\Phi(x) \Rightarrow \Psi(x)] \Rightarrow [\bigwedge_x \Phi(x) \Rightarrow \bigwedge_x \Psi(x)]$.

3.159. $\bigwedge_x [\Phi(x) \Leftrightarrow \Psi(x)] \Rightarrow [\bigwedge_x \Phi(x) \Leftrightarrow \bigwedge_x \Psi(x)]$.

Assuming that the propositional function Ψ does not contain x as a free variable, show that the following expressions are tautologies of the predicate calculus (Problems 3.160–3.166):

3.160. $(\bigwedge_x \Psi) \Leftrightarrow \Psi$.

3.161. $\bigwedge_x [\Phi(x) \vee \Psi] \Rightarrow \bigwedge_x \Phi(x) \vee \Psi$.

3.162. $\Psi \wedge \bigvee_x \Phi(x) \Rightarrow \bigvee_x [\Psi \wedge \Phi(x)]$.

3.163. $[\Psi \Rightarrow \bigwedge_x \Phi(x)] \Rightarrow \bigwedge_x [\Psi \Rightarrow \Phi(x)]$.

3.164. $[\Psi \Rightarrow \bigvee_x \Phi(x)] \Rightarrow \bigvee_x [\Psi \Rightarrow \Phi(x)]$.

3.165. $[\bigvee_x \Phi(x) \Rightarrow \Psi] \Rightarrow \bigwedge_x [\Phi(x) \Rightarrow \Psi]$.

3.166. $[\bigwedge_x \Phi(x) \Rightarrow \Psi] \Rightarrow \bigvee_x [\Phi(x) \Rightarrow \Psi]$.

Check whether the following formulas are tautologies (Problems 3.167–3.177):

3.167. $\bigwedge_x \Phi(x) \Rightarrow \bigwedge_z \Phi(z)$.

3.168. $\bigwedge_x [\Phi(x) \vee \Psi(x)] \Rightarrow \bigwedge_x \Phi(x) \vee \bigwedge_x \Psi(x)$.

3.169. $\bigwedge_y \bigvee_x \Phi(x, y) \Rightarrow \bigvee_x \bigwedge_y \Phi(x, y)$.

3.170. $\bigvee_x \Phi(x) \wedge \bigvee_x \Psi(x) \Rightarrow \bigvee_x [\Phi(x) \wedge \Psi(x)]$.

3.171. $\bigwedge_x [\Phi(x) \Leftrightarrow \bigwedge_x \Psi(x)] \Rightarrow \bigwedge_x [\Phi(x) \Leftrightarrow \Psi(x)]$.

3.172. $\bigwedge_x [\Phi(x) \Rightarrow \bigwedge_x \Psi(x)] \Rightarrow \bigwedge_x [\Phi(x) \Rightarrow \Psi(x)]$.

3.173. $\bigvee_x [\Phi(x) \Rightarrow \Psi(x)] \Rightarrow [\bigvee_x \Phi(x) \Rightarrow \bigvee_x \Psi(x)]$.

3.174. $\bigwedge_x \bigwedge_y \Phi(x, y) \Rightarrow \bigwedge_x \Phi(x, x)$.

3.175. $\bigvee_x \bigvee_y \Phi(x, y) \Rightarrow \bigvee_x \Phi(x, x)$.

3.176. $\bigwedge_x [\Phi(x) \Rightarrow \bigwedge_x \Phi(x)]$.

3.177. $\bigwedge_x [\bigvee_x \Phi(x) \Rightarrow \Phi(x)]$.

Let Ax_L denote the set of formulas of the form:

(a) $\bigwedge_x [\Phi(x) \Rightarrow \Psi(x)] \Rightarrow [\bigwedge_x \Phi(x) \Rightarrow \bigwedge_x \Psi(x)]$,

(b) $\bigwedge_x \Phi(x) \Rightarrow \Phi(x)$,

(c) $\bigwedge_x \bigwedge_y \Phi(x, y) \Rightarrow \bigwedge_y \bigwedge_x \Phi(x, y)$,

(d) $\bigwedge_x \bigwedge_y \Phi(x, y) \Rightarrow \bigwedge_x \Phi(x, x)$, providing Φ does not already contain a quantifier for the variable x such that y is in the range of that quantifier.

(e) $\Phi \Rightarrow \bigwedge_x \Phi$

if Φ does not contain x as a free variable,

(f) $\bigwedge_x [\Phi(x) \Rightarrow \Psi(x)] \Rightarrow [\bigvee_x \Phi(x) \Rightarrow \bigvee_x \Psi(x)]$,

(g) $\Phi(x) \Rightarrow \bigvee_x \Phi(x)$,

(h) $\bigvee_x \bigvee_y \Phi(x, y) \Rightarrow \bigvee_y \bigvee_x \Phi(x, y)$,

(i) $\bigvee_x \Phi(x, x) \Rightarrow \bigvee_x \bigvee_y \Phi(x, y)$,

providing Φ does not already contain a quantifier for the variable x that y is in the range of that quantifier,

(j) $\bigvee_x \Phi \Rightarrow \Phi$

if Φ does not contain x as a free variable.

Moreover, Ax_L contains all formulas which can be obtained from the tautologies of the propositional calculus by the following procedure: First substitute for the propositional variables the formulas of the predicate calculus; if the resulting formula has free variables, insert in front of that formula the quantifier $\bigwedge_v$ for every such variable v.

The set Ax_L is one of the possible sets of axioms for the predicate calculus.

3.178. Using the method employed in problems 3.151–3.166, prove that the formulas (b)–(j) are tautologies of the predicate calculus.

3.179. Using the inference rules (see Chapter I), deduce formulas 3.151–3.166 from the axioms.

3.180. Show that the formula

$$\bigwedge_x \Phi(x) \Rightarrow \bigvee_x \Phi(x).$$

is not a consequence of Ax_L.

3.181. Show that if the set Ax_L is completed by one sentence of the form

$$\bigvee_x [\Phi(x) \Leftrightarrow \Phi(x)],$$

then the sentence $\bigwedge_x \Phi(x) \Rightarrow \bigvee_x \Phi(x)$, can be derived.

Note: The set AxL together with the sentence $\bigvee_x [\Phi(x) \Leftrightarrow \Phi(x)]$ is called a *system of axioms of the predicate calculus over a nonempty domain.*

***3.182.** Prove that if, in a given formula Φ being a theorem of the predicate calculus (tautology), we replace all the quantifiers by the same quantifiers restricted to some propositional function $\Psi(x)$, then the resulting formula is again a tautology.

Find the proofs of the theorems given below and point out what inference rules and what tautologies of the predicate calculus were used in the consecutive steps (Problems 3.183–3.189):

3.183. Every natural number has at least one prime divisor.

3.184. For every two different real numbers there is a third one lying in between.

3.185. There is no largest real number.

3.186. Every real function which is uniformly continuous is continuous.

3.187. Every two non-zero natural numbers have a common multiple.

3.188. For every b and $a \neq 0$ the equation $ax+b=0$ has a solution.

3.189. In every triangle there is a point equidistant from all the sides.

A formula Φ is *elementary* (*atomic*) if it is neither of the form $\Phi_1 \cdot \Phi_2$ (where $\cdot$ is a propositional connective) nor of the form $Q\Phi_1$ (where Q is either negation or a quantifier). It follows that all the variables in the atomic formula are free.

A formula Φ of the predicate calculus is in the *disjunctive* (*conjunctive*) *normal form* iff it is of the form:

$$\Phi = Q^1 x_1 \ldots Q^n x_n \Phi'(x_1, \ldots, x_n),$$

where $Q^i x_i$ $(1 \leqslant i \leqslant n)$ is a quantifier (universal or existential) and the formula Φ' does not contain quantifiers and is in the disjunctive (conjunctive) normal form (cf. Problems 1.80 and 1.81).

A formula Φ' is a *disjunctive* (*conjunctive*) *normal form of a formula* Φ iff Φ' is in the disjunctive (conjunctive) normal form, Φ' has the same free variables as Φ and $\Phi \Leftrightarrow \Phi'$ is a tautology.

Find the disjunctive (conjunctive) forms of the following formulas, assuming that Φ, Ψ, Θ are atomic and contain only those variables which are displayed (Problems 3.190–3.196):

3.190. $\bigwedge_x \Phi(x) \vee \bigwedge_x \Psi(x)$.

3.191. $\bigvee_x \Phi(x) \wedge \bigvee_x \Psi(x)$.

3.192. $\bigvee_x \Phi(x) \Rightarrow \bigwedge_x \Psi(x)$.

3.193. $\bigvee_x \Phi(x) \Leftrightarrow \bigwedge_x \Psi(x, y)$.

3.194. $\sim \bigvee_x [\Phi(x) \wedge \bigvee_y \Psi(x, y)]$.

3.195. $\sim \bigwedge_x [\bigvee_y \Phi(x, y) \Rightarrow \sim \bigwedge_y \Psi(x, y, z) \vee \bigwedge_z \Theta(x, y, z)]$.

3.196. $\bigvee_\lambda \{\bigwedge_x \Phi(x, y) \vee \sim \bigvee_y [\Psi(x, y) \vee \bigwedge_z \Theta(z, x)]\}$.

3.197. Show that every formula has a conjunctive normal form.

3.198. Show that every formula has a disjunctive normal form.

3.199. Is the normal conjunctive (disjunctive) form for given formula unique?

3.200. Find the normal forms of the formulas of problems 3.151–3.174.

CHAPTER 4

RELATIONS, EQUIVALENCES

An *n-ary relation* is a set consisting of ordered n-tuples. Thus an n-ary relation is a set R such that, for some sets $A_1, \ldots, A_n$, $R \subset A_1 \times \ldots \times A_n$.

The ith *domain* of a given relation $R \subset A_1 \times \ldots \times A_n$ is the set

$$D_i(R) = \{x \in A_i : \bigvee_{a_1} \ldots \bigvee_{a_{i-1}} \bigvee_{a_{i+1}} \ldots \bigvee_{a_n} \langle a_1, \ldots, a_{i-1}, x, a_{i+1}, \ldots \ldots, a_n \rangle \in R.\}$$

In particular, if R is binary, then its first domain $D_1(R)$ is called simply the *domain* of R and denoted by $D(R)$, whereas its second domain $D_2(R)$ is called the *counterdomain* of R (or the *range* of R) and denoted by $D^*(R)$.

Thus, given $R \subset A \times B$,

$$D(R) = \{x \in A: \bigvee_{y \in B} \langle x, y \rangle \in R\},$$

$$D^*(R) = \{y \in B: \bigvee_{x \in A} \langle x, y \rangle \in R\}.$$

The set $D(R) \cup D^*(R)$ is called the *field* of relation R.

Instead of writing $\langle x_1, \ldots, x_n \rangle \in R$, we often write $R(x_1, \ldots, x_n)$, and in the case of binary relations the notation $x_1 R x_2$ is often used.

Let us now consider relations $R \subset X^2$ $(X^2 = X \times X)$. Some of those relations are classified as follows:

(a) *reflexive* in X, i.e. such that

$$\bigwedge_{x \in X} \langle x, x \rangle \in R$$

(or $\bigwedge_{x \in X} xRx$, $\bigwedge_{x \in X} R(x, x)$),

(b) *antireflexive* in X, i.e. such that

$$\bigwedge_{x \in X} \langle x, x \rangle \notin R$$

(or $\bigwedge_{x \in X} \sim xRx$, $\bigwedge_{x \in X} \sim R(x, x)$),

(c) *symmetric* in X, i.e. such that

$$\bigwedge_{x\in X}\bigwedge_{y\in X}\langle x, y\rangle \in R \Rightarrow \langle y, x\rangle \in R$$

(or $\bigwedge_{x\in X}\bigwedge_{y\in X} xRy \Rightarrow yRx$, $\bigwedge_{x\in X}\bigwedge_{y\in X} R(x, y) \Rightarrow R(y, x)$),

(d) *antisymmetric* in X (sometimes called *weakly antisymmetric* in X), i.e. such that

$$\bigwedge_{x\in X}\bigwedge_{y\in X}\langle x, y\rangle \in R \wedge \langle y, x\rangle \in R \Rightarrow x=y$$

(or $\bigwedge_{x\in X}\bigwedge_{y\in X} xRy \wedge yRx \Rightarrow x=y$, $\bigwedge_{x\in X}\bigwedge_{y\in X} R(x, y) \wedge R(y, x) \Rightarrow x=y$),

(e) *asymmetric* in X (sometimes called *antisymmetric* in X), i.e. such that

$$\bigwedge_{x\in X}\bigwedge_{y\in X}\langle x, y\rangle \in R \Rightarrow \langle y, x\rangle \notin R$$

(or $\bigwedge_{x\in X}\bigwedge_{y\in X} xRy \rightarrow \sim yRx$, $\bigwedge_{x\in X}\bigwedge_{y\in X} R(x, y) \Rightarrow \sim R(y, x)$),

(f) *transitive* in X, i.e. such that

$$\bigwedge_{x\in X}\bigwedge_{y\in X}\bigwedge_{z\in X}\langle x, y\rangle \in R \wedge \langle y, z\rangle \in R \Rightarrow \langle x, z\rangle \in R$$

(or $\bigwedge_{x\in X}\bigwedge_{y\in X}\bigwedge_{z\in X} xRy \wedge yRz \Rightarrow xRz$, $\bigwedge_{x\in X}\bigwedge_{y\in X}\bigwedge_{z\in X} R(x, y) \wedge R(y, z) \Rightarrow R(x, z)$),

(g) *connected* in X, i.e. such that

$$\bigwedge_{x\in X}\bigwedge_{y\in X}\langle x, y\rangle \in R \vee \langle y, x\rangle \in R \vee x=y$$

(or $\bigwedge_{x\in X}\bigwedge_{y\in X} xRy \vee yRx \vee x=y$, $\bigwedge_{x\in X}\bigwedge_{y\in X} R(x, y) \vee R(y, x) \vee x=y$).

The relation I_x defined as

$$\{\langle x, y\rangle \in X^2 \colon x=y\} = \{\langle x, x\rangle \colon x \in X\}$$

is called an *identity relation* on X.

If $R \subset X^2$ and $X_1 \subset X$, then the relation $R \cap X_1^2$ is called the *restriction of* R *to* X_1 and denoted by $R{\restriction}X_1$.

The relation $R \subset X^2$ is reflexive (antireflexive, symmetric, etc.) if $R{\restriction}X_1$ is reflexive (antireflexive, symmetric, etc.) on the field X_1 of R.

A relation $R \subset X^2$ which is reflexive, symmetric and transitive is called an *equivalence relation*.

A relation $R \subset X^2$ which is reflexive, antisymmetric and transitive is called a *(partial) ordering*.

Find the domains of the following relations (Problems 4.1–4.6).

4.1. $R = \{\langle a, b\rangle, \langle a, c\rangle, \langle b, c\rangle\}$.

4.2. $R = \{\langle a, a\rangle, \langle a, b\rangle, \langle a, c\rangle\}$.

4.3. $R = \{\langle a, b, c\rangle, \langle a, c, b\rangle, \langle a, d, b\rangle\}$.

4.4. $R(a, b) \Leftrightarrow (a \in \mathscr{N} \wedge b \in \mathscr{N} \wedge a < b)$.

4.5. $R(a, b, c) \Leftrightarrow (a \in \mathscr{Z} \wedge b \in \mathscr{Z} \wedge c \in \mathscr{N} \wedge a^2 + b^2 < 10 - c^2)$.

4.6. $R(a, b, c) \Leftrightarrow \left(a \in \mathscr{R} \wedge b \in \mathscr{R} \wedge c \in \mathscr{N} \ldots \bigvee_{x \in \mathscr{R} - \{0\}} \dfrac{a \cdot b}{x} = c + 1\right)$.

4.7. Prove that if, for a certain i, $D_i(R) = O$, then $R = O$.

Assuming that different letters denote different elements, find which of the properties a – g pertain to the following relations $R \subset X^2$ where $X = \{a, b, c, d\}$. If R does not possess any of these properties, check if there exists a nonempty $X_1 \subset X$ such that $R \restriction X_1$ already has a given property (in X_1) (Problems 4.8–4.11):

4.8. $R = \{\langle a, a\rangle, \langle b, b\rangle, \langle a, b\rangle\}$.

4.9. $R = \{\langle a, a\rangle, \langle b, b\rangle, \langle c, c\rangle, \langle d, d\rangle, \langle a, b\rangle, \langle b, a\rangle\}$.

4.10. $R = \{\langle a, b\rangle, \langle b, a\rangle, \langle c, a\rangle, \langle a, c\rangle, \langle c, d\rangle, \langle a, d\rangle\}$.

4.11. $R = \{\langle a, b\rangle, \langle a, c\rangle, \langle b, c\rangle, \langle c, c\rangle, \langle a, a\rangle, \langle b, b\rangle\}$.

4.12. Prove that, for every binary R, $R \subset D(R) \times D^*(R)$.

4.13. Prove that the intersection of any two relations reflexive in X is reflexive in X.

4.14. Show that R is reflexive iff $I_{D(R) \cup D^*(R)} \subset R$.

4.15. Prove that R is antireflexive iff $I_{D(R) \cup D^*(R)} \cap R = O$.

4.16. Show that the intersection and the union of an arbitrary family of relations antireflexive in X are again antireflexive in X.

4.17. Is the union of an arbitrary family of reflexive relations necessarily reflexive?

4.18. Prove that if the relation R is antisymmetric in X then $R - I_X$ is asymmetric in X. Is the converse true?

4.19. Show that if the relation R is asymmetric in X then $R \cup I_X$ is antisymmetric in X. Is the converse true?

4.20. Prove that if $R \subset X^2$ has any of the properties (a)–(g) then $R \restriction D(R) \cup D^*(R)$ has the same properties.

4.21. Prove that if R is asymmetric then it is antirefiexive.

4.22. Show that if $R \subset X^2$ is connected then each of the sets $D(R) - D^*(R)$ and $D^*(R) - D(R)$ is at most a one-element set.

If $R \subset X^2$, then any $R_2 \subset X^2$ such that $R_1 \subset R_2$ is called an *extension* of the relation R_1.

Check if every relation in $R \subset X^2$ can be extended to a relation R' such that (Problems 4.23–4.29):

4.23. R' is reflexive in X^2.

4.24. R' is antireflexive in X^2.

4.25. R' is symmetric in X^2.

4.26. R' is asymmetric in X^2.

4.27. R' is antisymmetric in X^2.

4.28. R' is transitive in X^2.

4.29. R' is connected in X^2.

Given a relation R, we define

$$R^{-1} = \{\langle x, y\rangle : \langle y, x\rangle \in R\}.$$

4.30. Show that R is symmetric iff $R \subset R^{-1}$

4.31. Prove that $R \subset R^{-1} \Leftrightarrow R = R^{-1}$.

4.32. Check if the following are true:

(a) $(R \cup S)^{-1} = R^{-1} \cup S^{-1}$,

(b) $(R \cap S)^{-1} = R^{-1} \cap S^{-1}$,

(c) $I_X^{-1} = I_X$,

(d) $(X^2)^{-1} = X^2$.

4.33. Show, that the relation R is connected in X iff $R \cup R^{-1} \cup I_X = X^2$.

Check if the following statements are true (Problems 4.34–4.38):

4.34. The union of two symmetric relations in X is symmetric in X.

4.35. The intersection of two transitive relations in X is transitive in X.

4.36. The intersection of two connected relations in X is connected in X.

4.37. The union of two connected relations in X is connected in X.

4.38. If R is transitive in X and $R \subset S \subset X^2$, then S is transitive in X.

The *relational product* of the relations R and S (denoted by $R \circ S$) is a relation $T \subset D(R) \times D^*(S)$ defined as follows:

$$\langle x, z \rangle \in T \Leftrightarrow \bigvee_{y} (\langle x, y \rangle \in R \wedge \langle y, z \rangle \in S).$$

4.39. Prove that R is transitive iff $R \circ R \subset R$.

4.40. Show that $I_X \circ I_Y \circ I_{X \cap Y}$.

4.41. Show that $R \circ (S \circ T) = (\circ S) \circ T$.

4.42. Prove that $(R \circ S)^{-1} = S^{-1} \circ R^{-1}$.

4.43. Show that $I_{D(R)} \subset R \circ R^{-1}$, $I_{D^*(R)} \subset R^{-1} \circ R$.

4.44. In this problem we deal with (binary) relations in the set of real numbers $\mathscr{R}$, i.e. the subsets of $\mathscr{R}^2$.

(a) What is the natural interpretation of such relations?

(b) What is the geometric meaning of the sets $D(R)$ and $D^*(R)$?

(c) What is the geometric meaning of the property of reflexivity?

(d) What is the geometric meaning of the property of symmetry?

(e) What is the geometric meaning of the property of connectedness?

(f) What is the geometric meaning of the property of antireflexivity?

(g) What is the geometric meaning of the property of asymmetry?

(h) What is the geometric meaning of the property of antisymmetry?

Check which of the properties (a)–(g), pp. 37–38, hold for the given relation R (Problems 4.45–4.81):

4.45. $R \subset \mathscr{Z}^2 \wedge \bigwedge_{x,y \in \mathscr{Z}} xRy \Leftrightarrow 3 \mid x - y$.

4.46. $R \subset \mathscr{N} \wedge \bigwedge_{x,y \in \mathscr{N}} xRy \Leftrightarrow 2 \mid x + y$.

4.47. $R \subset \mathscr{N}^2 \wedge \bigwedge_{x,y \in \mathscr{N}} xRy \Leftrightarrow x \neq 0 \wedge x \mid y$.

4.48. $R \subset (\mathscr{N} - \{0\})^2 \wedge \bigwedge_{x,y \in \mathscr{N} - \{0\}} xRy \Leftrightarrow (x \mid y \wedge x \neq y)$.

4.49. $R \subset \mathscr{Z}^2 \wedge \bigwedge_{x,y \in \mathscr{Z}} xRy \Leftrightarrow (x = 2 \wedge y = 3)$.

4.50. $R \subset \mathscr{Z}^2 \wedge \bigwedge_{x,y \in \mathscr{Z}} xRy \Leftrightarrow (x = 1 \wedge y = 1)$.

4.51. $R \subset \mathscr{R}^2 \wedge \bigwedge_{x,y \in \mathscr{R}} xRy \Leftrightarrow x^2 = y^2$.

4.52. $R \subset \mathscr{R}^2 \wedge \bigwedge_{x,y \in \mathscr{R}} xRy \Leftrightarrow x^2 \neq y^2$.

4.53. $R \subset \mathscr{R}^2 \wedge \bigwedge_{x,y \in \mathscr{R}} xRy \Leftrightarrow x^3 = y^3$.

4.54. $R\subset\mathscr{R}^2\wedge\bigwedge_{x,y\in\mathscr{R}} xRy\Leftrightarrow x^3=y^2$.

4.55. $R\subset\mathscr{C}^2\wedge\bigwedge_{x,y\in\mathscr{C}} xRy\Leftrightarrow|x|<|y|$.

4.56. $R\subset\mathscr{Z}^2\wedge\bigwedge_{x,y\in\mathscr{Z}} xRy\Leftrightarrow|x|+|y|=3$.

4.57. $R\subset\mathscr{Z}^2\wedge\bigwedge_{x,y\in\mathscr{Z}} xRy\Leftrightarrow|x|+|y|\neq 3$.

4.58. $R\subset\mathscr{Z}^2\wedge\bigwedge_{x,y\in\mathscr{Z}} xRy\Leftrightarrow|x|+|y|\neq 4$.

4.59. $R\subset\mathscr{N}^2\wedge\bigwedge_{x,y\in\mathscr{N}} xRy\Leftrightarrow(x\leqslant 5\wedge y\leqslant 5\wedge x=y)\vee(x>5\wedge y>5\wedge \wedge 2|x+y)$.

4.60. $R\subset\mathscr{N}^2\wedge\bigwedge_{x,y\in\mathscr{N}} xRy\Leftrightarrow(x>y\vee y>x)$.

4.61. $R\subset\mathscr{R}^2\wedge\bigwedge_{x,y\in\mathscr{R}} xRy\Leftrightarrow x-y\in\mathscr{Q}$.

4.62. $R\subset\mathscr{Q}^2\wedge\bigwedge_{x,y\in\mathscr{Q}} xRy\Leftrightarrow x-y\notin\mathscr{Z}$.

4.63. $R\subset\mathscr{R}^2\wedge\bigwedge_{x,y\in\mathscr{R}} xRy\Leftrightarrow x-y\notin\mathscr{N}$.

4.64. $R\subset\mathscr{R}^2\wedge\bigwedge_{x,y\in\mathscr{R}} xRy\Leftrightarrow e^x=2e^y$.

4.65. $R\subset\mathscr{C}^2\wedge\bigwedge_{x,y\in\mathscr{C}} xRy\Leftrightarrow \mathrm{Re}\,x=\mathrm{Im}\,y$.

4.66. $R\subset\mathscr{N}^2\wedge\bigwedge_{x,y\in\mathscr{N}} xRy\Leftrightarrow(3x=2y)$.

4.67. $R\subset\mathscr{Q}^2\wedge\bigwedge_{x,y\in\mathscr{Q}} xRy\Leftrightarrow y=x+2$.

4.68. $R\subset\mathscr{R}^2\wedge\bigwedge_{x,y\in\mathscr{R}} xRy\Leftrightarrow|x-2|=|y+2|$.

4.69. $R\subset\mathscr{C}^2\wedge\bigwedge_{x,y\in\mathscr{C}} xRy\Leftrightarrow\bigvee_{a,b\in\mathscr{N}} x-y=a+bi$.

4.70. $R\subset(\mathscr{Q}^2)^2\wedge\bigwedge_{x,y,z,t\in Q}\langle x,y\rangle R\langle z,t\rangle\Leftrightarrow xt=yz$.

4.71. $R\subset(\mathscr{N}^3)^2\wedge\bigwedge_{x,y,z,t,u,w\in\mathscr{N}}\langle x,y,z\rangle R\langle t,u,w\rangle\Leftrightarrow(x=t\wedge y=w\wedge \wedge z=u)$.

4.72. $R\subset\mathscr{N}^2\wedge\bigwedge_{x,y\in\mathscr{N}} xRy\Leftrightarrow x\cdot y=4$.

***4.73.** $R\subset(\mathscr{P}(\mathscr{N}))^2\wedge\bigwedge_{X,Y\in\mathscr{P}(\mathscr{N})} XRY\Leftrightarrow X\dot{-}Y$ is a finite set.

4.74. Let Par denote the subset of the set of natural numbers consisting of even numbers.

$$R\subset(\mathscr{P}(\mathscr{N}))^2\wedge\bigwedge_{X,Y\in\mathscr{P}(\mathscr{N})} XRY\Leftrightarrow X\cap\mathrm{Par}=Y\cap\mathrm{Par}.$$

4.75. $R \subset (\mathscr{P}(\mathscr{N}))^2 \wedge \bigwedge_{X,Y \in \mathscr{P}(\mathscr{N})} XRY \Leftrightarrow X \cap Y \subset \mathscr{N} - \text{Par}.$

4.76. Let X be a set. A family $I \subset \mathscr{P}(X)$ is called an *ideal* iff:

1° $\quad O \in I,$

2° $\quad Z \in I \wedge Y \in I \Leftrightarrow Z \cup Y \in I.$

Given an ideal $I \subset \mathscr{P}(X)$, we define $R \subset (\mathscr{P}(X))^2$ as follows:

$$ZRY \Leftrightarrow Z \dot{-} Y \in I.$$

4.77. Let T be the set of all convergent sequences with rational terms, $\{x_n\} R \{y_n\} \Leftrightarrow \lim_{n\to\infty} \{x_n\} > \lim_{n\to\infty} \{y_n\}.$

4.78. Let the field of the relation R be as in 4.77;

$$\{x_n\} R \{y_n\} \Leftrightarrow \lim_{n\to\infty} x_n \cdot y_n = 7.$$

4.79. Let the field of the relation R be as in 4.77;

$$\{x_n\} R \{y_n\} \Leftrightarrow \sum_{n=1}^{\infty} \frac{1}{2^n} x_n = \sum_{n=1}^{\infty} \frac{1}{2^n} y_n.$$

4.80. Given a set X consider the family of all binary relations on X and a relation R on that family defined as follows:

$$ARB \Leftrightarrow D(A) \subset D(B).$$

4.81. Consider sets X and Y and an ideal $I \subset P(X)$ such that $X \notin I$. In the set ${}^X Y$ of all functions with the domain X and with values in Y define a relation R as follows:

$$fRg \Leftrightarrow \{x \in X : f(x) \neq g(x)\} \in I.$$

A *partition* of the set X is any family $\{X_i : i \in I\}$ of nonempty subsets of X satisfying the following conditions:

(a) $\quad \bigwedge_{i,j \in I} (i \neq j \Leftrightarrow X_i \cap X_j = O),$

(b) $\quad \bigcup_{i \in I} X_i = X.$

Every equivalence $R \subset X^2$ determines a partition of the set X consisting of so-called *abstraction classes* (or *equivalence classes*), The abstraction class of an element x, $[x]_R$ is the set $\{y \colon yRx\}$.

4.82. Prove that whenever R is an equivalence then:

(a) $[x]_R = [y]_R \Leftrightarrow xRy$,

(b) $[x]_R \cap [y]_R \neq O \Leftrightarrow xRy$.

4.83. Prove that every partition $\mathscr{X} = \{X_i : i \notin I\}$ of a set X determines the following equivalence

$$xR_{\mathscr{X}}\, y \Leftrightarrow \bigvee_i (x \in X_i \wedge y \in X_i).$$

4.84. Show that if the family $\mathscr{X}$ does not satisfy condition b) of the definition of partition, then $R_{\mathscr{X}}$ is not an equivalence.

4.85. Prove that the condition a) is not a necessary and sufficient condition for the family $\mathscr{X}$ (satisfying condition b) to determine the equivalence (as in 4.83). Find the appropriate weakened version of a).

If the equivalence R determines the partition $\mathscr{X}$ into equivalence classes and $\mathscr{X}$ determines the equivalence R_1, then $R = R_1$. Similarly the partition $\mathscr{X}_1$ determined by the relation $R_{\mathscr{X}}$ is again equal to $\mathscr{X}$.

For a given set X and a relation $R \subset X^2$ check if R is an equivalence. If it is, find the equivalence classes of R (Problems 4.86–4.123).

4.86. $X = \mathrm{Par}$, $xRy \Leftrightarrow 3 \mid x - y$.

4.87. $X = \mathscr{C}$, $z_1 R z_2 \Leftrightarrow \mathrm{Re}\, z_1 = \mathrm{Re}\, z_2$.

4.88. $X = \mathscr{R}$, $xRy \Leftrightarrow x - y = 2$.

4.89. $X = \mathscr{N}$, $xRy \Leftrightarrow 2 \mid x + y$.

4.90. $X = \mathscr{Z}$, $xRy \Leftrightarrow x^2 \leqslant y^2$.

4.91. $X = \{1, 2, 3\}$, $xRy \Leftrightarrow x + y \neq 3$.

4.92. $X = \mathscr{R}[t]$, $xRy \Leftrightarrow \bigvee_{a,b,c} (x - y = at^2 + bt + c)$.

4.93. $X = \mathscr{Z}[t]$, $xRy \Leftrightarrow$ the difference $x - y$ has even coefficients.

4.94. $X = \mathscr{C}_\infty$, $xRy \Leftrightarrow \bigvee_{z \in \mathscr{R}} (z \neq 0 \wedge xz = y)$.

4.95. $X = \mathscr{R}$, $xRy \Leftrightarrow \bigvee_{a \in \mathscr{R}} (x + yi)^2 = ai$.

4.96. $X = \mathscr{C}_\infty[a, b]$, $fRg \Leftrightarrow \bigvee_{k,n \in \mathscr{N}} (f^{(n)} = g^{(k)})$ (remember $f^{(n)}$ is the nth derivative of f).

4.97. $X = \mathscr{Q}[t]$, $xRy \Leftrightarrow \bigvee_{a,b \in \mathscr{Q}} (x - y = at + b)$.

4.98. $X = \mathscr{C}$, $xRy \Leftrightarrow x + y \in \mathscr{R}$.

4.99. $X = \{1, 2, \ldots, 16\}$, $xRy \Leftrightarrow 4 \mid x^2 - y^2$.

4.100. X is the set of all convergent sequences with rational terms, $\{x_n\}_{n\in\mathcal{N}} R \{y_n\}_{n\in\mathcal{N}} \Leftrightarrow \lim x_n = \lim y_n$.

4.101. X as in 4.100,

$$\{x_n\}_{n\in\mathcal{N}} R \{y_n\}_{n\in\mathcal{N}} \Leftrightarrow \{x_n - y_n\}_{n\in\mathcal{N}} \text{ converges to } 0.$$

4.102. $X=\mathcal{N}$, k a fixed natural number greater than 2, $xRy \Leftrightarrow k|x+y$.

4.103. X and k as in 4.102, $xRy \Leftrightarrow k|x-y$.

4.104. X – the set of 2×2 matrices with real entries. Det A is the determinant of A; $ARB \Leftrightarrow \text{Det } A = \text{Det } B$.

4.105. X as in 4.104, $I=\begin{bmatrix}1 & 0\\ 0 & 1\end{bmatrix}$, $ARB \Leftrightarrow \bigvee_{k\in R} A-B=kI$.

4.106. $X=\mathscr{C}[t]$, $fRg \Leftrightarrow f-g \in \mathscr{R}[t]$.

4.107. $X=\mathscr{R}[t]$, $fRg \Leftrightarrow$ the degree of $f-g$ is odd.

4.108. $X=\mathscr{R}[t]$, $fRg \Leftrightarrow$ the degree of $f-g$ is even.

4.109. $X=\mathcal{N}-\{0\}$, $xRy \Leftrightarrow$ $x\cdot y$ is odd.

4.110. $X=\mathcal{N}-\{0\}$, $xRy \Leftrightarrow \bigvee_{t\in\mathcal{N}} xy=t^2$.

4.111. $X=\mathcal{N}\times(\mathcal{N}-\{0\})$, $\langle r,s\rangle R \langle t,u\rangle \Leftrightarrow ru=st$.

4.112. $X=\mathcal{N}\times\mathcal{N}$, $\langle r,s\rangle R \langle t,u\rangle \Leftrightarrow r+u=s+t$.

4.113. $X=\mathscr{Z}\times(\mathscr{Z}-\{0\})$, $\langle r,s\rangle R \langle t,u\rangle \Rightarrow ru=st$.

4.114. $X=\mathscr{R}$, $xRy \Leftrightarrow x-y \in \mathcal{N}$.

4.115. $X=\mathcal{N}$, $xRy \Leftrightarrow (x\in\text{Par} \wedge y\in\text{Par} \wedge x=y) \vee$
$\vee (x\notin\text{Par} \wedge y\notin\text{Par} \wedge 3\,|\,x-y)$.

4.116. $X=\mathcal{N}$, $xRy \Leftrightarrow (x\in\text{Par} \wedge y\in\text{Par} \wedge 3\,|\,x-y) \vee$
$\vee (x\notin\text{Par} \wedge y\notin\text{Par} \wedge 5\,|\,x-y)$.

4.117. $X=\mathscr{P}(Y)$ (where Y is a fixed set with at least two elements), $ARB \Leftrightarrow A\subset B \vee B\subset A$.

4.118. $X=\mathscr{P}(Y)$, a is a fixed element of the set Y, $ARB \Leftrightarrow a\notin(A\cup B)$.

4.119. $X=\mathscr{Z}$, $xRy \Leftrightarrow (|x|<5 \wedge |y|<5 \wedge x=y) \vee (|x|\geqslant 5 \wedge$
$\wedge |y|\geqslant 5 \wedge 2|x-y)$.

4.120. $X=\mathscr{P}(Y)$, C is a fixed subset of the set Y, $ARB \Leftrightarrow A \dot{-} B \subset C$.

4.121. $X=\mathscr{R}[t]-\{0\}$, $xRy \Leftrightarrow x\cdot y$ is of even degree.

4.122. $X=\mathscr{Q}[t]$, $fRg \Leftrightarrow fg$ is of odd degree.

4.123. $X=\mathcal{N}$, $xRy \Leftrightarrow \bigvee_{k,l\in\mathcal{N}} (k>0 \wedge l>0 \wedge x^k=y^l)$.

* **4.124.** Given sets X_1, X_2 and equivalences $R_1\subset X_1^2$ and $R_2\subset X_2^2$. In the set $X_1\times X_2$ a relation S is defined as follows:

$$\langle x_1, x_2\rangle S \langle y_1, y_2\rangle \Leftrightarrow [(x_1 R_1 y_1)\wedge(x_2 R_2 y_2)].$$

Is S an equivalence? If so, show the equivalence classes of S.

4.125. Given equivalences $R_1, R_2 \subset X^2$, check if:

(a) $R_1 \cap R_2$ is an equivalence,

(b) $R_1 \cup R_2$ is an equivalence,

(c) $\sim R_1 = X^2 - R_1$ is an equivalence.

If so, find dependence of the new partition on the partitions determined by R_1 and R_2, respectively.

4.126. Given family $\{R_t\}_{t \in T}$ of equivalences in a set X, check if:

(a) $\bigcap_{t \in T} R_t$ is an equivalence,

(b) $\bigcup_{t \in T} R_t$ is an equivalence.

For either relation, if it is not an equivalence, find an appropriate counterexample.

4.127. Given a mapping $f: X \to Y$, define in the set X a relation $\sim_f$ as follows:

$$x \sim_f y \Leftrightarrow f(x) = f(y).$$

(a) Prove that $\sim_f$ is an equivalence.

(b) Find a necessary and sufficient condition for the relation $\sim_f$ to be an identity.

4.128. Given a partition of the set $\mathscr{R}$ into half-closed segments $\{\langle x, x+1): x \in \mathscr{Z}\}$, find an equivalence for which it is the determined partition.

4.129. Given the partition of a set $\mathscr{Z}$ into the sets of even and odd numbers, find an equivalence for which it is the determined partition.

4.130. Divide a Cartesian plane $\mathscr{R}^2$ into five sets as follows: open quarts and the set theoretical union of axes. Find an equivalence for which it is the determined partition.

4.131. Divide a Cartesian plane $\mathscr{R}^2$ into circles with centers at the origin. Find an equivalence for which it is the determined partition.

4.132. Given partitions $\mathscr{A} = \{A_t\}_{t \in T}$, $\mathscr{B} = \{B_s\}_{s \in S}$ of a set X, assume

$$\bigwedge_{t \in T} \bigvee_{s \in S} A_t \subset B_s.$$

What is the relationship between the corresponding equivalences?

4.133. Given equivalences R_1 and R_2 on the set X such that $R_1 \subset R_2$, i.e.

$$\bigwedge_{x,y} \langle x, y \rangle \in R_1 \Rightarrow \langle x, y \rangle \in R_2 .$$

What is the relationship between determined partitions?

A binary relation can be represented in a useful and instructive way by means of so-called *diagrams*. They are produced as follows: Let $R \subset A \times B$ and assume that A, B are finite. The elements of the sets A and B are represented by points on the Cartesian plane and we assume that the elements of A are denoted by dots $(\cdot)$ whereas the elements of B by crosses $(+)$. In particular, when $A = B$ all the points are denoted by dots. If $R(a, b)$ holds, then we draw an arrow from a to b. Some properties of R can easily be reconstructed from its diagram.

4.134. Draw the diagram of the relation $R \subset A^2$ (where $A = \{0, 1, 2\}$) defined as follows: $xRy \Leftrightarrow x < y$.

4.135. Draw the diagram of $R \subset A^2$ where $A = \{1, 2, ..., 10\}$ and $xRy \Leftrightarrow x | y \wedge x \neq y$.

4.136. Draw the diagram of $R \subset A^2$ where $A = \{1, 2, 3, 4\}$ and $xRy \Leftrightarrow 2 | x + y$.

Find the properties of the diagrams of (Problems 4.137–4.143):

4.137. A reflexive relation.

4.138. A transitive relation.

4.139. A weakly antisymmetric relation.

4.140. A symmetric relation.

4.141. A connected relation.

4.142. An antireflexive relation.

4.143. An asymmetric relation.

If $R \subset A^2$ is a reflexive and transitive relation, then we can simplify the diagram by erasing all arrows having the same origin and end-point. Also, given arrows from a to b and from b to c, we erase the arrow from a to c, etc.

The relation xRy can be deduced from the diagram as the possibility of passing from x to y according to the arrows.

In problems 4.144–4.150 we discuss only such simplified diagrams of reflexive and transitive relations.

Find what properties hold for the diagrams of (Problems 4.144–4.146):

4.144. A symmetric relation.

4.145. A weakly antisymmetric relation.

4.146. A weakly antisymmetric and connected relation.

Find the properties of the point corresponding to (Problems 4.147–4.150):

4.147. A minimal element of partial ordering.

4.148. A maximal element of partial ordering.

4.149. The maximum element of partial ordering.

4.150. The minimum element of partial ordering.

CHAPTER 5

FUNCTIONS

A relation $R \subset X \times Y$ is called a *function* if it satisfies the following condition:

$$\bigwedge_{x \in X} \bigwedge_{y \in Y} \bigwedge_{z \in Y} (\langle x, y \rangle \in R \wedge \langle x, z \rangle \in R \Rightarrow y = z).$$

Traditionally, we use the letters f, g, h (possibly with indices) to denote functions and write $f(x) = y$ instead of $\langle x, y \rangle \in f$.

The *domain* (the *set of arguments*) of the function f is the set Df defined by the condition:

$$x \in Df \Leftrightarrow \bigvee_{y} (\langle x, y \rangle \in f),$$

or equivalently

$$x \in Df \Leftrightarrow \bigvee_{y} (f(x) = y).$$

The *counterdomain* (*codomain*, the *set of values*) of the function f is the set Rf defined by condition:

$$y \in Rf \Leftrightarrow \bigvee_{x} (\langle x, y \rangle \in f),$$

or equivalently

$$y \in Rf \Leftrightarrow \bigvee_{x} (f(x) = y).$$

A *mapping* of a set X into set Y is a function f such that $Df = X$ and $Rf \subset Y$. This situation (f maps X into Y) is usually denoted by $f\colon X \to Y$. The set of all mappings of X into Y is denoted by $\overset{X}{\smile}Y$.

Thus, if $f \subset X \times Y$ is a function, then f maps Df into Y.

If $Df = X$, then we say that f is *total* on X. Otherwise, we say that f is a *partial* mapping on X. Thus every function $f \subset X \times Y$ is a partial mapping on X, and every partial function is total on its domain. The set of all partial functions on X with values in Y is denoted by $\overset{X}{\smile}Y$. Thus

$$f \in \overset{X}{\smile}Y \Leftrightarrow \bigvee_{X_1 \subset X} f \in {}^{X_1}Y.$$

Given mappings $f: X\to Y$ and $g: Y\to Z$, we define the mapping $g\circ f: X\to Z$ as follows:

$$\langle x, z\rangle \in g\circ f \Leftrightarrow \bigvee_{y} (\langle x, y\rangle \in f \wedge \langle y, z\rangle \in g)$$
$$\Leftrightarrow \bigvee_{y} [y=f(x) \wedge z=g(y)].$$

The mapping $g\circ f$ is called the *superposition* of f and g.

The operation of superposition is associative, i.e. $h\circ(g\circ f)=(h\circ g)\circ \circ f$. Let us note, however, that it is not commutative.

If the set X is included in Df, then the *image* of the set X (denoted by $f{*}X$) is the set of values of the function f on elements of, X i.e.

$$f{*}X=\{y: \bigvee_{x\in X} f(x)=y\}^{1}.$$

If f maps X into Y (i.e. $f: X\to Y$), then $f^{-1} * Z$, for $Z\subset Y$, is defined as $\{x: f(x)\in Z\}$. The set $f^{-1} * Z$ is called the *counterimage* of Z by f.

5.1. Prove that, according to our definition, if $f: X\to Y$, $A\subset X$, $A\neq O$ then $f * A\neq O$.

5.2. Show an example of nonempty sets X, Y, A and a mapping $f: X\to Y$ such that $A\subset Y$, $A\neq O$ but $f^{-1} * A=O$.

The notion of a mapping (function) is restricted in different ways.

An *injection* (*one-to-one mapping, one-to-one function*) is a mapping (function) satisfying the condition:

$$\bigwedge_{x}\bigwedge_{y} [f(x)=f(y)\Rightarrow x=y]$$

or equivalently

$$\bigwedge_{x}\bigwedge_{y} [x\neq y\Rightarrow f(x)\neq f(y)].$$

We then write $f: X\xrightarrow{1-1} Y$.

Another important class of mappings consists of so called *surjections* (*mappings "onto"*). They are mappings $f: X\to Y$ satisfying the condition $Rf=Y$. Thus $f: X\to Y$ is a surjection iff

$$\bigwedge_{y\in Y}\bigvee_{x\in X} (f(x)=y).$$

We then write $f: X\xrightarrow[\text{onto}]{} Y$.

Let us note that f always maps its domain onto set of values. A mapping that is simultaneously an injection and a surjection is called a *bijection*. In this case we write $f: X\xrightarrow[\text{onto}]{1\text{-}1} Y$. If f is one-to-one, then the relation f^{-1} is also a function. We call it an *inverse function*. If $f\colon X \xrightarrow[\text{onto}]{1-1} Y$ then $f^{-1}\colon Y \xrightarrow[\text{onto}]{1-1} X$.

A function of n variables, $f(x_1, \ldots, x_n)$ is a function defined on a subset of the Cartesian product $X_1 \times \ldots \times X_n$.

5.3. Prove that the relation I_X is a mapping of X into X which is both an injection and a surjection and thus $I_X\colon X \xrightarrow[\text{onto}]{1-1} X$.

5.4. Check if the relation $R \subset \mathscr{R} \times \mathscr{R}$ defined by $\langle x, y\rangle \in R \Leftrightarrow x^2 = y^2$ is a function.

5.5. Check if the relation $R \subset \mathscr{R}^+ \times \mathscr{R}^+$ defined by $\langle x, y\rangle \in R \Leftrightarrow x^2 = y^2$ is a function.

5.6. Check if the relation $R \subset \mathscr{N} \times \mathscr{Z}$ defined by $\langle x, y\rangle \in R \Leftrightarrow x^2 = y^3$ is a function.

5.7. Check if the relation $R \subset \mathscr{N} \times \mathscr{Z}$ defined by $\langle x, y\rangle \in R \Leftrightarrow x^3 = y^2$ is a function.

5.8. Check if the relation $R \subset \mathrm{C} \times \mathrm{C}$ defined by $\langle x, y\rangle \in R \Leftrightarrow \operatorname{Im} x = \operatorname{Re} y$ is a function.

5.9. In this problem we investigate relations $R \subset \mathscr{R} \times \mathscr{R}$ (thus naturally interpreted as subsets of a Cartesian plane).

(a) Give the geometrical condition for the relation R to be a function.

(b) What is the geometrical sense of the fact that R^{-1} is a function?

(c) What is the geometrical sense of the fact that R is a one-to-one function?

(d) What is the geometrical interpretation of the following sentence: For every relation R there exists a relation S such that $S \subset R$, $DS = DR$ and S is a function.

For a given mapping $f\colon \mathscr{R} \to \mathscr{R}$ check if f is an injection and if f is a surjection. If f is not a surjection, find Rf. If f is not an injection, find x_1, x_2 such that $x_1 \neq x_2$ but $f(x_1) = f(x_2)$ (Problems 5.10–5.24):

5.10. $f(x) = 2^x$.

5.11. $f(x) = x^3$.

5.12. $f(x) = E[x]$.

5.13. $f(x) = \begin{cases} \dfrac{2x+1}{x-1} & \text{if } x \neq 1, \\ 0 & \text{if } x = 1. \end{cases}$

5.14. $f(x) = 2^x + x$.

5.15. $f(x) = 3^x - 2^x$.

5.16. $f(x)=\dfrac{2x}{x^2+1}$.

5.17. $f(x)=x^3-x^2$.

5.18. $f(x)=\begin{cases} x\ln|x| & \text{if} \quad x\neq 0, \\ 0 & \text{if} \quad x=0. \end{cases}$

5.19. $f(x)=\begin{cases} \sqrt{x+1} & \text{if} \quad x\in\mathcal{R}^+, \\ 2x & \text{if} \quad x\in\mathcal{R}-\mathcal{R}^+. \end{cases}$

5.20. $f(x)=x^2$.

5.21. $f(x)=x^4-5x^2+4$.

5.22. $f(x)=x2^{x-1}$.

5.23. $f(x)=\sin x$.

5.24. $f(x)=\begin{cases} \dfrac{x}{x+1} & \text{if } x\neq -1, \\ 1 & \text{if } x=-1. \end{cases}$

5.25. Let $f: X\to X$ be an injection of X into X and g a mapping of X into X. Let $f\circ g=f$. Show that $g=I_X$. Show that the assumption that f is an injection cannot be omitted.

5.26. Let $f: X\to X$ be a surjection of X onto X and g a mapping of X into X. Let $g\circ f=f$. Show that $g=I_X$. Show that the assumption that f is a surjection cannot be omitted.

5.27. Prove that if $f: X\to Y$ and $g: Y\to X$ and $g\circ f=I_X$ then f is an injection and g is a surjection.

Let $f: X\to Y$. A mapping $g: Y\to X$ is called *left* (*right*) *inverse* for f if $g\circ f=I_X$ ($f\circ g=I_X$).

5.28. Prove that if f maps X onto Y and X is nonempty then:

(a) f possesses a left inverse iff f is an injection,

(b) f possesses a right inverse iff f is a surjection.

5.29. Let $f: X\to Y$, $g_1: Y\to X$, $g_2: Y\to X$ be mappings such that

$$f\circ g_1=f\circ g_2=I_Y,\ g_1\circ f=g_2\circ f=I_X.$$

Prove that $g_1=g_2$.

5.30. Let $f: X\to Y$ and $g: Y\to Z$ be injections. Prove that $g\circ f: X\to Z$ is also an injection.

5.31. Assume that $f: X\to Y$ and $g: Y\to Z$ are mappings such that $g\circ f$ is an injection and f is a surjection. Show that g is an injection.

Find an example showing that if f is not a surjection then g need not be an injection.

5.32. Assume that $f: X \to Y$ and $g: Y \to Z$ are mappings such that $g \circ f$ is a surjection and g is an injection. Show that f is a surjection. Find an example showing that if g is not an injection then f need not be a surjection.

5.33. Give an example of a set X and a mapping $f: X \to X$ which is a surjection and is not an injection. Can such an example be found among finite sets?

5.34. Give an example of a set X and a mapping $f: X \to X$ which is an injection but not a surjection. Can such an example be found among finite sets?

5.35. Prove that, whenever X is a finite set, $f: X \to X$, then f is an injection iff f is a surjection.

5.36. Prove that $f: X \to Y$ is a surjection iff the counterimage of every one-element set $\{y\}$ ($y \in Y$) is nonempty.

5.37. Prove that $f: X \to Y$ is an injection iff the counterimage of every one-element set $\{y\}$ ($y \in Y$) is at most a one-element set.

5.38. Let $a \neq b$, $X = \{a, b, \{a, b\}\}$, $Y = \{a, b\}$, $f: X \to Y$ be a mapping defined as follows:

$$f(a) = f(b) = a, \quad f(\{a, b\}) = b.$$

Find $f * \{a, b\}$. Interprete the result.

5.39. Let $f: \mathcal{N} \to \mathcal{N}$ be defined by $f(n) = n^2 + 1$. Is f a surjection? Is f an injection?

5.40. Let $f: \mathcal{R} \to \mathcal{R}^+$ be a mapping defined by $f(x) = 2^x$. Prove that there is a mapping inverse to f. How can it be expressed?

5.41. Let $f: \mathcal{R} \to \mathcal{R}$ be a mapping defined by $f(x) = 2^x$. Investigate if the mapping inverse to f exists. Interpret the answer in detail.

The *image of the element* $x \in X$ is the image of the set $\{x\}$.

5.42. Let $X = \mathcal{R}_n[t]$, n being a given natural number $\geqslant 4$. Let φ: $X \to X$ be a mapping defined as follows:

$$\varphi(x(t)) = x(2t).$$

(a) Check if φ is an injection.

(b) Check if φ is a surjection.

(c) Find the image of the set of constants (i.e. of $\mathcal{R}_0(t]$).

(d) Find the counterimage of $\mathcal{R}_2[t]$.

5.43. Let $X=C_\infty\langle 0, 1\rangle$ and let $\varphi: X\to X$ be a mapping defined by $\varphi(f)=f'-1$.

(a) Check if φ is a surjection.
(b) Check if φ is an injection.
(c) Find the image of the set of constant functions.
(d) Find the image of $\mathscr{R}[t]$.
(e) Find the counterimage of $\mathscr{Q}[t]$.
(f) Find the counterimage of the set $\{\sin x, \cos x\}$.
(g) Find $\varphi^{-1} * (\varphi * \{e^x\})$.

5.44. Let $f: \mathscr{R}\to\mathscr{R}$ be defined by $f(x)=x^2-3x+2$.

(a) Find $f*\langle 0, 1\rangle$.
(b) Find $f*\langle -2, -1\rangle$.
(c) Find $f^{-1}*(-\infty, -6\rangle$.
(d) Find $f^{-1}*\{-3, -4\}$.
(e) Find $f*\{1, 2\}$.

5.45. Let $f: \mathscr{R}\to\mathscr{R}$ be defined by $f(x)=\sin x+1$.

(a) Find $f*\langle 0, \frac{3}{2}\pi\rangle$.
(b) Find $f*\{0, \pi\}$.
(c) Find $f*\{\frac{1}{2}\pi, \frac{1}{4}\pi, \frac{1}{6}\pi\}$.
(d) Find $f^{-1}*(\frac{1}{2}, \infty)$.
(e) Find $f^{-1}*(-\infty, 1\rangle$.
(f) Find $f^{-1}*\{0\}$.

5.46. Let $f: \mathscr{R}[t]\to\mathscr{R}[t]$ be the mapping defined by $f(\varphi(t))=\varphi^2(t)$.

(a) Find the image of the set of constant polynomials (i.e. $f*\mathscr{R}_0[t]$).
(b) Find $f^{-1}*\mathscr{R}_4[t]$.
(c) Find $f^{-1}*\{t^2+2t+1\}$.
(d) Find $f^{-1}*\{t^2+2t+2\}$.
(e) Find $f^{-1}*(f*\{(t-1), (t^2-1)\})$.

5.47. Let $f: \mathscr{R}[t]\to\mathscr{C}$ be the mapping defined by $f(\varphi)=\varphi(i)$.

(a) Is f an injection?
(b) Is f a surjection?
(c) Find the counterimage of the number 0 (i.e. $f^{-1}*\{0\}$).
(d) Find $f^{-1}*\mathscr{R}$.
(e) Find $f^{-1}*(f*\mathscr{R}[t])$.

5.48. Let $f: \mathscr{R}[t]\times(\mathscr{N}-\{0\})\to\mathscr{R}[t]$ be the mapping defined by $f\langle\varphi, n\rangle=n\varphi$.

(a) Is f a surjection?

(b) Find $f*(\mathscr{R}_3[t]\times\{2\})$.

(c) Find $f^{-1}*\{0\}$.

(d) Find the counterimage of the polynomial t^2+t+1.

5.49. Let X be the set of all roots of unity, i.e. of all complex numbers t such that for a certain $n\in\mathscr{N}-\{0\}$, $t^n=1$. Let k be a fixed natural number >1, and let φ: $X\to X$ be a mapping defined by $\varphi(t)=t^k$.

(a) Is φ a surjection?

(b) Is φ an injection?

(c) Find $\varphi^{-1}*\{1\}$.

(d) Find the image of the set of roots of degree $2k$.

(e) Find the counterimage of the set $\{1, -1, i, -i\}$.

5.50. Let X be the set of all $n\times n$ matrices with real entries, $n>1$, and let φ: $X\to\mathscr{R}$ be the mapping defined by $\varphi(T)=\text{Det } T$ (where Det T is the determinant of T).

(a) Is φ a surjection?

(b) Is φ an injection?

(c) Find the image of the set of all matrices T such that $t_{11}=0$.

5.51. Let X be a vector space $\mathscr{R}^3$ and let a mapping φ: $X\to X$ be defined by $\varphi(x)=x\times[0, 0, 1]$ (vectorial product).

(a) Is φ a surjection?

(b) Is φ an injection?

(c) Find $\varphi^{-1}*\{[0,0,0]\}$.

(d) Find the image of the set of all vectors of the form $[0, t, 0]$($t\in\mathscr{R}$).

(e) Find $\varphi^{-1}*\{[0,0,t]: t\in\mathscr{R}-\{0\}]$.

5.52. Let X be a vector space $\mathscr{R}^n$ where $n>0$ and let the mapping φ: $X\to\mathscr{R}$ be defined by:

$$\varphi([x_0, \ldots, x_{n-1}])=\sum_{k=0}^{n-1} x_k^2.$$

(a) Is φ a surjection?

(b) Is φ an injection?

(c) Find $\varphi^{-1}*\{-1\}$.

(d) Find $\varphi^{-1}*\{0\}$.

(e) Find $\varphi^{-1}*\{1\}$.

5.53. Let φ: $\mathscr{R}\to\mathscr{L}$ be a mapping defined by $\varphi(x)=E[x]+1$.

(a) Is φ a surjection?

(b) Find $\varphi^{-1}*\{0\}$.

(c) Find $\varphi^{-1}*(\mathscr{Z}-\mathscr{N})$.
(d) Find $\varphi*\mathscr{R}^{+}$.
(e) Find $\varphi*\{\sqrt{2}, 2, 2\sqrt{2}\}$.
5.54. Let $\varphi\colon \mathscr{N}^2\to\mathscr{N}$ be a mapping defined by $\varphi(\langle n, k\rangle)=n+k+1$.
(a) Is φ a surjection?
(b) Is φ an injection?
(c) Find $\varphi*(\mathscr{N}\times\{1\})$.
(d) Find $\varphi^{-1}*\{0\}$.
(e) Find $\varphi^{-1}*\text{Par}$.
5.55. Let $\varphi\colon \mathscr{N}^2\to\mathscr{N}$ be a mapping defined by $\varphi(\langle n, k\rangle)=nk$.
(a) Is f a surjection?
(b) Find $\varphi*(\mathscr{N}\times\{2\})$.
(c) Find $\varphi^{-1}*\{0\}$.
(d) Find $\varphi^{-1}*\text{Par}$.
(e) Find $\varphi^{-1}*\{2^n : n\in\mathscr{N}\}$.
5.56. Let $\varphi\colon \mathscr{Z}^2\to\mathscr{Z}$ be a mapping defined by $\varphi(\langle n, k\rangle)=n^2k$.
(a) Is φ a surjection?
(b) Is φ an injection?
(c) Find $\varphi^{-1}*\{0\}$.
(d) Find $\varphi^{-1}*\{1\}$.
(e) Find $\varphi^{-1}*\mathscr{N}$.
(f) Find $\varphi*(\mathscr{Z}\times\{1\})$.
5.57. Let $f\colon \mathscr{R}^2\to\mathscr{R}^2$ be a mapping defined by $\varphi(\langle x, y\rangle)$
$=\langle x+y, x-y\rangle$.
(a) Is φ a surjection?
(b) Is φ an injection?
(c) Find the image of the line $L\colon y=x+1$.
(d) Find $\varphi*\{\langle 0, 0\rangle\}$.
5.58. Let $f\colon \mathscr{N}^2\to\mathscr{N}$ be a mapping defined by $f(\langle x, y\rangle)=x^2+y^2$.
(a) Is f a surjection?
(b) Is f an injection?
(c) Find $f^{-1}*\{0\}$.
(d) Find $f^{-1}*\{24\}$.
5.59. Let $f\colon \mathscr{N}^2\to\mathscr{N}$ be a mapping defined by $f(\langle x, y\rangle)=|x^2-y^2|$.
(a) Is f a surjection?
(b) Is f an injection?
(c) Find $f^{-1}*\{0\}$.

(d) Find $f*[(\mathscr{N}-\mathrm{Par})\times\mathrm{Par}]$.

(e) Find $f*(\mathscr{N}\times\{0\})$.

5.60. Let f: $\mathscr{N}^2\to\mathscr{N}$ be a mapping defined by $f(\langle x, y\rangle)=\max(x, y)$.

(a) Is f a surjection?

(b) Is f an injection?

(c) Find $f^{-1}*\{0\}$.

(d) Find $f^{-1}*\{k\}$.

(e) Find $f*(\mathrm{Par}\times\mathscr{N})$.

5.61. Let f: $\mathscr{C}\to\mathscr{C}$ be a mapping defined by $f(x+yi)=x-yi$.

(a) Is f an injection?

(b) Is f a surjection?

(c) Find $f*\mathscr{R}$.

(d) Find $f^{-1}*\{x+i\colon x\in\mathscr{R}^+\}$. Let X be an arbitrary subset of the set $\mathscr{C}$. Find $f^{-1}*(f*X)$.

5.62. Let f: $\mathscr{R}\to\mathscr{R}$ be a mapping defined by $f(x)=x^2+x-2$.

(a) Find $f*\mathscr{R}$.

(b) Find $f*\mathscr{R}^+$.

(c) Find $f^{-1}*(\mathscr{R}-\mathscr{R}^+)$.

(d) Find $f^{-1}*\langle -1, 0\rangle$.

(e) Find $f^{-1}*\{0\}$.

5.63. Let f: $\mathscr{R}\to\mathscr{R}$ be a mapping defined by $f(x)=x^2-5x+4$.

(a) Find $f*\mathscr{R}$.

(b) Find $f*(\mathscr{R}-\mathscr{R}^+)$.

(c) Find $f^{-1}*\mathscr{R}^+$.

(d) Find $f^{-1}*\langle 2, 3\rangle$.

(e) Find $f^{-1}*\{0\}$.

5.64. Let f: $\mathscr{R}\to\mathscr{R}$ be a mapping defined by $f(x)=2^{|x|}$.

(a) Find $f*\mathscr{R}$.

(b) Find $f^{-1}*\langle 0, 1\rangle$.

(c) Find $f*\langle 0, 1)$.

(d) Find $f^{-1}*\langle 2, 3\rangle$.

(e) Find $f^{-1}*\{4\}$.

5.65. Let f: $\mathscr{R}\to\langle 0, 1\rangle$ be a mapping defined by $f(x)=x-E(x)$.

(a) Is f a surjection?

(b) Find $f^{-1}*\{0\}$.

(c) Find $f^{-1}*\langle 0, 1)$.

(d) Find $f^{-1}*(0, 1)$.

5.66. Let f: $\mathscr{R} \to \mathscr{R}$ be a mapping defined by $f(x) = x^3 - x^2 - x + 1$.
(a) Is f a surjection?
(b) Find $f * \langle 3, \infty)$.
(c) Find $f^{-1} * \mathscr{R}$.
(d) Find $f^{-1} * \{0\}$.
(e) Find $f * \{-1, 2\}$.
5.67. Let f: $Q \to Q$ be a mapping defined by $f(x) = x^2 + 1$.
(a) Is f a surjection?
(b) Is $\langle 1, 2\rangle \cap Q$ included in $f * \mathscr{Q}$?
(c) Find $f^{-1} * \{\frac{13}{9}\}$.
(d) Find $f^{-1} * \{1\}$.
(e) Find $f^{-1} * \langle \frac{10}{9}, 2\rangle$.
5.68. Let f: $\mathscr{R} \to \mathscr{R}^+$ be a mapping defined by $f(x) = |x^2 - 5x + 6|$.
(a) Is f a surjection?
(b) Find $f^{-1} * \{0\}$.
(c) Find $f * (2, \infty)$.
(d) Find $f^{-1} * (0, 1)$.
(e) Find $f * \{0, 1, 2, 3\}$.
5.69. Let f: $\mathscr{R} \to \mathscr{R}$ be a mapping defined by $f(x) = -x^2 + 7x - 12$.
(a) Is f a surjection?
(b) Find $f^{-1} * \{0\}$.
(c) Find $f^{-1} * \langle -1, 1\rangle$.
(d) Find $f * \{1, 2, 3, 4\}$.
(e) Find $f * \mathscr{R}$.

Let f be a mapping of the set X in the set $Y(f\colon X \to Y)$. Let A, B be subsets of X (Problems 5.70–5.74):

5.70. Prove that $(f * A) \cup (f * B) = f * (A \cup B)$.

5.71. Prove that $f * (A \cap B) \subset (f * A) \cap (f * B)$. Find an example showing that the inclusion cannot generally be replaced by equality.

5.72. Prove that $(f * A) - (f * B) \subset f * (A - B)$. Find an example showing that the inclusion cannot generally be replaced by equality.

5.73. Prove that $A \subset B \to f * A \subset f * B$.

5.74. Prove that $A \subset f^{-1} * (f * A)$. Find an example showing that the inclusion cannot generally be replaced by equality.

Let f be the mapping of the set X into the set Y. Let A, B be subsets of Y. Show that (Problems 5.75–5.79):

5.75. $(f^{-1} * A) \cup (f^{-1} * B) = f^{-1} * (A \cup B)$.

5.76. $(f^{-1}*A)\cap(f^{-1}*B)=f^{-1}*(A\cap B)$.

5.77. $f^{-1}*(A-B)=(f^{-1}*A)-(f^{-1}*B)$.

5.78. $A\subset B\Rightarrow f^{-1}*A\subset f^{-1}*B$.

5.79. $f*(f^{-1}*A)=A\cap(f*X)$.

5.80. Prove that for any family $\{A_t\}_{t\in T}$ of subsets of the set X.

(a) $f*(\bigcup_{t\in T} A_t)=\bigcup_{t\in T}(f*A_t)$.

(b) $f*(\bigcap_{t\in T} A_t)\subset\bigcap_{t\in T}(f*A_t)$.

Show that the inclusion in part b) cannot be replaced by equality.

5.81. Prove that for any family $\{B_t\}_{t\in T}$ of subsets of the set Y:

(a) $f^{-1}*(\bigcup_{t\in T} B_t)=\bigcup_{t\in T}(f^{-1}*B_t)$.

(b) $f^{-1}*(\bigcap_{t\in T} B_t)=\bigcap_{t\in T}(f^{-1}*B_t)$.

5.82. Show that in problems 5.71, 5.72, 5.74, 5.80b, inclusion can be replaced by equality if and only if f is an injection.

5.83. Show that the counterimage of an arbitrary nonempty set is nonempty if and only if f is a surjection.

5.84. Prove that if f: $X\to Y$ then $f*(f^{-1}*A)=A$ (for every $A\subset Y$) iff f is a surjection.

5.85. Prove that if f: $X\to Y$, $A\subset X$, $B\subset Y$, then:

(a) $f*(A\cap f^{-1}*B)=(f*A)\cap B$,

(b) $A\cap f^{-1}*B\subset f^{-1}*[(f*A)\cap B]$.

5.86. Prove that the notation $f^{-1}*B$ is consistent, i.e. if f: $X\to Y$ and there exists an inverse mapping g: $Y\to X$, then $g*B=f^{-1}*B$.

5.87. Show that if $f\subset A\times B$ then Df is the least set X such that $f\subset X\times B$ and Rf is the least set Y such that $f\subset A\times Y$.

5.88. Consider mappings f_i: $\mathcal{N}\times\mathcal{N}\to\mathcal{N}$ $(i=1,2)$ defined as follows:

$$f_1(\langle n,k\rangle)=2^n(2k+1)-1,$$

$$f_2(\langle n,k\rangle)=\tfrac{1}{2}[(n+k+1)(n+k)]+n.$$

Prove that both f_1 and f_2 are injections.

In problems 5.89–5.92 find the superposition $g\circ f$ and the images $(g\circ f)*A$, $(g\circ f)*B$.

5.89. $f:\mathcal{R}\to\mathcal{R}$ is defined by $f(x)=x+1$, $g:\mathcal{R}\to\mathcal{Z}$ is defined by $g(x)=E[x-\frac{1}{3}]$, $A=\langle 0,1)$, $B=\mathcal{R}^+$.

5.90. $f: \mathscr{R}^2 \to \mathscr{C}$ is defined by $f(\langle x, y\rangle) = x + iy$, $g: \mathscr{C} \to \mathscr{R}$ defined by $g(x) = |x| + 1$, $A = (0, 1) \times \langle 0, 1\rangle$, $R = \{\langle x, y\rangle : x^2 + y^2 \leqslant 2\}$.

5.91. $f: \mathscr{N} \to \mathscr{R}$ is defined by $f(n) = \sqrt{n}$, $g: \mathscr{R} \to \mathscr{R}$ defined by $g(x) = x^4 - x^2$, $A = \{2, 4, 6, 8, 10\}$, $B = \{0, 1\}$.

5.92. $f: \mathscr{Z} \to \mathscr{Z}$ is defined by $f(k) = k^2$, $g: \mathscr{Z} \to \mathscr{R}$ is defined by $g(x) = e^x$, $A = \mathscr{N}$, $B = \{2k : k \in \mathscr{N}\}$.

5.93. Prove that:

(a) $O \in \overset{X}{\smile}Y$,

(b) ${}^XY \subseteq \overset{X}{\smile}Y$.

5.94. Prove that $\overset{X}{\smile}Y$ is closed under the operation of intersection.

5.95. Prove that $\overset{X}{\smile}Y$ is closed under the operation of sum iff it is closed under the generation of complement (to ${}^XY$).

5.96. Let $X = \{1, 2, 3\}$, $Y = \{2, 3, 4\}$. Check if the following relations are partial functions of X into Y.

(a) $\{\langle 1, 3\rangle, \langle 2, 4\rangle, \langle 3, 3\rangle\}$,

(b) $\{\langle 1, 4\rangle, \langle 2, 5\rangle, \langle 3, 3\rangle\}$,

(c) $\{\langle 3, 2\rangle, \langle 3, 4\rangle, \langle 2, 2\rangle\}$,

(d) $\{\langle 1, 2\rangle, \langle 1, 4\rangle, \langle 4, 2\rangle\}$,

(e) $\{\langle 1, 3\rangle, \langle 2, 4\rangle\}$,

(f) $\{\langle 1, 4\rangle, \langle 2, 1\rangle\}$.

NOTE

[1] Note that this symbol is different from that used by H. Rasiowa in her *Introduction to modern mathematics*. For the reason for the change see problem 5.38.

CHAPTER 6

GENERALIZED SET-THEORETICAL OPERATIONS

Let f be a function defined on a set T such that, for every $t \in T$, $f(t)$ is a set A_t. If T is a set of pairs, then instead of writing $A_{<m,n>}$ we traditionally write $A_{m,n}$. The set $f * T$ is then called *indexed family of sets* and usually denoted by $\{A_t\}_{t \in T}$. For indexed families of sets the operations of generalized union $(\bigcup_{t \in T} A_t)$ and of generalized intersection $(\bigcap_{t \in T} A_t)$ are defined as follows:

$$x \in \bigcup_{t \in T} A_t \Leftrightarrow \bigvee_{t \in T} x \in A_t, \qquad x \in \bigcap_{t \in T} A_t \Leftrightarrow \bigwedge_{t \in T} x \in A_t.$$

If $T = \mathcal{N}$, then instead of $\bigcup_{n \in \mathcal{N}} A_n$ and $\bigcap_{n \in \mathcal{N}} A_n$ we often use the notation $\bigcup_{n=0}^{\infty} A_n$ and $\bigcap_{n=0}^{\infty} A_n$ When the set T of indices is known we simplify the notation by using $\bigcap_t A_t$ instead of $\bigcap_{t \in T} A_t$ (and analogously for union $\bigcup_t A_t$).

The *generalized Cartesian product* of the family $\{A_t\}_{t \in T}$, $\mathrm{P}_{t \in T} A_t$ is defined as follows: $\mathrm{P}_{t \in T} A_t$ consists of all functions f such that $D(f) = T$ and $R(f) \subset \bigcup_{t \in T} A_t$ and for all $t, f(t) \in A_t$.

6.1. Prove that if $T = \{n_1, \ldots, n_k\}$ and $\{A_t\}_{t \in T}$ is an indexed family of sets, then:

(a) $\bigcup_{t \in T} A_t = A_{n_1} \cup \ldots \cup A_{n_k}$,

(b) $\bigcap_{t \in T} A_t = A_{n_1} \cap \ldots \cap A_{n_k}$.

Find the union and the intersection of the family $\{A_t\}$ for $t \in \mathcal{N}$, $t \in \mathcal{R}^+$, where $A_t \subset R$ is defined as follows (Problems 6.2–6.15):

6.2. $A_t = \{x : t \leqslant x < t+1\}$.

6.3. $A_t = \{x : -t \leqslant x \leqslant t\}$.

6.4. $A_t = \{x : t \leqslant x\}$.

6.5. $A_t=\left\{x:0\leqslant x\leqslant\frac{1}{t+1}\right\}$.

6.6. $A_t=\left\{x:-\frac{1}{t+1}<x\leqslant\frac{1}{t+1}\right\}$.

6.7. $A_t=\{x:\sqrt{t}\leqslant x\leqslant\sqrt{2t}\}$.

6.8. $A_t=\left\{x:\frac{t}{t+1}\leqslant x<\frac{t+1}{t+2}\right\}$.

6.9. $A_t=\left\{x:1-\frac{1}{t+1}<x<2+\frac{1}{t+1}\right\}$.

6.10. $A_t=\left\{x:-\frac{1}{t+1}<x<1-\frac{1}{t+1}\right\}$.

6.11. $A_t=\{x:(t+1)^{\frac{1}{t+1}+1}<x<(t+2)^{\frac{1}{t+1}+1}\}$.

6.12. $A_t=\left\{x:-10-\frac{1}{t+1}<x<2t^2-6t+1\right\}$.

6.13. $A_t=\{x:t^2<x<(t+1)^2\}$

6.14. $A_t=\{x:-(t+1)^2<x<2t^2-5t+5\}$.

6.15. $A_t=\{x:\sin x=t\}$.

Find the union and the intersection of $\{A_t\}_{t\in T}$, $A_t\leqslant R$ (Problems 6.16–6.23):

6.16. $A_t=\{x:\operatorname{tg}x=t\}$.

6.17. $A_t=\{\langle x, y\rangle:x^2+y^2\leqslant t^2\}$.

6.18. $A_t=\{\langle x, y\rangle:x^2+y^2\geqslant t^2\}$.

6.19. $A_t=\{\langle x, y\rangle:x=t\cdot y)$.

6.20. $A_t=\{\langle x, y\rangle:x=t\cdot y^2\}$.

6.21. $A_t=\{\langle x, y\rangle:x+t=y\}$.

6.22. $A_t=\{\langle x, y\rangle:x<t\cdot y\}$.

6.23. $A_t=\{\langle x, y\rangle:x^2<t^2\cdot y^2\}$.

Assume $A_{m,n}\subset\mathscr{R}$ (Problems 6.24–6.27):

6.24. $A_{n,m}=\{x:n^2\leqslant x<m^2\}$, $n, m\in\mathscr{N}$, find

$$\bigcup_{n,m}A_{n,m},\quad \bigcap_{n,m}A_{n,m},\quad \bigcup_{n}\bigcap_{m}A_{n,m},\quad \bigcap_{n}\bigcup_{m}A_{n,m}.$$

6.25. $A_{n,m}=\{x:n\leqslant x<m\}$ $(n, m\in\mathscr{N})$, find

$$\bigcup_{n,m}A_{n,m},\quad \bigcap_{n,m}A_{n,m},\quad \bigcup_{n}\bigcap_{m}A_{n,m},\quad \bigcap_{n}\bigcup_{m}A_{n,m}.$$

6.26. $A_{n,m}=\{x : n^2 \leqslant x < m^2+(n+1)^2\}$ $(n, m \in \mathcal{N})$, find

$$\bigcup_n \bigcap_m A_{n,m}, \quad \bigcap_n \bigcup_m A_{n,m}, \quad \bigcup_m \bigcap_n A_{n,m}, \quad \bigcap_m \bigcup_n A_{n,m},$$

6.27. $A_{n,m}=\{x : n^m \leqslant x\}$ $(n, m \in \mathcal{N}, n \neq 0)$, find

$$\bigcup_n \bigcap_m A_{n,m}, \quad -\bigcup_n \bigcap_m A_{n,m}, \quad \bigcap_n \bigcup_m A_{n,m}, \quad -\bigcap_n \bigcup_m A_{n,m},$$

$$\bigcap_m \bigcup_n A_{n,m}, \quad \bigcap_m \bigcup_n -A_{n,m}, \quad \bigcup_m \bigcap_n A_{n,m}, \quad \bigcup_m \bigcap_n -A_{n,m}.$$

6.28. Prove that $\bigcap_{t \in T} A_t$ is the largest set (with respect to inclusion) included in all sets $A_t (t \in T)$.

6.29. Prove that $\bigcup_{t \in T} A_t$ is the least set (with respect to inclusion) containing all sets $A_t (t \in T)$.

6.30. Show that for an arbitrary space X

(a) $\quad -\bigcup_{t \in T} A_t = \bigcap_{t \in T} -A_t,$

(b) $\quad -\bigcap_{t \in T} A_t = \bigcup_{t \in T} -A_t.$

Prove that (Problems 6.31–6.41):

6.31. $\bigcap_{t \in T} (A_t \cap B_t) = \bigcap_{t \in T} A_t \cap \bigcap_{t \in T} B_t.$

6.32. $\bigcup_{t \in T} (A_t \cup B_t) = \bigcup_{t \in T} A_t \cup \bigcup_{t \in T} B_t.$

6.33. $\bigcap_{t \in T} (A_t \cup B) = B \cup \bigcap_{t \in T} A_t.$

6.34. $\bigcup_{t \in T} (A_t \cap B) = B \cap \bigcup_{t \in T} A_t.$

6.35. $\bigcup_{t \in T} (A_t \cap B_t) \subset \bigcup_{t \in T} A_t \cap \bigcup_{t \in T} B_t.$

6.36. $\bigcap_{t \in T} A_t \cup \bigcap_{t \in T} B_t \subset \bigcap_{t \in T} (A_t \cup B_t).$

6.37. $\bigcup_{t \in T} \bigcup_{s \in S} A_{t,s} = \bigcup_{s \in S} \bigcup_{t \in T} A_{t,s}.$

6.38. $\bigcap_{t \in T} \bigcap_{s \in S} A_{t,s} = \bigcap_{s \in S} \bigcap_{t \in T} A_{t,s}.$

6.39. $\bigcup_{t \in T} A_t \cap \bigcup_{s \in S} B_s = \bigcup_{t \in T} \bigcup_{s \in S} (A_t \cap B_s).$

6.40. $\bigcap_{t \in T} A_t \cup \bigcap_{s \in S} B_s = \bigcap_{t \in T} \bigcap_{s \in S} (A_t \cup B_s).$

6.41. $\bigcup_{t \in T} \bigcap_{s \in S} A_{t,s} \subset \bigcap_{s \in S} \bigcup_{t \in T} A_{t,s}.$

Check if the following holds (Problems 6.42–6.44):

6.42. $\bigcap_{t\in T}(A_t\cup B_t)=\bigcap_{t\in T}A_t\cup\bigcap_{t\in T}B_t$.

6.43. $\bigcup_{t\in T}(A_t\cap B_t)=\bigcup_{t\in T}A_t\cap\bigcup_{t\in T}B_t$.

6.44. $\bigcup_{t\in T}\bigcap_{s\in S}A_{t,s}=\bigcap_{s\in S}\bigcup_{t\in T}A_{t,s}$.

Prove that (Problems 6.45–6.46):

6.45. $\bigcup_{s\in S}(A_s\cap B_s)\subset\bigcup_{s,t\in S}(A_s\cap B_t)=\bigcup_{s\in S}A_s\cap\bigcup_{t\in S}B_t$.

Can the inclusion be replaced here by equality?

6.46. $[(\bigcap_{t\in T}A_t)\cup(\bigcap_{t\in T}B_t)_3=\bigcap_{t,s\in T}(A_t\cup B_s)\subset\bigcap_{t\in T}(A_t\cup B_t)$.

Show that the inclusion cannot be replaced here by equality.

6.47. Check if the following is true for arbitrary families:

$$\bigcup_{t\in T}(A_t-B_t)\subset[\bigcup_{t\in T}A_t-\bigcap_{t\in T}B_t].$$

Can the inclusion be replaced here by equality?

6.48. Show that

$$[(-\bigcup_{t\in T}A_t)\cup(\bigcap_{t\in T}B_t)]\subset\bigcap_{t\in T}(-A_t\cup B_t).$$

Can the inclusion be replaced here by equality?

6.49. Assume $A_1\supset A_2\supset\ldots\supset A_n\supset\ldots$ and $B_1\supset B_2\supset\ldots\supset B_n\supset\ldots$. Prove that

$$\bigcap_{n=1}^{\infty}(A_n\cup B_n)=(\bigcap_{n=1}^{\infty}A_n)\cup(\bigcap_{n=1}^{\infty}B_n).$$

6.50. Assume that for all n $A_n\subset A_0$. Prove that

$$A_0=(A_0-A_1)\cup(A_1-A_2)\cup\ldots\cup\bigcap_{n=0}^{\infty}A_n.$$

6.51. Assume $A_0\supset A_1\supset A_2\supset\ldots$ Prove that

$$(A_1-A_2)\cup(A_3-A_4)\cup\ldots\cup\bigcap_{n=0}^{\infty}A_n=A_0-\bigcup_{n=0}^{\infty}(A_{2n}-A_{2n+1}).$$

6.52. Prove the following generalized commutativity theorem: If $f\colon T\xrightarrow[\text{onto}]{1-1}T$ and $\{A_t\}_{t\in T}$ is an indexed family, then:

(a) $\bigcup_{t\in T}A_t=\bigcup_{t\in T}A_{f(t)}$,

(b) $\bigcap_{t\in T}A_t=\bigcap_{t\in T}A_{f(t)}$.

6.53. Prove the following generalized associativity property: If $T=\bigcup_{s\in S} T_s$, then

(a) $\bigcup_{t\in T} A_t=\bigcup_{s\in S}\bigcup_{t\in T_s} A_t$,

(b) $\bigcap_{t\in T} A_t=\bigcap_{s\in S}\bigcap_{r\in T_s} A_t$.

6.54. Prove the following generalized distributivity principle:

Let K be the set of all mappings of S into T (i.e. $K={}^S T$). For an arbitrary indexed family $\{A_{s,t}\}_{s\in S,\,t\in T}$ the following holds:

(a) $\bigcap_{s\in S}\bigcup_{t\in T} A_{s,t}=\bigcup_{f\in K}\bigcap_{s\in S} A_{s,f(s)}$,

(b) $\bigcup_{s\in S}\bigcap_{t\in T} A_{s,t}=\bigcap_{f\in K}\bigcup_{s\in S} A_{s,f(s)}$.

6.55. Prove that if $T=\bigcup_{s\in S} T_s$ and K is a family of subsets of T such that $A\in K$ implies $A\cap T_s$ is nonempty, then:

(a) $\bigcap_{s\in S}\bigcup_{t\in T_s} A_t=\bigcup_{Y\in K}\bigcap_{t\in Y} A_t$,

(b) $\bigcup_{s\in S}\bigcap_{t\in T_s} A_t=\bigcap_{Y\in K}\bigcup_{t\in Y} A_t$.

6.56. Prove that:

(a) $(\bigcup_{t\in S} A_t)\times(\bigcup_{s\in S} B_s)=\bigcup_{t\in T}\bigcup_{s\in T}(A_t\times B_s)$,

(b) $(\bigcap_{t\in T} A_t)\times(\bigcap_{s\in S} B_s)=\bigcap_{t\in T}\bigcap_{s\in S}(A_t\times B_s)$.

Find $\mathrm{P}_{t\in T} A_t$ for the following T and A_t (Problems 6.57–6.65):

6.57. $T=\mathcal{N}$, $A_t=\{1\}$.

6.58. $T=\mathcal{N}$, $A_t=\{0,1\}$.

6.59. $T=\mathcal{N}$, $A_t=\{0,1,\ldots,t\}$.

6.60. $T=\mathcal{R}$, $A_t=\langle 0,1\rangle$.

6.61. $T=\mathcal{R}$, $A_t=\langle 0,t\rangle$.

6.62. $T=\mathcal{R}^+$, $A_t=\langle -t,t\rangle$.

6.63. $T=\mathcal{R}$, $A_t=\{t\}$.

6.64. $T=\mathcal{R}$, $A_t=\{-t\}$.

6.65. $T=\mathcal{R}$, $A_t=\{t^2+1\}$.

6.66. Prove that if T is finite and, for all $t\in T$, A_t is nonempty, then $\mathcal{P}_{t\in T} A_t\neq O$.

6.67. Prove that if $T\neq O$ and, for all $t\in T$, $A_t\neq 0$, then $\mathrm{P}_{t\in T} A_t\neq O$.

CHAPTER 7

CARDINAL NUMBERS

The sets X and Y are called *equipollent* (*equinumerous*) if there is a mapping $f\colon Y \xrightarrow[\text{onto}]{1-1} Y$. We then write

$$X \sim Y \quad \text{or} \quad X \text{ eq } Y.$$

It is easy to show (Problems 5.30, 5.31) that, for arbitrary X, Y and Z, if $X \sim Y$ and $Y \sim Z$ then $X \sim Z$; further (Problem 5.28), if $X \sim Y$, then $Y \sim X$ and finally $X \sim X$.

To every set we adjoin an object called its *cardinal number* (or *cardinality*). Cardinal numbers are usually denoted by letters of the German or Hebrew alphabet (possibly with indices). If the cardinal number $\mathfrak{m}$ is adjoint to X, then we write $\overline{\overline{X}} = \mathfrak{m}$. We adjoin the same cardinal number to equipollent sets. Thus: $\overline{\overline{X}} = \overline{\overline{Y}} \Leftrightarrow X \sim Y$. The cardinal number $\overline{\overline{X}}$ is called the *cardinality* of X (sometimes: the *power* of X). If X and Y are finite sets, then $X \sim Y$ if and only if the number of elements of X is equal to the number of elements of Y.

The notion of equipollence is considered as a formalization of the intuitive notion of having the same number of elements. Thus, if X is a finite set, then $\overline{\overline{X}}$ is just the number of its elements.

For infinite sets new numbers are introduced. In particular the cardinality of the set of natural numbers is denoted by $\aleph_0$ (aleph-zero) and the cardinality of the set of real numbers is denoted by $\mathfrak{c}$ (continuum).

A set X is *finite* if it is empty or if there exists a natural number $k > 0$ such that $X \sim \{0, \ldots, k-1\}$. The set $\{0, \ldots, k-1\} = \{n : n < k\}$ is also called a *proper initial segment of the set* $\mathcal{N}$ (cf. p. 83, Problem 8.64).

Sets which are either finite or of cardinality $\aleph_0$ are called *denumerable* (or *countable*).

We say that the cardinal number $\mathfrak{n}$ is *greater than* or *equal to* $\mathfrak{m}$ (which is denoted by $\mathfrak{m} \leqslant \mathfrak{n}$) if there exist sets X and Y such that:

1° $X \subset Y$.

2° $\overline{\overline{X}} = \mathfrak{m}$, $\overline{\overline{Y}} = \mathfrak{n}$.

We write $\mathfrak{m} < \mathfrak{n}$ to abbreviate $\mathfrak{m} \leqslant \mathfrak{n}$ and $\mathfrak{m} \neq \mathfrak{n}$. One can show that for arbitrary cardinal numbers $\mathfrak{m}$ and $\mathfrak{n}$ the following holds:

1° $\mathfrak{m} \leqslant \mathfrak{m}$,

2° $\mathfrak{m}_1 \leqslant \mathfrak{m}_2 \wedge \mathfrak{m}_2 \leqslant \mathfrak{m}_3 \Rightarrow \mathfrak{m}_1 \leqslant \mathfrak{m}_3$,

3° $\mathfrak{m}_1 \leqslant \mathfrak{m}_2 \wedge \mathfrak{m}_2 \leqslant \mathfrak{m}_1 \Rightarrow \mathfrak{m}_1 = \mathfrak{m}_2$

(Cantor–Bernstein theorem).

In some considerations related to the cardinality questions we need the so called *axiom of choice*, which can be formulated as follows:

If $A = \{A_t\}_{t \in T}$ is a family of sets such that

1° $A_t \neq 0$ for every $t \in T$,

2° $A_{t_1} \cap A_{t_2} = O$ for every $t_1, t_2 \in T$, $t_1 \neq t_2$,

3° $T \neq O$,

then there exists a set S such that $\overline{\overline{S \cap A_t}} = 1$ for every $t \in T$.

Prove that the following sets A and B are equipollent (Problems 7.1–7.3):

7.1. $A = \{1, 2\}$, $B = \{5, 7\}$.

7.2. $A = \{x \in \mathcal{N} : x < 7\}$, $B = \{x \in \mathcal{N} : 1 < x^2 < 70\}$.

7.3. $A = \{x \in \mathcal{R} : x^2 - 2x + 1 = 0\}$, $B = \{O\}$.

7.4. Prove that for all sets A, B and C we have

(a) $A \sim A$,

(b) $A \sim B \Rightarrow B \sim A$,

(c) $A \sim B \wedge B \sim C \Rightarrow A \sim C$.

7.5. Prove that if A and B are finite sets then $A \sim B$ if and only if A and B have the same number of elements.

Prove that for arbitrary sets A, B, C, A_1, A_2, B_1, B_2 (Problems 7.6–7.9):

7.6. If $A_1 \sim B_1$ and $A_2 \sim B_2$ then $A_1 \times A_2 \sim B_1 \times B_2$.

7.7. If $A_1 \cap A_2 = O$, $B_1 \cap B_2 = O$ and $A_1 \sim B_1$, $A_2 \sim B_2$ then $A_1 \cup A_2 \sim B_1 \cup B_2$.

7.8. If $A \sim B$ and $C \cap (A \cup B) = O$ then $A \cup C \sim B \cup C$.

7.9. If $A \sim B$ then $\mathcal{P}(A) \sim \mathcal{P}(B)$.

Prove that the following sets are denumerable. Point out those which have cardinality $\aleph_0$ (Problems 7.10–7.13):

7.10. $\{x \in \mathcal{N} : 10 | x\}$.

7.11. $\{x \in \mathcal{R} : \bigvee_{y \in \mathcal{N}} x = \ln y\}$.

7.12. $\{x \in \mathcal{N} : \bigvee_{y \in \mathcal{R}} x = \sin y\}$.

7.13. $\{x \in \mathcal{N} : \bigvee_{y \in \mathcal{R}} x = \operatorname{tg} y\}$.

7.14. Prove that the set A is denumerable if and only if its elements form the range of a sequence (finite or infinite).

7.15. Prove that a set A is of cardinality $\aleph_0$ ($\overline{\overline{A}} = \aleph_0$) if and only if its elements form the range of an infinite sequence with different entries.

7.16. Prove that if A is denumerable and $B \subset A$ then B is also denumerable.

7.17. Prove that if A and B are denumerable, then $A \cup B$, $A \cap B$, $A - B$ and $A \dot{-} B$ are also denumerable.

7.18. Prove that if A and B are denumerable then $A \times B$ is also denumerable. Under what assumptions has $A \times B$ cardinality $\aleph_0$?

7.19. Prove that if $\overline{\overline{X}} \geqslant \aleph_0$ then there exists an $X_1 \subset X$, $\overline{\overline{X}}_1 = \aleph_0$, such that $X - X_1 \sim X$.

7.20. Prove that if $\{A_n\}_{n \in \mathcal{N}}$ is a family of denumerable sets then $\bigcup_{n \in \mathcal{N}} A_n$ is also denumerable.

7.21. Prove that if $\{A_n\}_{n \in \mathcal{N}}$ is a family of finite sets then $\bigcup_{n \in \mathcal{N}} \underset{k<n}{P} A_k$ is a denumerable set. When has it cardinality $\aleph_0$?

Deduce from the above that the set of all finite sequences of natural numbers is denumerable.

7.22. Prove that the set of integers is denumerable.

7.23. Prove that the set of rational numbers is denumerable.

7.24. Prove that for each n the set of polynomials of degree n with rational coefficients is denumerable.

7.25. Prove that for all n, the set

$$\{x \in \mathcal{R} : \bigvee_{a_0 \in \mathcal{Q}} \ldots \bigvee_{a_n \in \mathcal{Q}} a_0 x^n + \ldots + a_{n-1} x + a_n = 0\}$$

is denumerable.

7.26. Prove that the set of algebraic numbers, i.e. of the roots of polynomials with rational coefficients, is denumerable.

7.27. Prove that every set of pairwise disjoint intervals on the real line is denumerable.

7.28. Prove that the set of (local) extrema of a continuous real function is denumerable.

7.29. Prove that if $f(x)$ is a monotone real function then the set of its discontinuity points is denumerable.

Prove that the following sets are of cardinality $\aleph_0$ (Problems 7.30–7.36):

7.30. The set of intervals (of the real line) with rational extrema.

7.31. The set of circles on the plane with centers at the points with both coordinates rational and with a rational radius.

7.32. The set of equilateral triangles (in the plane) having the center of gravity at a point with rational coordinates and one of the vertices with rational coordinates.

7.33. Any set consisting of pairwise disjoint cubes in the three-dimensional space.

7.34. Any set consisting of pairwise disjoint discs in the plane.

7.35. The set of all (finite dimensional) matrices with rational entries.

7.36. The set $\{x \in \mathscr{R} : \bigvee_{n \in \mathscr{N} - \{0\}} x^n \in \mathscr{Q}\}$

7.37. Let A be a denumerable set. Prove that the set of all finite sequences whose terms are elements of A is denumerable. When is that set of cardinality $\aleph_0$?

7.38. Find the cardinality of the set of all sequences with integer terms convergent to 0.

7.39. Find the cardinality of the set of all eventually constant sequences with rational terms.

7.40. Prove that a nonempty set A is denumerable if and only if there exists a function $g: N \xrightarrow[\text{onto}]{} A$.

7.41. Prove that $\overline{\overline{A}} = \aleph_0$ if and only if there exist functions f and g such that $f: A \xrightarrow[\text{onto}]{} N$ and $g: N \xrightarrow[\text{onto}]{} A$.

7.42. Prove that the set of points of the real line is of cardinality $\mathbf{c}$.

Without the use of the Cantor–Bernstein theorem prove that $A \sim B$ (Problems 7.43–7.50):

7.43. A, B are two open intervals of the real line.

7.44. A, B are two closed intervals.

7.45. A is an open interval, B an interval with one extreme.

7.46. A, B are circles in the plane.

7.47. A, B are closed discs in the plane.

7.48. A is the real line, B an open interval.

7.49. A is the real line, B a closed interval.

7.50. A is a closed halfline, B the real line.

7.51. Prove that the set of the points of the plane has cardinality $\mathfrak{c}$.

7.52. Prove that if A and B are sets of cardinality $\mathfrak{c}$ then $A \cup B$ and $A \times B$ also have cardinality $\mathfrak{c}$.

7.53. Prove that if A is a denumerable set and B has cardinality $\mathfrak{c}$ then $A \cup B$, $B - A$, and $B \dot{-} A$ are of cardinality $\mathfrak{c}$.

7.54. What can be said of the cardinality of the set $A \times B$ if $\bar{\bar{A}} = \mathfrak{c}$ and B is denumerable?

7.55. Prove that if $\{A_t\}_{t \in \mathscr{R}}$ is a family of cardinality $\mathfrak{c}$ and A_t are nonempty, pairwise disjoint and denumerable, then $\bigcup_{t \in R} A_t$ is of cardinality $\mathfrak{c}$.

7.56. Prove that if the cardinality of each $A_t (t \in \mathscr{R})$ is $\mathfrak{c}$ then the set $\bigcup_{t \in R} A_t$ is also of cardinality $\mathfrak{c}$.

Check if the following sets are of cardinality $\mathfrak{c}$ (Problems 7.57–7.74):

7.57. $\{x \in \mathscr{R} : -1 < x < 1\}$.

7.58. $\{x \in \mathscr{R} : \bigvee_{n \in \mathscr{N}} x^n \in \mathscr{Q}\}$.

7.59. $\{x \in \mathscr{R} : 0 \leqslant x\}$.

7.60. $\{x \in \mathscr{R} : x < 0\}$.

7.61. $\{\langle x, y \rangle : x \in \mathscr{R} \wedge y \in \mathscr{R} \wedge x^2 = y\}$.

7.62. $\{\langle x, y \rangle : x \in \mathscr{R} \wedge y \in \mathscr{R} \wedge x = y\}$.

7.63. $\{\langle x, y \rangle : x^2 + y^2 = 1 \wedge x \in \mathscr{R} \wedge y \in \mathscr{R}\}$.

7.64. $\{\langle x, y \rangle : x \in \mathscr{R} \wedge y \in \mathscr{R} \,\&\, y = f(x)\}$ where f is a fixed real function.

7.65. $\{\langle x, y \rangle : x \in \mathscr{R} \wedge y \in \mathscr{R} \wedge \bigvee_{w \in Q} x - y = q\}$.

7.66. $\{\langle x, y \rangle : x \in \mathscr{R} \wedge y \in \mathscr{R} \wedge x < 1\}$.

7.67. $\{\langle x, y \rangle : x \in \mathscr{R} \wedge y \in \mathscr{R} \wedge x^2 = 4\}$.

7.68. $\{\langle x, y \rangle : x \in \mathscr{R} \wedge y \in \mathscr{Q} \wedge x^2 = 4\}$.

7.69. $\{\langle x, y \rangle : x \in \mathscr{R} \wedge y \in \mathscr{R} \wedge x < 0 \wedge y < 0\}$.

7.70. $\{\langle x, y \rangle : x \in \mathscr{R} \wedge y \in \mathscr{R} \wedge 0 \leqslant x < 1 \wedge 0 \leqslant y < 1\}$.

7.71. $\{\langle x, y\rangle \in \mathcal{R}^2 : x \cdot y < 1\}$.

7.72. $\{\langle x, y\rangle \in \mathcal{R}^2 : x^2 + y^2 < 1\}$.

7.73. $\{\langle x, y, z\rangle \in \mathcal{R}^3 : x = y = z\}$.

7.74. $\{\langle x, y, z\rangle \in \mathcal{R}^3 : 0 \leqslant x < 1 \wedge 0 \leqslant y < 1 \wedge 0 \leqslant z < 1\}$.

7.75. Prove that, for all $n > 0$, the set $\mathcal{R}^n$ is of cardinality $\mathfrak{c}$.

7.76. Prove that if for all $n \in \mathcal{N}$, $A_n = \{0, 1\}$, then $\mathop{\mathcal{P}}_{n \in \mathcal{N}} A_n$ is of cardinality $\mathfrak{c}$.

7.77. Given a family of denumerable sets A_n such that almost all A_n's (i.e. all but a finite number of A_n's) are finite, prove that $\overline{\overline{\mathop{\mathcal{P}}_{n \in N} A_n}} \leqslant \mathfrak{c}$.

7.78. Prove that if $X \subset Y$ then $\overline{\overline{X}} \leqslant \overline{\overline{Y}}$.

7.79. Prove that for all cardinal numbers $\mathfrak{m}_1$, $\mathfrak{m}_2$ and $\mathfrak{m}_3$:

(a) $\mathfrak{m}_1 \leqslant \mathfrak{m}_2$.

(b) $\mathfrak{m}_1 \leqslant \mathfrak{m}_2 \wedge \mathfrak{m}_2 \leqslant \mathfrak{m}_3 \Rightarrow \mathfrak{m}_1 \leqslant \mathfrak{m}_3$.

7.80. Prove that $\overline{\overline{X}} \leqslant \overline{\overline{Y}}$ if and only if there exists a $Y_1 \subset Y$ such that $X \sim Y_1$.

7.81. Prove that $\overline{\overline{X}} \leqslant \overline{\overline{Y}}$ if and only if there exists an f such that f: $X \xrightarrow{1\text{-}1} Y$.

7.82. Prove that, for nonempty, $\overline{\overline{X}} \leqslant \overline{\overline{Y}}$ if and only if there exists an f such that f: $Y \xrightarrow[\text{onto}]{} X$.

7.83. Prove that if X is such that $\overline{\overline{X}} > n$ for every $n \in \mathcal{N}$, then $\overline{\overline{X \cup \{a\}}} = \overline{\overline{X}}$ for every a.

7.84. Prove that if $\overline{\overline{X}} \geqslant \aleph_0$ then, for every a, $\overline{\overline{X \cup \{a\}}} = \overline{\overline{X}}$.

7.85. Prove that if $\{A_n\}_{n \in \mathcal{N}}$ is a family of nonempty sets such that $\{n \in \mathcal{N} : 2 \leqslant \overline{\overline{A}}_n \leqslant \aleph_0\}$ is not finite, then $\overline{\overline{\mathop{\mathcal{P}}_{n \in \mathcal{N}} A_n}} \geqslant \mathfrak{c}$. Find when $\overline{\overline{\mathop{\mathcal{P}}_{n \in \mathcal{N}} A_n}} = \mathfrak{c}$.

Recall that ${}^A B$ is the set of all mappings f: $A \to B$. Thus $f \in {}^A B \Leftrightarrow f$: $A \to B$. The set ${}^A B$ is sometimes denoted by B^A. In particular, when $B = \{0, 1\}$ then, instead of ${}^A B$ (or B^A), we use 2^A.

7.86. Prove that $2^A \sim \mathcal{P}(A)$.

7.87. Prove that if A is finite, $\overline{\overline{A}} \geqslant 2$ and $\overline{\overline{B}} \geqslant \aleph_0$, then ${}^B A \sim 2^B$.

7.88. Prove that if $A_1 \sim B_1$ and $A_2 \sim B_2$ then ${}^{A_1} A_2 \sim {}^{B_1} B_2$.

7.89. Prove that if $A \cap B = O$ then for all C

$$ {}^{(A \cup B)} C \sim {}^A C \times {}^B C. $$

7.90. Prove that ${}^C(A \times B) \sim {}^C A \times {}^C B$.

7.91. Prove that $\overline{\overline{{}^{\mathcal{N}} \mathcal{N}}} = \mathfrak{c}$.

7.92. Prove that if $\{A_t\}_{t\in\mathscr{R}}$ is a family of sets such that, for all $t\in\mathscr{R}$, $\overline{\overline{A_t}}\geqslant 2$ then $\overline{\overline{\underset{t\in T}{\mathscr{P}} A_t}}\geqslant\mathfrak{c}$.

7.93. Does there exist a set X such that $\overline{\overline{\mathscr{P}(X)}}=\aleph_0$?

7.94. Prove that if $\overline{\overline{X}}=\aleph_0$ then $\overline{\overline{\mathscr{P}(X)}}=\mathfrak{c}$.

7.95. Prove that there is no set X and no function f such that f: $X\xrightarrow[\text{onto}]{}\mathscr{P}(X)$.

7.96. Prove that $\overline{\overline{X}}<\overline{\overline{\mathscr{P}(X)}}$.

7.97. Prove that there is no set containing all sets.

7.98. Prove that there are infinitely many cardinal numbers.

7.99. Prove that if $A_1\subset B\subset A_2$ and $A_1\sim A_2$ then $A_1\sim B$ and $B\sim A_2$.

7.100. Prove that if there exists an injection f: $A\rightarrow A$ such that $f\times A\subset B$ and $B\subset A$ then $A\sim B$.

CHAPTER 8

ORDERINGS

A (*partial*) *ordering relation* in the set A is any relation $R \subset A^2$ which satisfies the following conditions:

(a) *Reflexivity*

$$\bigwedge_{x \in A} (\langle x, x \rangle \in R) \qquad (\bigwedge_{x \in A} (xRx)).$$

(b) *Antisymmetry*

$$\bigwedge_{x \in A} \bigwedge_{y \in A} [(\langle x, y \rangle \in R \wedge \langle y, x \rangle \in R) \Rightarrow x = y]$$

$$(\bigwedge_{x \in A} \bigwedge_{y \in A} [(xRy \wedge yRx) \Rightarrow x = y]).$$

(c) *Transitivity*

$$\bigwedge_{x \in A} \bigwedge_{y \in A} \bigwedge_{z \in A} [(\langle x, y \rangle \in R \wedge \langle y, z \rangle \in R) \Rightarrow \langle x, z) \rangle \in R]$$

$$(\bigwedge_{x \in A} \bigwedge_{y \in A} \bigwedge_{z \in A} [(xRy \wedge yRz) \Rightarrow xRz]).$$

If the relation R satisfies in addition the condition of connectedness

(d) *Connectedness*

$$\bigwedge_{x \in A} \bigwedge_{y \in A} (\langle x, y \rangle \in R \vee \langle y, x \rangle \in R \vee x = y)$$

$$(\bigwedge_{x \in A} \bigwedge_{y \in A} (xRy \vee yRx \vee x = y)),$$

then we say that R is a *linear ordering*.

In older literature the term linear ordering was used to denote a relation which is asymmetric, transitive and connected (see Problems 4.18 and 4.19).

Well-orderings are linear orderings satisfying – in addition – the following condition e:

(e) $$\bigwedge_{X\subset A}\{X\neq O\Rightarrow\bigvee_{y}[y\in X\wedge\bigwedge_{z}(z\in X\Rightarrow\langle y,z\rangle\in R)\}]$$

$$(\bigwedge_{X\subset A}[X\neq O\Rightarrow\bigvee_{y\in X}\bigwedge_{z\in X}(yRz)]).$$

A *partially* (*linearly*), *well-ordered set* is a pair $\langle A, R\rangle$ such that R (partially) orders (linearly orders, well-orders) the set A.

Some of the elements of A may play a special role in $\langle A, R\rangle$:

(a) An element $a\in A$ is called *maximal* in $\langle A, R\rangle$ if

$$\bigwedge_{x\in A}(aRx\Rightarrow x=a).$$

(b) An element $a\in A$ is called *minimal* in $\langle A, R\rangle$ if

$$\bigwedge_{x\in A}(xRa\Rightarrow x=a).$$

(c) An element $a\in A$ is called *largest* in $\langle A, R\rangle$ if

$$\bigwedge_{x\in A}(xRa).$$

(d) An element $a\in A$ is called *smallest* in $\langle A, R\rangle$ if

$$\bigwedge_{x\in A}(aRx).$$

A *chain* in the ordered set $\langle A, R\rangle$ is any subset B of A such that $\langle B, R\cap B^2\rangle$ is a linearly ordered set.

Note that $R\cap B^2$ is simply the restriction of R to the set B.

An *antichain* in the ordered set $\langle A, R\rangle$ is any subset B of A such that:

$$\bigwedge_{x,y\in B}[x\neq y\Rightarrow\sim xRy\wedge\sim yRx].$$

Thus if B is a chain, then any two elements of B are comparable (in the sense of R) whereas if it is an antichain then any two elements of B are incomparable.

Every one-element set is both a chain and an antichain.

If X is a subset of A then an *upper bound* of X (in $\langle A, R\rangle$) is any element $a\in A$ such that $\bigwedge_{x\in X}(xRa)$. The notion of *lower bound* of $X\subset A$ is defined similarly.

KURATOWSKI–ZORN LEMMA. Let $\langle A, R\rangle$ be an ordered set and assume that every chain in A possesses an upper bound. Then A possesses a maximal element. Moreover, for every $x \in A$ there exists a maximal element b such that xRb.

8.1. Let $R \subset (\mathcal{N} - \{0\})^2$ be defined as follows:

$$xRy \Rightarrow \bigvee_{z \in \mathcal{N}} (xz = y)$$

(we usually use the symbol $x|y$ to denote this relation).

(a) Show that $|$ is an ordering.

(b) Find the minimal elements of $|$.

(c) Prove that the set $\langle \mathcal{N} - \{0\}, | \rangle$ has no maximal element.

(d) Find the general form of a chain in $\langle \mathcal{N} - \{0\}, | \rangle$.

(e) Find the general form of an antichain in $\langle \mathcal{N} - \{0\}, | \rangle$.

(f) Prove that the inclusion relation in the set of all chains of the set $\langle \mathcal{N} - \{0\}, | \rangle$ is an ordering and find the form of all maximal and minimal chains.

8.2. Let X be a set and $\mathcal{P}(X)$ the family of its subsets. Prove that the relation $\subset$ orders $\mathcal{P}(X)$. Find the largest element and the smallest element in the set $\langle \mathcal{P}(X), \subset \rangle$.

8.3. Assume that the set X contains at least two elements. Consider the set $U = \mathcal{P}(X) - \{0\} - \{X\}$, and the relation $\subset$ restricted to the elements of that set. Check if:

(a) The set $\langle U, \subset \rangle$ possesses a largest element.

(b) The set $\langle U, \subset \rangle$ possesses a least element.

(c) Find the form of the maximal and minimal elements of $\langle U, \subset \rangle$.

(d) What is the cardinality of the set consisting of the maximal and minimal elements of U? Consider the case where X is a two-element set.

8.4. Prove that if $\langle X, R\rangle$ is an ordered set and $Y \subset X$ then the set $\langle Y, R \cap Y^2 \rangle$ is also an ordered set. Is the above statement true if $\langle X, R\rangle$ is linearly (well-) ordered? What can be said about the minimal, maximal, etc. elements in $\langle X, R\rangle$ and $\langle Y, R \cap Y^2\rangle$?

8.5. The set $\langle X, R^{-1}\rangle$ is called the *dual set* of $\langle X, R\rangle$.

(a) Prove that if $\langle X, R\rangle$ is an ordered set then $\langle X, R^{-1}\rangle$ is also an ordered set.

(b) Is an analogous statement true for linearly ordered sets?

(c) Is an analogous statement true for well-ordered sets?

(d) What is the connection between the minimal, maximal, etc. elements in $\langle X, R\rangle$ and its dual?

8.6. (a) Prove that if in the ordered set $\langle X, R\rangle$ there is a largest element, then it is the unique maximal element in $\langle X, R\rangle$.

(b) Prove that if in the ordered set $\langle X, R\rangle$ there is a least element then it is the unique minimal element in $\langle X, R\rangle$.

(c) What connection between points (a) and (b) is suggested in problem 8.5?

8.7. Given $X \neq O$, is there a relation $R \subset X^2$ such that simultaneously:

(a) $\langle X, R\rangle$ is an ordered set?

(b) R is an equivalence in X?

8.8. Given X such that $\overline{\overline{X}} \geqslant 2$, is there a relation $R \subset X^2$ such that simultaneously:

(a) $\langle X, R\rangle$ is a linearly ordered set?

(b) R is an equivalence in X?

8.9. Assume that $\langle X, R\rangle$ is a well-ordered set. Find a necessary and sufficient condition for $\langle X, R^{-1}\rangle$ to be a well-ordered set.

8.10. Prove that if X is a finite set and $\langle X, R\rangle$ is a linearly ordered set, then it is also a well-ordered set.

8.11. Find an example of an ordered set $\langle X, R\rangle$ such that $\langle X, R\rangle$ possesses exactly one maximal element but not a largest element.

8.12. Find an example of an ordered set $\langle X, R\rangle$ such that $\langle X, R\rangle$ possesses exactly one minimal element but not a least element.

8.13. (a) Prove that if the set X is finite then in the ordered set $\langle X, R\rangle$ there exists at least one maximal element.

(b) Prove an analogous result for the minimal elements.

8.14. Prove that if X is a finite set and there is a unique maximal (minimal) element in the ordered set $\langle X, R\rangle$ then it is the largest (least) element of the set $\langle X, R\rangle$. Can the assumption of finiteness of the set X be dropped?

8.15. Show that the ordered set $\langle X, R\rangle$ is linearly ordered if and only if every antichain in $\langle X, R\rangle$ is a one-element set.

8.16. Show that, in a linearly ordered set, a maximal (minimal) element – if it exists – is the largest (least) one.

8.17. What is the meaning of the fact that A is a chain in the ordered set $\langle A, R\rangle$?

8.18. Prove that in a linearly ordered set every finite nonempty subset possesses both a least and a largest element.

8.19. Prove that in an ordered set every finite nonempty subset possesses both minimal and maximal elements.

8.20. For every $n \in \mathcal{N}$ construct an example of an ordered set with exactly n minimal elements. Make an analogous construction for maximal elements.

8.21. In the set T of all finite sequence of natural numbers (i.e. in the set $\bigcup_{k \in \mathcal{N}} \mathcal{N}^k$) define a relation R as follows: $xRy \Leftrightarrow$ the sequence x is an initial segment (not necessarily proper) of the sequence y.

(a) Show that R is an ordering in T.

(b) Is there a largest element in $\langle T, R \rangle$?

(c) Is there a minimal element in $\langle T, R \rangle$?

(d) What is the cardinality of the set of all chains in $\langle T, R \rangle$?

8.22. Consider the set $T = \{2^n \colon n \in \mathcal{N}\} \cup \{3\}$ with the divisibility relation (cf. Problem 8.1).

(a) Are there maximal (minimal) elements in $\langle T, | \rangle$?

(b) Is there a smallest element in $\langle T, | \rangle$?

(c) Draw the diagram of our relation.

8.23. Consider the set $T = \{2, 3, \ldots, 15\}$ with the divisibility relation restricted to it.

(a) Draw the diagram of our relation.

(b) Find all the maximal, minimal, etc. elements.

(c) Find all chains of length 3.

8.24. Let T be the set of all functions with the domain equal to $\langle 0, 1 \rangle$ and with values in $\mathcal{R}^+$ ($T = {}^{\langle 0,1 \rangle}\mathcal{R}^+$). Define a relation R as follows:

$$fRg \Leftrightarrow \bigwedge_x [f(x) \leqslant g(x)].$$

(a) Prove that R is an ordering relation in T.

(b) Find the maximal, minimal, etc. elements of $\langle T, R \rangle$.

8.25. Given an ordered set $\langle X, S \rangle$ and a set $T \neq O$, in the set F of all functions from T to X (i.e. ${}^T X$), define a relation R as follows:

$$fRg \Leftrightarrow \bigvee_{t \in T} [f(t) S g(t)].$$

(a) Prove that R is an ordering relation in T.

(b) Find a necessary and sufficient condition for $\langle F, R\rangle$ to possess a minimal (maximal) element.

(c) Find a necessary and sufficient condition for $\langle F, R\rangle$ to possess a least (largest) element.

8.26. Let A_t for $t \in \{0, 1, 2, 3, 4, 5\}$ be a family of sets defined as follows:

$$A_t = \{z \in \mathscr{C} : k_t + 1 \leqslant \operatorname{im} z \leqslant t + 2\},$$

where

$$k_t = \begin{cases} 0 & \text{for} \quad t = 0, 2, 4, \\ +1 & \text{for} \quad t = 1, \\ -1 & \text{for} \quad t = 3, 5. \end{cases}$$

(a) Considering this family as partially ordered by the inclusion relation, find the elements which are maximal, minimal, etc.

(b) Do there exist largest and smallest elements in our family?

(c) Draw the diagram of this ordering.

8.27. For $t \in \{1, 2, 3, 4, 5, 6\}$, let A_t be a family of sets defined as follows:

$$A_t = \{z \in \mathscr{C} : |z - k_t| \leqslant t\},$$

where

$$k_t = \begin{cases} 2 & \text{for} \quad t = 1, \\ 0 & \text{for} \quad t = 2, 3, 6, \\ -1 & \text{for} \quad t = 4, \\ 1 & \text{for} \quad t = 5. \end{cases}$$

Consider this family as ordered by the inclusion relation.

(a) Find the elements of our family which are maximal, minimal, etc.

(b) Draw the diagram of this partial ordering.

8.28. For $t \in \{0, 1, 2, 3, 4, 5, 6\}$, let A_t be a family of sets defined as follows:

$$A_t = \{x \in \mathscr{R} : k_t \leqslant x \leqslant t + 1\},$$

where

$$k_t = \begin{cases} 0 & \text{for} \quad t = 0, 2, \\ -1 & \text{for} \quad t = 1, 3, \\ -2 & \text{for} \quad t = 4, 6, \\ -3 & \text{for} \quad t = 5. \end{cases}$$

(a) Considering this family as ordered by the inclusion relation, find the maximal, minimal, etc. elements.

(b) Draw the diagram of this partial ordering.

8.29. For $t \in \{0, 1, 2, 3, 4\}$, let A_t be a family of subsets of $\mathcal{Q}$ defined as follows:

$$A_t = \{x \in \mathcal{Q} : t < x < t^2\}.$$

(a) Considering this family as ordered by the inclusion relation, find the elements which are maximal, minimal, etc.

(b) Draw the diagram of this partial ordering.

8.30. Let $\alpha(n)$ be the number of prime divisors of a natural number n. In the set $\mathcal{N} - \{0, 1\}$ we introduce a relation R as follows:

$$xRy \Leftrightarrow [\alpha(x) < \alpha(y)] \vee [\alpha(x) = \alpha(y) \wedge x \leqslant y].$$

(a) Prove that $\langle \mathcal{N} - \{0, 1\}, R\rangle$ is an ordered set.

(b) Is it a linearly ordered set?

(c) Find the elements which are maximal, minimal, etc.

8.31. Let $\beta(n)$ be the number of all divisors of a natural number n. In the set $\mathcal{N} - \{0\}$ we introduce a relation R as follows:

$$xRy \Leftrightarrow [\beta(x) < \beta(y)] \vee [\beta(x) = \beta(y) \wedge x \leqslant y].$$

(a) Prove that $\langle \mathcal{N} - \{0\}, R\rangle$ is an ordered set and find the elements which are maximal, minimal, etc.

(b) Is it a well-ordered set?

8.32. Define in the set $\mathcal{N} - \{0\}$ a relation R as follows:

$$xRv \Leftrightarrow (2 | x \wedge 2 | y \wedge y \leqslant x) \vee (2 | x \wedge \sim 2 | y) \vee (\sim 2 | x \wedge \sim 2 | y \wedge$$
$$\wedge x \leqslant y).$$

(a) Prove that the set $\langle \mathcal{N} - \{0\}, R\rangle$ is linearly ordered.

(b) Is it a well-ordered set?

8.33. Define in the set ${}^{\mathcal{N}}\mathcal{N}$ a relation R as follows:

$$\underline{a} R \underline{b} \Leftrightarrow \bigwedge_n (a_n \leqslant b_n).$$

(a) Prove that $\langle {}^{\mathcal{N}}\mathcal{N}, R\rangle$ is an ordered set. Find the elements which are maximal, minimal, etc.

(b) Is it a linearly ordered set?

8.34. Define a relation S in the set ${}^{\mathcal{N}}\mathcal{N}$ as follows:

$$\underline{a}S\underline{b} \Leftrightarrow \bigvee_k \bigwedge_{n<k} (a_n = b_n \wedge a_k < b_k) \vee (\underline{a} = \underline{b}).$$

(a) Prove that the set $\langle {}^{\mathcal{N}}\mathcal{N}, S\rangle$ is linearly ordered.

(b) Prove that $\bigwedge_{a,b} (\underline{a}R\underline{b} \Rightarrow \underline{a}S\underline{b})$, where R is the relation introduced in Problem 8.33.

8.35. Define a relation R in the set $\mathscr{C}$ as follows:

$$xRy \Leftrightarrow (\operatorname{Re} y \leqslant \operatorname{Re} x) \wedge (\operatorname{Im} y \leqslant \operatorname{Im} x).$$

(a) Prove that $\langle \mathscr{C}, R\rangle$ is an ordered set.

(b) Is it a linearly ordered set?

8.36. Define a relation S in the set $\mathscr{C}$ as follows:

$$xSy \Leftrightarrow (\operatorname{Re} y < \operatorname{Re} x) \vee [(\operatorname{Re} y = \operatorname{Re} x) \wedge (\operatorname{Im} y \leqslant \operatorname{Im} x)].$$

(a) Prove that $\langle \mathscr{C}, S\rangle$ is a linearly ordered set.

(b) Is it a well-ordered set?

(c) Show that $\bigwedge_{x,y} (xRy \Rightarrow xSy)$, where R is the relation introduced in 8.35.

An element $x \in X$ is called an (*immediate*) *successor* in the ordered set $\langle X, R\rangle$, of an element $y \in X$ if the following condition is satisfied:

$$yRx \wedge \bigwedge_z [(yRz \wedge zRx) \Rightarrow (y = z \vee z = x)].$$

8.37. Define the notion of a predecessor in such a way that: x is a predecessor of y in the set $\langle X, R\rangle$ if and only if x is a successor of y in the dual set $\langle X, R^{-1}\rangle$ (cf. Problem 8.5).

8.38. Prove that in a well-ordered set every element (with the exception of the largest one—provided it exists) possesses a successor. Is it true that every element (with the exception of the least one) must possess a predecessor?

8.39. In the linearly ordered set $\langle X, R\rangle$ every element possesses a successor. Is $\langle X, R\rangle$ a well-ordered set?

8.40. (a) Give an example of an ordered set in which there is a least element, and every element possesses two successors and at most one predecessor.

(b) What can be said about the cardinality of the set of all chains in such a set?

We say that ordered sets $\langle X, R\rangle$ and $\langle Y, S\rangle$ are *similar* if there exists a bijection f of X and Y such that

$$\bigwedge_{x,y}[xRy \Leftrightarrow f(x)\, Sf(y)].$$

In this case we write: $\langle X, R\rangle \approx \langle Y, S\rangle$.

8.41. Check that $\langle X, R\rangle \approx \langle Y, S\rangle$ implies $X \sim Y$.

8.42. (a) Check that if $\langle X, R\rangle$ is a linearly ordered set and $\langle X, R\rangle \approx \langle Y, S\rangle$, then $\langle Y, S\rangle$ is also a linearly ordered set.

(b) Show an analogous property of well-ordered sets.

8.43. Show that similarity relationship has the properties of reflexivity, symmetry, and transitivity. Why are we unable to apply the abstraction principle in this case?

8.44. Show that if f is a similarity function of $\langle X, R\rangle$ and $\langle Y, S\rangle$ then

(a) f preserves the elements which are maximal, minimal, etc. (i.e. the image of a maximal element is maximal etc.).

(b) f preserves chains and antichains – in other words, if T is a chain (an antichain) in $\langle X, R\rangle$ then $f * T$ is a chain (an antichain) in $\langle Y, S\rangle$.

(c) f preserves successors and predecessors.

Every ordered set X R has a *relational type*, $TR(\langle X, R\rangle)$, which satisfies the condition:

$$TR(\langle X, R\rangle) = TR(\langle Y, S\rangle) \Leftrightarrow \langle X, R\rangle \approx \langle Y, S\rangle.$$

The type of a linearly ordered set is called an *order type*, and the type of a well-ordered set is called an *ordinal number*. If the type of the set $\langle X, R\rangle$ is α then the type of its dual set $\langle X, R^{-1}\rangle$ is denoted by α^*.

Several order types have traditional names: ω – the type of the set $\langle \mathcal{N}, \leqslant\rangle$, η – the type of the set $\langle \mathcal{Q}, \leqslant\rangle$, λ – the type of the set $\langle \mathcal{R}, \leqslant\rangle$.

An ordered set $\langle X, R\rangle$ is *dense* if

$$\bigwedge_{x,y}[(xRy \wedge x \neq y) \Rightarrow \bigvee_{z}(x \neq z \wedge z \neq y \wedge xRz \wedge zRy)].$$

8.45. Prove that, for every type α, $\alpha^{**} = \alpha$.

8.46. Prove that:

(a) The set $(\mathcal{Q}, \leqslant\rangle$ is dense.

(b) The set $\langle \mathcal{R}, \leqslant\rangle$ is dense.

8.47. Show that "being dense" is invariant under similarity, i.e.

if the set $\langle X, R\rangle$ is dense and $\langle X, R\rangle \approx \langle Y, S\rangle$ then the set $\langle Y, S\rangle$ is also dense. Deduce from this that the notation of "dense order type" is sound.

8.48. Show that if $X \geqslant 2$ and the set $\langle X, R\rangle$ is well-ordered then it is not dense.

8.49. Is the set $\langle X, R\rangle$ considered
(a) in 8.33
(b) in 8.34
dense?

8.50. Is the set $\langle X, R\rangle$ considered
(a) in 8.35
(b) in 8.36
dense?

8.51. Assume that $\langle X, R\rangle$ is a dense ordered set, $x \in X$, $y \in X$. Is it possible that y is a successor of x?

8.52. Show that η is a unique denumerable dense order type without the least and the largest elements.

In other words, if $\langle X, R\rangle$ is a linearly ordered set, denumerable and dense, without the least and the largest elements, then $\langle X, R\rangle \approx \approx \langle \mathscr{Q}, \leqslant\rangle$.

8.53. Show that the set $\langle \mathscr{Z} - \mathscr{N}, \leqslant\rangle$ is of type ω^*.

8.54. Show that $\eta^* = \eta$, i.e. that $\langle \mathscr{Q}, \leqslant\rangle \approx \langle \mathscr{Q}, \geqslant\rangle$.

8.55. Show that $\lambda^* = \lambda$.

8.56. Show that every linearly ordered set $\langle X, R\rangle$ such that:
a) its every element possesses both a successor and a predecessor,
b) whenever xRy then the set $\{z : xRz \wedge zRy\}$ is finite, is similar to the set $\langle \mathscr{Z}, \leqslant\rangle$.

8.57. Is the set $\langle \{1 - 1/n : n \in \mathscr{N}^+\} \cup \{1\}, \leqslant\rangle$ similar to the set $\langle \mathscr{N}, \leqslant\rangle$? What will be the answer if the set of natural numbers is completed by an element ∞ greater than all natural numbers?

A subset $Y \subset X$ is *dense* in the linearly ordered set $\langle X, R\rangle$ if the following holds:

$$\bigwedge_{x,y} [xRy \wedge x \neq y \Rightarrow \bigvee_{z} (z \in Y \wedge xRz \wedge zRy)].$$

8.58. Prove that:
(a) X is dense in $\langle X, R\rangle$.
(b) $\mathscr{Q}$ is dense in $\langle \mathscr{R}, \leqslant\rangle$.

8.59. Prove that every linearly ordered set $\langle X, R\rangle$ which possesses a dense denumerable subset is similar to $\langle Y, \leqslant\restriction Y\rangle$ for a certain $Y\subset\mathscr{R}$.

8.60. Show that the following holds for linearly ordered sets: $\langle X, R\rangle$ is similar to $\langle Y, S\rangle$ if and only if there exists a bijection f of X and Y such that

$$\bigwedge_{x,y\in X}[(xRy\Rightarrow f(x)\,Sf(y)].$$

8.61. Prove that the set $\langle\mathscr{N}, \leqslant\rangle$ is well-ordered, whereas the set $\langle\mathscr{N}, \geqslant\rangle$ is not well-ordered.

8.62. Prove that if $\langle X, R\rangle$ is a well-ordered set and $X_1\subset X$ then $\langle X_1, R\cap X_1^2\rangle$ is also a well-ordered set.

8.63. Prove that every relation $R\subset X^2$ satisfying conditions (b) and (e) satisfies also conditions (a), (c), (d) (see p. 76). Show that condition b (antisymmetry) cannot be omitted.

An *initial segment* of the linearly ordered set $\langle X, R\rangle$ is any subset $Y\subset X$ satisfying the condition: $\bigwedge_{x,y}(x\in Y\wedge yRx\Rightarrow y\in Y)$.

8.64. (a) What are the initial segments of $\langle\mathscr{N}, \leqslant\rangle$?

(b) What are the initial segments of $\langle\mathscr{Q}, \leqslant\rangle$?

8.65. Let Y be a nonempty initial segment of $\langle\mathscr{Q}, \leqslant\rangle$ without a largest element. Prove that $\langle Y, \leqslant\restriction Y\rangle\approx\langle\mathscr{Q}, \leqslant\rangle$.

8.66. Define $O_R(x)=\{y: yRx\wedge y\neq x\}$. Prove that $O_R(x)$ is an initial segment of $\langle X, R\rangle$.

8.67. Show the following principle of transfinite induction: If $\langle X, R\rangle$ is a well-ordered set, and Φ a property (i.e. a subset of the set X) such that

$$\bigwedge_{x\in X}\{\bigwedge_{y}[y\in O_R(x)\Rightarrow\Phi(y)]\Rightarrow\Phi(x)\},$$

then $\bigwedge_{x\in X}\Phi(x)$.

8.68. Prove that no proper initial segment of a well-ordered set is similar to the set itself. In other words if $\langle X, R\rangle$ is a well-ordered set, Y is an initial segment of $\langle X, R\rangle$ and $X\neq Y$, then $\sim(\langle Y, R\cap Y^2\rangle\approx\approx\langle X, R\rangle)$.

8.69. Assume that $\langle X, R\rangle$ is a well-ordered set and f is an injection satisfying the condition $xRy\Rightarrow f(x)Rf(y)$. Show that $\bigwedge_{x} xRf(x)$.

8.70. Show the Kuratowski–Zorn Lemma using the axiom of choice.

8.71. Show that for every relation $R \subset X^2$ such that $\langle X, R\rangle$ is an ordered set, there exists a relation $S \subset X^2$ such that $R \subset S$ and $\langle X, S\rangle$ is a linearly ordered set.

8.72. Show the following strong version of Problem 8.71. For every maximal element a of the ordered set $\langle X, R\rangle$ there exists a relation S such that $R \subset S$, the set $\langle X, S\rangle$ is linearly ordered and a is the largest element of $\langle X, S\rangle$.

8.73. Formulate and prove a theorem analogous to that of 8.72 for the case of minimal elements.

8.74. Prove (using the Kuratowski–Zorn Lemma) the following version of the axiom of choice:

For every family $\{X_t\}_{t \in T}$ of nonempty sets there exists a function $f: T \to \bigcup_{t \in T} X_t$ such that for all $t \in T$, $f(t) \in X_t$.

8.75. Prove (using the Kuratowski–Zorn Lemma) the following version of the axiom of choice:

For every family $\{X_t\}_{t \in T}$ of nonempty pairwise disjoint sets there exists a set W such that $\overline{\overline{W \cap X_t}} = 1$.

8.76. Show that every ring with unity possesses a maximal ideal.

8.77. Show that every linear space possesses an algebaric basis.

8.78. Show that every antichain can be extended to a maximal antichain.

8.79. Consider the set $\langle \mathscr{P}(X), \subset\rangle$ and the set ${}^X\{0, 1\}$ with the relation R defined as follows:

$$fRg \Leftrightarrow \bigwedge_{x \in X} [f(x) \leqslant g(x)].$$

Show that

$$\langle \mathscr{P}(X), \leqslant\rangle \approx \langle {}^X\{0, 1\}, R\rangle.$$

8.80. Assume that $\{\langle X_t, R_t\rangle\}_{t \in T}$ is a family of nonempty ordered sets. Consider the set

$$\langle \mathop{\mathrm{P}}_{t \in T} X_t, R\rangle, \quad \text{where} \quad fRg \Leftrightarrow \bigwedge_t [f(t) R_t g(t)].$$

(a) Show that it is an ordered set.

(b) Assume that, for all t, $\langle X_t, R_t\rangle$ is a linearly ordered set. Is then $\langle \mathop{\mathrm{P}}_{t \in T} X_t, R\rangle$ a linearly ordered set?

CHAPTER 9

SETS OF PROBLEMS

9.I.1. Let $\varphi\colon \mathscr{R}[x] \to \mathscr{R}[x]$ be a mapping defined by $\varphi(f)=f(2x^2)$:

(a) Find the counterimage of the set of all constant polynomials.

(b) Find the image of the set of polynomials of degree at most two.

(c) Find the counterimage of the polynomial x^3+1 and the cardinality of that counterimage.

(d) Find the counterimage of the polynomial x^4+x^2+1.

2. Find the cardinality of the set $\mathscr{R}_2[x]$, and construct a bijection of that set and one of the following sets: $\mathscr{N}$, $\mathscr{Q}$, $\mathscr{R}$, $\mathscr{R}^2$.

3. Prove that:

$$\bigcup_{s\in S}(A_s\cap B_s)\subset\bigcup_{s\in S}\ \bigcup_{t\in S}(A_s\cap B_t).$$

Provide an example showing that inclusion cannot be replaced by equality here.

4. Using – besides logical symbols – the symbols for arithmetic operations, absolute value, $\mathscr{N}$ and $\mathscr{R}$, – write a formal expression of the following: There exist sequences of reals which are bounded but not convergent.

5. In the set $Z=\{3, 6, 7, 8, 12, 14, 21, 32, 66, 128, 135\}$ define a relation R by: $xRy=x|y$.

(a) Prove that $\langle Z, R\rangle$ is an ordered set.

(b) Draw the diagram of the relation R.

(c) Show the maximal, minimal, etc. elements.

(d) Find a relation S such that $R\subset S$ and $\langle X, S\rangle$ is a linearly ordered set.

6. In the set $\mathscr{Q}[x]$ introduce a relation R as follows:

$$\varphi_1 R\varphi_2\Leftrightarrow\bigvee_{a,b,c}[\varphi_1(x)-\varphi_2(x)=ax^2+bx+c].$$

(a) Show that R is an equivalance relation.

(b) Find the cardinality of the equivalence class of the polynomial $\varphi(x) \equiv 0$.

(c) Find the cardinality of the collection of all abstraction classes.

9.II.1. Let P be the set of all formulas of the propositional calculus. Introduce a relation $\sim$ as follows: $\Phi_1 \sim \Phi_2$ if and only if the expression $\Phi_1 \Leftrightarrow \Phi_2$ is a tautology.

(a) Prove that $\sim$ is an equivalence.

(b) What formulas belong to the class of $\sim[p \wedge \sim p]$?

(c) What formulas belong to the class of $\sim[p \vee \sim p]$?

(d) In the set of equivalence classes $P/\sim$ we introduce a relation $\leqslant$ as follows: $[\Phi_1] \leqslant [\Phi_2]$ if and only if the expression $\Phi_1 \Rightarrow \Phi_2$ is a tautology.

Check that the definition of $\leqslant$ is sound, i.e. if $\Phi_1 \sim \Psi_1$, $\Phi_2 \sim \Psi_2$ and $\Phi_1 \Rightarrow \Phi_2$ is a tautology then also $\Psi_1 \Rightarrow \Psi_2$ is a tautology.

(e) Construct in $P/\sim$ a chain of cardinality $\aleph_0$.

(f) Construct in $P/\sim$ an antichain of cardinality $\aleph_0$.

2. Using—besides the logical symbols—the symbols $\mathscr{P}$, $\times$, $\langle\ \rangle$, $\in$, write a formal statement expressing the following: The set of all subsets of the set $\mathscr{N}$ is equipollent to the set $\mathscr{R}$.

3. Given: an n-element set X. What is the cardinality of the family of the relations reflexive in X?

4. Give an example of an infinite ordered set which:

(a) has a least element,

(b) has a successor for every element,

(c) has a predecessor for every element, apart from the least one,

(d) is not similar to the set $\langle \mathscr{N}, \leqslant \rangle$.

5. Construct a function mapping the set $\langle 0, 1 \rangle$ to $\mathscr{N}$ such that the counterimage of every natural number has cardinality $\mathfrak{c}$.

6. Prove that the mapping f: $\mathscr{N} \times \mathscr{N} \to \mathscr{N}$ defined by $f(x, y) = \binom{x+y+1}{2} + x$ is a bijection.

9.III.1. In the set of functions ${}^{\langle 0, 1\rangle}\langle 0, 1 \rangle$ introduce a relation R as follows:

$$f R g \Leftrightarrow f \mid \mathscr{Q} \cap \langle 0, 1 \rangle = g \mid \mathscr{Q} \cap \langle 0, 1 \rangle:$$

(a) Prove that R is an equivalence relation.

(b) Find the cardinality of the equivalence class of the function $f(x) \equiv 0$.

(c) Find the cardinality of the collection of all equivalence classes.

(d) Are any two equivalence classes equipollent? If so, find a bijection of the class of the function $f(x) = x$ and the class of the function $f(x) = 1 - x$.

2. Using—besides the logical symbols—the symbols $\subset, \in, \langle\ \rangle, -$, write a formal statement expressing the following: There exists a subset X of the set $\mathcal{N}$ such that $\mathcal{N} - X$ is equipollent to $\mathcal{N}$.

3. Given an n-element set X, what is the cardinality of the family of the relations symmetric in X?

4. Prove that any at least two-element set, linearly ordered and dense, has cardinality at least $\aleph_0$.

5. Construct a function mapping the segment $\langle 0, 1\rangle$ to $\mathcal{N}$ such that the counterimage of every even number has cardinality continuum, but the counterimage of every odd number has cardinality 1.

6. Prove that the mapping $f: \mathcal{N} \times \mathcal{N} \to \mathcal{N}$ defined by $f(x, y) = 2^x(2y+1) - 1$ is a bijection.

9.IV.1. Let $f: X \to Y$. f determines the following relation $\sim_f$ in X:

$$x_1 \sim_f x_2 \Leftrightarrow f(x_1) = f(x_2).$$

(a) Prove that $\sim_f$ is an equivalence relation.

(b) Prove that f can be represented as $g \circ h$ where h is a surjection and f is an injection, moreover $h: X \to X/\sim$, $g: X/\sim \to Y$.

(c) Prove that g is a surjection if and only if f is a surjection, and h is an injection if and only if f is an injection. (The decomposition $f = g \circ h$ is called the *canonical decomposition* of the function f.)

2. In the set of all straight lines in the plane we introduce a relation R as follows: $L_1 R L_2 \Leftrightarrow \bigvee_{w \in Q}$ (The angle between L_1 and L_2 is equal to $w \cdot \pi$ or $L_1 \| L_2$).

(a) Prove that R is an equivalence.

(b) Find the cardinality of the equivalence class of the axis $0x$.

(c) Find the cardinality of the set of equivalence classes.

3. An ordered set $\langle X, R\rangle$ is called a *tree* if and only if, for every x, the set $O_R(x) = \{y : yRx\}$ is well-ordered by the relation R restricted to its elements. We define the type $t(x)$ of the element x as the ordinal number of the set $O_R(x)$.

(a) Prove that the relation $\sim_R$ defined by $x_1 \sim_R x_2 \Leftrightarrow t(x_1)=t(x_2)$, is an equivalence relation.

(b) In the set of equivalence classes of the relation $\sim_R$ introduce a relation S as follows: $[x_1]\, S\, [x_2] \Leftrightarrow t(x_1) \leqslant t(x_2)$. Prove that $\leqslant$ is a well-ordering.

(c) Show that every equivalence class of $\sim_R$ is an antichain.

4. Given families of sets $\{A_t\}_{t\in T}$ and $\{B_t\}_{t\in T}$ and sets $U=\bigcup_{t\in T}(A_t-B_t)$ and $V=\bigcup_{t\in T}A_t-\bigcup_{t\in T}B_t$:

(a) prove that $V\subset U$,

(b) find families $\{A_n\}_{n\in\mathcal{N}}$, $\{B_n\}_{n\in\mathcal{N}}$ such that it is not true that $U\subset V$ and in addition $\bigwedge_{n\in\mathcal{N}} \overline{\overline{A_n}}=\overline{\overline{B_n}}=\aleph_0$.

5. Using—besides logical symbols—the symbols $\times$, $\{,\}$, $\langle\,\rangle$, $\in$, write the formal expression of the following: The set of all zero-one sequences is equipollent to the set $\mathscr{R}$.

6. For a given set $X\subset\mathcal{N}$ such that $\overline{\overline{\mathcal{N}-X}}\neq\aleph_0$ construct a mapping $f\colon\mathcal{N}\to\mathcal{N}$ with the following properties:

(a) the counterimage of every element of X has two elements.

(b) the counterimage of every element of $\mathcal{N}-X$ is infinite.

9.V.1. Let $R\subset X^2$, $S\subset Y^2$. Define the Cartesian product $R\times S$ $\subset(X\times Y)^2$ as follows:

$$\langle x_1, y_1\rangle (R\times S)\langle x_2, y_2\rangle \Leftrightarrow x_1 R x_2 \wedge y_1 S y_2 .$$

(a) Prove that a product of equivalence relations is an equivalence relation.

(b) Let $\overline{\overline{[x]_R}}=\mathfrak{m}$, $\overline{\overline{[y]_S}}=\mathfrak{n}$. Find the cardinality of $[\langle x, y\rangle]_{R\times S}$.

(c) Prove that a product of orderings is an ordering.

2. Prove that no two of the sets

$$A,\ 2^A,\ 2^{2^A},\ \ldots,\ 2^{2^{\cdot^{\cdot^{\cdot^{2^A}}}}}\ldots$$

are equipollent. Derive from this that there exist infinitely many infinite cardinal numbers.

3. In the set $\mathcal{N}\times\mathcal{N}$ introduce a relation R as follows:

$$\langle k, l\rangle R \langle m, n\rangle \Leftrightarrow \max(k, l) = \max(m, n).$$

(a) Prove that R is an equivalence.

(b) Find the general form of its equivalence class.

(c) Are there two different but equipollent equivalence classes of R?

4. Given the tree T (cf. the set IV, problem 3) of all zero-one finite sequences, $tTs \Leftrightarrow t$ is an initial segment of s,

(a) What is the cardinality of the set of chains in T?

(b) Construct an antichain in T of cardinality $\aleph_0$.

5. Given families $\{A_t\}_{t\in T}$, $\{B_t\}_{t\in T}$ of subsets of the set X and the sets $U = \bigcup_{t\in T} A_t \cup \bigcap_{\in T} B_t$, $V = -\bigcup_{t\in T}(A_t - B_t)$.

(a) prove that $V \subset U$.

(b) find families $\{A_n\}_{n\in\mathcal{N}}$, $\{B_n\}_{n\in\mathcal{N}}$ of infinite subsets of the set $\mathcal{R}$ such that $U\subset V$ is false.

6. Given the tree $\langle X, R\rangle$ and the antichain $U\subset X$, define a relation S in the set X as follows: $xSy \Leftrightarrow \bigvee_{u\in U}(uRx \wedge uRy) \vee (x=y)$.

(a) Prove that S is an equivalence relation.

(b) In the set X/S introduce a relation $\leqslant$ as follows: $[x]\leqslant[y] \Leftrightarrow xRy$. Prove that this definition is sound, i.e. does not depend on the choice of representatives of equivalence classes.

(c) Prove that each class $[u]$ for $u\in U$ is maximal in $\leqslant$.

(d) Prove that if $[x]$ is a maximal element of $\leqslant$ and $\bigwedge_{u\in U}(u\notin[x])$, then x is a maximal element of R.

(e) Prove that if $[x]$ is minimal in $\leqslant$ then x is minimal in R.

9.VI.1. In the set ${}^{\mathcal{N}}\mathcal{N}$ (of all sequences with natural terms) define a relation R as follows:

$$\{a_n\}_{n\in\mathcal{N}} R \{b_n\}_{n\in\mathcal{N}} \Leftrightarrow \bigwedge_k (a_{2k}=b_{2k}).$$

(a) Prove that R is an equivalence.

(b) Find the cardinality of the class $[\{2n\}_{n\in\mathcal{N}}]$.

(c) Prove that any two equivalence classes are equipollent and have the cardinality of the continuum.

(d) Prove that the cardinality of the set of equivalence classes is the continuum.

2. Define a mapping $f: {}^{\mathcal{N}}\mathcal{N} \to {}^{\mathcal{N}}\mathcal{N}$ as follows:

$$f(\{a_n\}_{n\in\mathcal{N}}) = \{a_{2n}\}_{n\in\mathcal{N}}.$$

(a) Investigate the relation $\sim_f$ (cf. Set IV, Problem 1).
(b) Find a set X such that $X \not\subset f*X$.
(c) Find a set X such that $X = f*X$.

3. In the set ${}^{\mathcal{N}}\mathcal{N}$ introduce a relation $\leqslant$ as follows:

$$\{a_n\} \leqslant \{b_n\} \Leftrightarrow \bigvee_k [\bigwedge_m (m<k \Rightarrow (a_m = b_m) \wedge b_k \leqslant a_k) \vee \{a_n\} = \{b_n\}].$$

Prove that $\leqslant$ is a linear ordering.

4. A set A is called *Dedekind-infinite* (D.i.) if and only if it is equipollent to a proper subset of it.

(a) Prove that every D.i. set contains a subset of cardinality $\aleph_0$.
(b) Prove that if A contains a subset of cardinality $\aleph_0$ then it is D.i.

5. A set A is *Dedekind-finite* (D.f.) if and only if it is not equipollent to any of its proper subsets.

(a) Prove that A is D.f. if and only if A does not contain a subset of cardinality $\aleph_0$.
(b) Prove that the union $A \cup B$ of D.f. sets is again D.f.

6. Using the logical symbols, write a formal expression of the following: The set ${}^{\mathcal{N}}\mathcal{N}$ is linearly orderable.

9.VII.1. In the set $\mathcal{N}[x]$ introduce a relation R as follows:

$$(a_n x^n + \ldots + a_0) R (b_m x^m + \ldots + b_0) \Leftrightarrow n \leqslant m \wedge \bigwedge_{k\leqslant n} (a_k \leqslant b_k).$$

(a) Prove that R is an ordering.
(b) Construct in $\langle \mathcal{N}[x], R\rangle$ a chain of cardinality $\aleph_0$.
(c) Construct in $\langle \mathcal{N}[x], R\rangle$ an antichain of cardinality $\aleph_0$.
(d) Find the maximal, minimal, etc. elements of $\langle \mathcal{N}[x], R\rangle$.

2. In the set $\mathscr{Z}[x]$ introduce a relation R as follows: $fRg \Leftrightarrow$ all the coefficients of $f-g$ are divisible by 3.

(a) Prove that R is an equivalence relation.
(b) Find the cardinality of the equivalence class of $f(x) \equiv 3x$.

3. Given families of sets $\{A_t\}_{t\in T}$ and $\{B_t\}_{t\in T}$ and the sets

$$U = \bigcap_{t\in T} (A_t - B_t), \quad V = \bigcap_{t\in T} A_t - \bigcap_{t\in T} B_t:$$

(a) Prove that $U \subset V$.

(b) Is it true that $V \subset U$? If it is not, find families $\{A_n\}_{n\in\mathcal{N}}$ and $\{B_n\}_{n\in\mathcal{N}}$ of subsets of $\mathcal{N}$ such that $U \neq V$.

4. Using – besides the logical symbols – the symbols $\mathcal{N}$, $\mathcal{R}$, $\in$, $\langle\rangle$, $||$, write a formal expression of the following: Every bounded sequence of real numbers contains a convergent subsequence.

5. Given a mapping φ: $\mathcal{Z}[x]\times\mathcal{Z}[x]\to\mathcal{Z}[x]$ defined as follows: $\varphi(f,g)=f-g$.

(a) Find the image of the set $\{\langle x, x^2\rangle, \langle x^2, x^3\rangle, \langle x^2+1, x^3+1\rangle\}$.

(b) Find the counterimage of the polynomial 0.

(c) Find the image of the set $\mathcal{Z}_3[x]\times\mathcal{Z}_3[x]$.

(d) Find the counterimage of the set $\mathcal{Z}[x]$.

6. In the set $A=(\mathcal{R}[x]-\{0\})\times\mathcal{N}$ define the relation R as follows: $\langle f, n_1\rangle R\langle g, n_2\rangle \Leftrightarrow f|g \wedge n_1|n_2$.

(a) Prove that R is an ordering.

(b) What is the cardinality of the set A?

(c) Construct an antichain of the cardinality of the continuum.

(d) What is the cardinality of the maximal chains in $\langle A, R\rangle$?

9.VIII.1. Prove that whenever f: $A\to\mathcal{P}(A)$, then

$$f^{-1}(\{x\in A : x\notin f(x)\})=O.$$

(this is known as Cantor's Diagonal Lemma).

2. Given an ordered set $\langle X, R\rangle$ such that every chain and every antichain in $\langle X, R\rangle$ are finite, prove that X is finite.

3. Find the cardinality of the set:

(a) of all equivalencies in $\mathcal{N}$.

(b) of all linear orderings in $\mathcal{N}$.

4. Let f be a mapping of a set A into itself such that $f(x)\neq x \wedge \wedge \bigwedge_{n\in\mathcal{N}-\{0,1\}} (f^n(x)\neq x)$. Define in A a relation R as follows: $xRy \Leftrightarrow \bigvee_{u\in\mathcal{N}-\{0\}} f^n(x)=y$.

(a) Prove that $\langle A, R\rangle$ is an ordered set.

(b) Find the maximal elements of $\langle A, R\rangle$.

(c) What does it mean that $\langle A, R\rangle$ possesses a largest element?

(d) Find all the maximal chains.

(e) Prove that if $x\in A-(f*A)$ then x is minimal in $\langle A, R\rangle$.

5. Let f and A be as in Problem 4. Prove that the relation S defined as

$$xSy \Leftrightarrow \bigvee_{k, l \in \mathcal{N} - (0)} f^k(x) = f^l(y)$$

is an equivalence relation.

6. Let $R \subset A^2$ be a relation which is both reflexive and transitive. Define a relation I in A as follows:

$$xIy \Leftrightarrow (xRy \wedge yRx).$$

(a) Prove that I is an equivalence relation.

(b) In the set of equivalence classes of I introduce a relation R_1 as follows: $[a]_I \, R_1 \, [b]_I \Leftrightarrow aRb$. Prove that this definition is sound, i.e. does not depend on the choice of representatives.

7. Using – besides the logical symbols – the symbols $\mathcal{N}$, R, $\in$, $<$, $\leqslant$, $| \, |$, write a formal expression of the following: The sequences $\{a_n\}_{n \in \mathcal{N}}$ and $\{b_n\}_{n \in \mathcal{N}}$ have the same limit.

CHAPTER 10

CARDINAL AND ORDINAL ARITHMETIC

Von Neumann ordinals (or simply *ordinals*) are sets A with the following properties 1–3:

$$(1)\qquad \bigwedge_{X} (X \in A \Rightarrow X \subset A),$$

$$(2)\qquad \bigwedge_{X,\,Y} [X, Y \in A \Rightarrow (X = Y \vee X \in Y \vee Y \in X)],$$

$$(3)\qquad \bigwedge_{X \subset A} [X \neq 0 \Rightarrow \bigvee_{Y \in X} (Y \cap X = 0)].$$

The least ordinal is the empty set O. Every ordinal consists of all smaller ordinals, the ordering is just the inclusion relation. In fact, the inclusion relation is a well-ordering of ordinals.

The basic property of (von Neumann) ordinals is the following one: For every well-ordered set $\langle X, R\rangle$ there exists an ordinal A such that $\langle X, R\rangle \approx \langle A, \subset\rangle$. This number is unique and is called the *ordinal number* of $\langle X, R\rangle$.[1] If the sets A and B are finite then for all the linear orderings R and S:

$$\langle A, R\rangle \approx \langle B, S\rangle \Leftrightarrow \overline{\overline{A}} = \overline{\overline{B}}.$$

This is the reason why – for finite sets – ordinals and cardinals are identified.

Traditionally, we use letters of the Greek alphabet, $\alpha, \beta, \ldots, \xi, \ldots$, to denote ordinals. If α is the ordinal number of the set $\langle X, R\rangle$ then we write $\alpha = \overline{\langle X, R\rangle}$. The *cardinal* of the ordinal ξ is the cardinality of (any) set ordered in type ξ. The cardinal of the ordinal number ξ is denoted by $\overline{\xi}$. If $\overline{\langle A, R\rangle} = \alpha$ and $\overline{\langle B, S\rangle} = \beta$ and $A \cap B = O$ then $\alpha + \beta = \overline{\langle A \cup B, R \cup S \cup (A \times B)\rangle}$.

Intuitively, all the elements of A precede all the elements of B, in the sets A and B themselves the order is not changed.

10 1. Prove that the definition of the addition of ordinals is sound.

10.2. Find α, β, γ such that $\beta+\alpha=\gamma+\alpha$ but $\beta\neq\gamma$.

10.3. Prove that $\alpha+\beta=\alpha+\gamma\Rightarrow\beta=\gamma$.

10.4. Prove that $\alpha+(\beta+\gamma)=(\alpha+\beta)+\gamma$.

10.5. Give examples of sets ordered in types:

(a) $\omega+1$,

(b) $\omega+\omega$,

(c) $\omega+\omega+1$.

10.6. Prove that $1+\omega=2+\omega=\ldots=n+\omega=\ldots=\omega$.

10.7. Prove that $\omega+\omega\neq\omega$.

10.8. Find α and β such that $\alpha+\beta\neq\beta+\alpha$.

Multiplication of ordinals: If $\overline{\langle A, R\rangle}=\alpha$, $\overline{\langle B, S\rangle}=\beta$, then $\alpha\cdot\beta=\overline{\langle A\times B, T\rangle}$, where T is the following relation:

$$\langle x, y\rangle T\langle x_1, y_1\rangle\Leftrightarrow(ySy_1\wedge y\neq y_1)\vee(y=y_1\wedge xRx_1).$$

10.9. Prove that the above definition is sound, i.e. does not depend on the choice of representatives of types α and β.

10.10. Prove that $\alpha\cdot\beta=\alpha\cdot\gamma\Rightarrow\beta=\gamma$.

10.11. Find α, β, γ such that $\beta\cdot\alpha=\gamma\cdot\alpha$ and $\beta\neq\gamma$.

10.12. Give examples of sets ordered in types:

(a) $\omega\cdot\omega$,

(5) $\omega\cdot\omega+\omega$,

(c) $\omega\cdot\omega+\omega+3$,

(d) $\omega\cdot\omega\cdot\omega+\omega\cdot\omega\cdot2+\omega$.

10.13. Prove that $\alpha\cdot(\beta\cdot\gamma)=(\alpha\cdot\beta)\cdot\gamma$.

10.14. Prove that $2\cdot\omega=\omega$.

10.15. Find α and β such that $\alpha\cdot\beta\neq\beta\cdot\alpha$.

10.16. Find infinite ordinals α and β such that $\alpha\neq\beta$ and $\alpha\cdot\beta=\beta\cdot\alpha$.

10.17. Prove that $\alpha\cdot(\beta+\gamma)=\alpha\cdot\beta+\alpha\cdot\gamma$.

10.18. Find α, β, γ such that $(\alpha+\beta)\cdot\gamma\neq\alpha\cdot\gamma+\beta\cdot\gamma$.

We introduce the *ordering relationship* between ordinals as follows:

(a) $\alpha\leqslant\beta\Leftrightarrow A$ set of type α is similar to an initial segment of a set of type β.

(b) $\alpha<\beta\Leftrightarrow\alpha\leqslant\beta\wedge\alpha\neq\beta$.

10.19. Prove that "a set" in the above definition can be changed in both places to "every set" without changing the meaning of the definition.

10.20. Prove that $\alpha \leqslant \beta \Leftrightarrow$. Every set of type α is similar to an initial set of a set of type β.

10.21. Prove that for arbitrary α and β:

$$\alpha \neq \beta \Rightarrow (\alpha < \beta \vee \beta < \alpha).$$

10.22. Prove that for arbitrary α, β, and γ:

$$\alpha \leqslant \beta \wedge \beta \leqslant \gamma \Rightarrow \alpha \leqslant \gamma.$$

10.23. Prove that for arbitrary α, β, and γ:

$$\alpha \leqslant \beta \wedge \beta < \gamma \Rightarrow \alpha < \gamma.$$

10.24. Prove that for arbitrary α, β, and γ:

$$\alpha < \beta \Rightarrow \gamma + \alpha < \gamma + \beta.$$

10.25. Prove that for arbitrary α, β, and γ:

$$\alpha \leqslant \beta \Rightarrow \alpha + \gamma \leqslant \beta + \gamma.$$

10.26. Find ordinals α, β, γ such that:

$$\alpha + \gamma \leqslant \beta + \gamma \quad \text{but} \quad \alpha > \beta$$

10.27. Prove that for arbitrary α, β, and γ:

$$\alpha + \beta < \alpha + \gamma \Rightarrow \beta < \gamma.$$

10.28. Prove that for arbitrary α and β:

$$\alpha \leqslant \beta \wedge \beta \leqslant \alpha \Rightarrow \alpha = \beta.$$

10.29. Prove that for arbitrary α and β:

(a) $\alpha \leqslant \beta \Leftrightarrow \alpha \subset \beta$,

(b) $\alpha < \beta \Leftrightarrow \alpha \in \beta$.

10.30. Prove that for arbitrary α and β: $\alpha \leqslant \beta \Leftrightarrow$ there is a unique γ such that $\alpha + \gamma = \beta$.

10.31. Prove the following "division with a remainder" theorem:

$$\bigwedge_{\beta > 0} \bigwedge_{\alpha} \bigvee_{\gamma, \rho} [(\alpha = \beta \cdot \gamma + \rho) \wedge (\rho < \beta)].$$

10.32. Prove that the ordinals γ and ρ from 10.31 are uniquely determined by α and β.

The *expotentation* of ordinals is defined by induction ($\alpha \geqslant 1$) as follows:

$$\alpha^0 = 1, \quad \alpha^{\beta+1} = \alpha^\beta \cdot \alpha, \quad \alpha^\lambda = \bigcup_{\xi < \lambda} \alpha^\xi.$$

10.33. Prove that for arbitrary α, β, and γ:

$$\alpha \leqslant \beta \Rightarrow \alpha^\gamma \leqslant \beta^\gamma.$$

10.34. Prove that for arbitrary α, β, and γ:

$$\alpha \leqslant \beta \Rightarrow \gamma^\alpha \leqslant \gamma^\beta.$$

10.35. Prove the following decomposition theorem of Cantor: For every ordinal $\mu > 1$ and $\alpha \geqslant 1$ there exists a unique decomposition

$$\alpha = \mu^{\alpha_0} \cdot \nu_1 + \ldots + \mu^{\alpha_n} \cdot \nu_n,$$

where

$$\alpha_0 > \alpha_1 > \ldots > \alpha_n, \quad 0 < \nu_i < \mu \quad (i = 0, 1, \ldots, n).$$

An ordinal β is a *limit* iff it does not possess a predecessor, i.e. there is no α such that $\beta = \alpha + 1$.

An ordinal β is *initial* iff $\bigwedge_\alpha (\alpha < \beta \Leftrightarrow \overline{\alpha} < \overline{\beta})$.

10.36. What is the meaning of the sign $<$ in the predecessor and in the successor of the above definition of an initial ordinal?

10.37. Prove that if α is a limit and $\beta < \alpha$ then $\beta + 1 < \alpha$.

10.38. Prove that if α is initial and $\beta < \alpha$ then $\beta \cdot \beta < \alpha$, $\beta \cdot \beta \cdot \beta < \alpha$, etc.

Infinite initial ordinals are well-ordered by $\subset$ and, so we can enumerate them with consecutive ordinals. The α initial ordinal is denoted by ω_α.

The cardinality of the set ω_α is denoted by $\aleph_\alpha$ (*aleph-alpha*). Those cardinals are called *alephs.*

10.39. Prove that for every infinite α there exists a β such that $\overline{\alpha} = \aleph_\beta$.

10.40. Find a well-ordering of the set $\omega_\alpha \times \omega_\alpha$ in type ω_α.

10.41. Prove that, for every α; $\aleph_\alpha \cdot \aleph_\alpha = \aleph_\alpha$.

Let $\overline{\overline{A}} = \mathfrak{m}$, $\overline{\overline{B}} = \mathfrak{n}$ and $A \cap B = O$. The sum $\mathfrak{m} + \mathfrak{n}$ (the product $\mathfrak{m} \cdot \mathfrak{n}$, the power $\mathfrak{m}^\mathfrak{n}$) is the cardinality of the set $A \cup B (A \times B, {}^B A)$.

10.42. Prove that for every α:

$$\aleph_\alpha + 1 = \aleph_\alpha + \aleph_0 = \aleph_\alpha \cdot 2 = \aleph_\alpha \cdot \aleph_0 = \aleph_\alpha \cdot \aleph_\alpha = \aleph_\alpha.$$

10.43. Prove that for every α and β:

$$\aleph_\alpha + \aleph_\beta = \aleph_\alpha \cdot \aleph_\beta = \aleph_{\max(\alpha, \beta)}.$$

10.44. Prove the following statements of ordinal arithmetic:

(a) $\alpha^{\beta+\gamma} = \alpha^\beta \cdot \alpha^\gamma$,

(b) $((\alpha)^\beta)^\gamma = \alpha^{\beta\gamma}$.

10.45. Prove that for every $\gamma > 1$ and ξ, $\xi \leqslant \gamma^\xi$ holds.

Function whose domain is an ordinal is called a *sequence.*

Let $\{\varphi_\xi\}_{\xi<\alpha}$ be an increasing sequence of ordinals (i.e. $\xi < \eta \Rightarrow \varphi_\xi < \varphi_\eta$). Define the limit of the sequence $\{\varphi_\xi\}_{\xi<\alpha}$ as follows:

$$\lim_{\xi<\alpha} \varphi_\xi = \bigcup_{\xi<\alpha} \varphi_\xi.$$

10.46. Prove that the limit of a sequence of ordinals is an ordinal itself.

10.47. Prove that the limit of the sequence of ordinals $\{\varphi_\xi\}_{\xi<\alpha}$ is the least ordinal β such that

$$\bigwedge_{\xi<\alpha} \varphi_\xi \leqslant \beta.$$

10.48. A function Φ defined on ordinals and with ordinal values is *continuous* if

$$\Phi(\lim_{\xi<\alpha} \varphi_\xi) = \bigcup_{\xi<\alpha} \Phi(\varphi_\xi).$$

Prove that every continuous, increasing function φ (i.e. a function satisfying in addition the condition $\xi < \eta \Rightarrow \varphi(\xi) < \varphi(\eta)$) possesses a critical point, i.e. a point such that $\xi = \varphi(\xi)$.

10.49. What is the relationship between the cardinals

(a) $\overline{\overline{\alpha+\beta}}$ and $\bar{\bar{\alpha}} + \bar{\bar{\beta}}$.

(b) $\bar{\bar{\alpha}} \cdot \bar{\bar{\beta}}$ and $\overline{\overline{\alpha \cdot \beta}}$.

(c) $\overline{\overline{\alpha^\beta}}$ and $\bar{\bar{\alpha}}^{\bar{\bar{\beta}}}$?

10.50. Prove that there is no set consisting of all ordinals.

10.51. Prove that there is no set consisting of all cardinals.

10.52. Prove (without using the axiom of choice) the equivalence of the following statements:

(a) ZERMELO'S THEOREM: For every set X there exists a relation $R \subset X^2$ such that $\langle X, R \rangle$ is a well-ordered set.

(b) All infinite cardinals are alephs.

10.53. Prove that the axiom of choice is equivalent to Zermelo's theorem.

Given a cardinal number $\mathfrak{m}$, define $Z(\mathfrak{m})$ as the set of all ordinals α such that $\bar{\alpha} \leqslant \mathfrak{m}$.

10.54. Prove that $Z(\mathfrak{m})$ is an initial ordinal.

10.55. Define $\aleph(\mathfrak{m}) = \overline{Z(\mathfrak{m})}$. Prove that $\sim \aleph(\mathfrak{m}) \leqslant \mathfrak{m}$.

10.56. Prove that $\mathfrak{m} < \mathfrak{m} + \aleph(\mathfrak{m})$.

10.57. Prove (without the axiom of choice) that $\aleph(\mathfrak{m}) \leqslant 2^{2^{2^{\mathfrak{m}}}}$.

10.58. Prove (without the axiom of choice) that $\aleph(\mathfrak{m}) \leqslant 2^{2^{\mathfrak{m}^2}}$.

10.59. Prove (using the axiom of choice) that $\aleph(\mathfrak{m}) \leqslant 2^{\mathfrak{m}}$.

10.60. Prove (without the axiom of choice) that if $\mathfrak{m}$ is an infinite cardinal and $\mathfrak{m} + \aleph(\mathfrak{m}) = \mathfrak{m} \cdot \aleph(\mathfrak{m})$ then $\mathfrak{m}$ itself is an aleph.

10.61. Prove that for every cardinal number $\mathfrak{m}$ there exists a cardinal number $\mathfrak{n}$ such that $\mathfrak{m} \leqslant \mathfrak{n}$ and $\mathfrak{n}^2 = \mathfrak{n}$.

10.62. Prove that if, for every infinite cardinal number $\mathfrak{n}$, $\mathfrak{n}^2 = \mathfrak{n}$ then all infinite cardinal numbers are alephs.

10.63. Prove the following lemma of Sierpiński: If $\mathfrak{m} + \mathfrak{n} = 2^{2\mathfrak{m}}$ then $\mathfrak{n} \geqslant 2^{\mathfrak{m}}$.

10.64. The *generalized continuum hypothesis* is the following statement: For all infinite $\mathfrak{m}$ and $\mathfrak{n}$, $\sim(\mathfrak{m} < \mathfrak{n} < 2^{\mathfrak{m}})$. Prove that the generalized continuum hypothesis implies the axiom of choice.

10.65. Prove that, for every α, $\aleph_\alpha^{\aleph_\alpha} = 2^{\aleph_\alpha}$.

10.66. Prove that, if for all $n \in \mathcal{N}$, $2^{\aleph_n} = \aleph_{\omega+1}$, then $2^{\aleph_\omega} = \aleph_{\omega+1}$

NOTE

[1] Thus in the case of well-orderings, the notion of relational type can be eliminated (cf. p. 81, Chapter VIII).

CHAPTER 11

FORMAL SYSTEMS AND THEIR PROPERTIES

Let T_1, T_2, T_3 be pairwise disjoint sets. Their elements are called – respectively – *predicate symbols, function symbols,* and *variables.*

We assume that the set T_3 is infinite. Moreover, we are given a function ϑ on the set $T_1 \cup T_2$ with the values in the set of natural numbers such that whenever $R \in T_1$ then $\vartheta R \neq O$.

Intuitiviely, the function ϑ carries the information on the arity of the relation or of the function described by the corresponding relational symbol or by the corresponding function symbol.

The set of terms is defined as the least set $\mathscr{T}$ such that

(1) $T_3 \subset \mathscr{T}$.

(2) If $F \in T_2$, $\vartheta F = n$ and $t_1, \ldots, t_n \in \mathscr{T}$ then $F(t_1, \ldots, t_n) \in \mathscr{T}$.

From the above definition it follows that whenever $\vartheta F = O$ then $F \in T$. Such function symbols are called *constants.*

The *set of formulas* is the smallest set $\mathscr{F}$ such that:

(1) If $R \in T_1$, $\vartheta R = n$ and $t_1, \ldots, t_n \in \mathscr{T}$ then $R(t_1, \ldots, t_n) \in \mathscr{F}$.

(2) If $t_1, t_2 \in \mathscr{T}$ then $t_1 = t_2 \in \mathscr{F}$.

(3) If Φ, $\Psi \in \mathscr{F}$ then $\Phi \vee \Psi \in \mathscr{F}$, $\sim\Phi \in \mathscr{F}$, $\Phi_1 \Psi \in \mathscr{F}$, $\Phi \Rightarrow \Psi \in \mathscr{F}$ and for every $x \in T_3$, $\bigwedge_x \Phi \in \mathscr{F}$ and $\bigvee_x \Phi \in \mathscr{F}$.

The formalized language is a quadruple $\mathbf{J} = \langle T_1, T_2, T_3, \vartheta \rangle$, and the sets $\mathscr{T}_\mathbf{J}$ and $\mathscr{F}_\mathbf{J}$ are, respectively, its set of terms and its set of formulas.

Note: Often, when $R \in T_1$ and $\vartheta(R) = 2$, we write $x_1 R x_2$ instead of $R(x_1\ x_2)$.

$\mathbf{L}$ is the set of all the formulas arising from the tautologies of the propositional calculus or the tautologies of the predicate calculus by the substitution of arbitrary propositional functions for the propositional variables or the predicate variables.

In addition, we include in the set $\mathbf{L}$ all the expressions which follow immediately from the basic properties of the equality relation, for instance: $x_1 = x_1$, $[x = y \wedge \Phi(x)] \Rightarrow \Phi(y)$ etc. Finally we accept all statements following from the fact that a given symbol is a function symbol, for instance the expressions

$$z = x \Rightarrow f(z) = f(x) \quad \text{or} \quad \Phi[f(x), y] \Rightarrow \bigvee_z \Phi(z, y).$$

The set **L** is called *logic with equality*, or simply *logic*.

Let **J** be a language and let $\mathscr{F}_\mathbf{J}$ be the set of its formulas: A sequence $\langle \Phi_0, \ldots, \Phi_n \rangle$ is called a *proof of the formula* Ψ from the set $X \subset \mathscr{F}_\mathbf{J}$ if

(a) $\Phi_n = \Psi$.

(b) For every $m \leqslant n$ three cases are possible: $\Phi_m \in X \cup (\mathbf{L} \cap \mathscr{F}_\mathbf{J})$, there exist $k, s < m$ such that $\Phi_k = (\Phi_s \Rightarrow \Phi_m)$ or Φ_m is of the form $\bigwedge_x \Psi(x)$ and $\Psi(x)$ is Φ_k for a certain $k < m$.

If there exists a proof of Ψ from X then Ψ is called a *consequence of the set* X. The set of all consequences of the set X is denoted by $\mathrm{Cn}\, X$.

A *formal system* (*theory*) in the language **J** is a set of formulas S such that $S = \mathrm{Cn}\, S$.

A system S is called *consistent* if $S \neq \mathscr{F}_\mathbf{J}$.

A system S is called *complete* if, for every $\Phi \in \mathscr{F}_\mathbf{J}$, $\Phi \in S$ or $\sim\Phi \in S$.

A set $X \subset \mathscr{F}_\mathbf{J}$ is a *set of axioms* (*axiomatics*) for the system S if $S = \mathrm{Cn}\, X$.

The above definitions imply that the language in which S is built is determined by the system. Therefore we often omit the definition of the language while defining the system itself.

11.1. Prove that for a given language $\mathbf{J} = \langle T_1, T_2, T_3, \vartheta \rangle$:

(a) $\overline{\overline{\mathscr{T}_\mathbf{J}}} = \max(\overline{\overline{T_2}}, \overline{\overline{T_3}})$,

(b) $\overline{\overline{\mathscr{F}_\mathbf{J}}} = \max(\overline{\overline{\mathscr{T}_\mathbf{J}}}, \overline{\overline{T_1}})$.

Let the language **J** be fixed and consider the operation $\mathrm{Cn}\, X$ on the sets of formulas of **J** Problems 11.2–11.16):

Note: In the following problems the letters of the Latin alphabet are used to denote the sets of formulas whereas the letters of the Greek alphabet are used to denote the formulas.

11.2. $X \subset Y \Rightarrow \mathrm{Cn}\, X \subset \mathrm{Cn}\, Y$.

11.3. $X \subset \mathrm{Cn}\, Y \wedge Y \subset \mathrm{Cn}\, Z \Rightarrow X \subset \mathrm{Cn}\, Z$.

11.4. $\mathrm{Cn}\,\mathrm{Cn}\, X = \mathrm{Cn}\, X$.

11.5. $\mathrm{Cn}\, O = \mathbf{L} \cap \mathscr{F}_\mathbf{J}$.

11.6. $\mathrm{Cn}\{\Phi, \Psi, \Theta\} = \mathrm{Cn}\{\Phi \wedge \Psi, \Theta\} = \mathrm{Cn}\{\Phi \wedge \Psi \wedge \Theta\}$.

11.7. $\mathrm{Cn}\{\Phi\} \cap \mathrm{Cn}\{\Psi\} = \mathrm{Cn}\{\Phi \vee \Psi\}$.

11.8. $(\Phi \Rightarrow \Psi) \in \mathrm{Cn}\, X \Leftrightarrow \Psi \in \mathrm{Cn}(X \cup \{\Phi\})$.

11.9. $(\Phi \vee \Psi) \in \mathrm{Cn}\, X \Leftrightarrow \Psi \in \mathrm{Cn}\,[X \cup (\sim \Phi)]$.

11.10. $\mathrm{Cn}\,[\Phi(x)] = \mathrm{Cn}\,[\bigwedge_x \Phi(x)]$.

11.11. $\Psi(x) \in \mathrm{Cn}\, X \Rightarrow \bigvee_x \Psi(x) \in \mathrm{Cn}\, X$.

11.12. $\Psi \in \mathrm{Cn}(X \cup \{\Phi\}) \wedge \Psi \in \mathrm{Cn}(X \cup \{\Theta\}) \Rightarrow \Psi \in \mathrm{Cn}(X \cup \{\Phi \vee \Theta\})$.

11.13. $\mathrm{Cn}(X \cup Y) = \mathrm{Cn}(\mathrm{Cn}\, X \cup \mathrm{Cn}\, Y) = \mathrm{Cn}(X \cup \mathrm{Cn}\, Y)$.

11.14. $\mathrm{Cn}\, X = \bigcup_{Y \in \mathrm{Fin}\, X} \mathrm{Cn}\, Y$, where $\mathrm{Fin}\, X = \{Y \subseteq X \colon \overline{\overline{Y}} < \aleph_0\}$.

11.15. $\mathrm{Cn}(X - \mathbf{L}) = \mathrm{Cn}\, X$.

11.16. $\Phi \vee \Psi \in \mathrm{Cn}\, X \wedge \Theta \in \mathrm{Cn}(Y \cup \{\Phi\}) \Rightarrow \Psi \vee \Theta \in \mathrm{Cn}(X \cup Y)$.

11.17. Prove that, for every set of formulas X, Cn X is a system. Is $\mathscr{F}_{\mathbf{J}}$ a system?

11.18. Are the following sets of formulas systems?

(a) $\{\Phi, \Psi\}$,

(b) O,

(c) $\mathbf{L} \cap \mathscr{F}_{\mathbf{J}}$.

11.19. Is it true that if $X \subset \mathscr{F}_{\mathbf{J}}$ is a theory then every Y such that $X \subset Y \subset \mathscr{F}_{\mathbf{J}}$ is also a theory? What about Y such that $Y \subset X$? Give appropriate examples.

11.20. Let $\mathbf{J} = \langle T_1, T_2, T_3, \vartheta \rangle$ be a language and let $a_1, a_2 \in T_2$ be such that $\vartheta(a_1) = \vartheta(a_2) = O$ (i.e. a_1 and a_2 are constants). Prove that $\Phi(a_1) \in \mathrm{Cn}\, O \Leftrightarrow \Phi(a_2) \in \mathrm{Cn}\, O$ for all $\Phi \in \mathscr{F}_{\mathbf{J}}$.

11.21. Let $S_\alpha (\alpha < \mu)$ be a family of systems such that $\alpha < \beta \Rightarrow S_\alpha \subset S_\beta$. Prove that $\bigcup_{\alpha < \mu} S_\alpha$ is also a system.

11.22. Let $S_{\mathbf{J}}$ be the set of all systems of a language $\mathbf{J}$. Prove that:

(a) $\mathrm{Cn}\, X = \bigcap_{X \subset Y \in S_{\mathbf{J}}} Y$.

(b) If $O \neq K \subset S_{\mathbf{J}}$, then $\bigcap_{X \in K} X \in S_{\mathbf{J}}$.

11.23. Prove that if $X, Y \in S_{\mathbf{J}}$, then

$$X \cup Y \in S_{\mathbf{J}} \Leftrightarrow (X \cup Y = X) \vee (X \cup Y = Y).$$

11.24. Prove that if $X \in S_{\mathbf{J}}$, $\Phi, \Psi \in X$ then

$$X \cup \{\Phi \vee \Psi\} = X \cup \{\Phi \wedge \Psi\} = X \quad \text{and} \quad X = X \cup \{\Theta \Rightarrow \Psi\}$$

for any formula θ.

11.25. Prove that if X is a system in a language $\mathbf{J}$ then X is a filter in the Lindenbaum algebra of $\mathbf{J}$ (cf. Problem 12.106).

Ax_S is the family of all sets of axioms for a system S. Thus: $U \in \mathrm{Ax}_S \Leftrightarrow \mathrm{Cn}\, U = S$.

11.26. Prove that if $X \in \mathrm{Ax}_S$ then $X - \mathbf{L} \in \mathrm{Ax}_S$.

11.27. Prove that whenever $X \in \mathrm{Ax}_S$ then, for every $Y \subset S$, $X \cup Y \in \mathrm{Ax}_S$.

11.28. Prove that if $X \in \mathrm{Ax}_{S_1}$ and $Y \in \mathrm{Ax}_{S_2}$ then $X_1 \cup X_2 \in \mathrm{Ax}_{\mathrm{Cn}(S_1 \cup S_2)}$.

11.29. What can be said about S is $O \in \mathrm{Ax}_S$?

11.30. Prove that if $X_1 \in \mathrm{Ax}_S$ and $X_2 \in \mathrm{Ax}_S$ then

$$\bigwedge_{\Phi \in \mathscr{F}} (\Phi \in X_1 \Rightarrow \Phi \in \mathrm{Cn}\, X_2) \wedge (\Phi \in X_2 \Rightarrow \Phi \in \mathrm{Cn}\, X_1).$$

11.31. Prove that if X is a consistent system and $Y \subset X$ then $\mathrm{Cn}\, Y \neq \mathscr{F}$.

11.32. Prove that if $X \in S_{\mathbf{J}}$ and $Y \in S_{\mathbf{J}}$ and both X and Y are consistent then $X \cap Y$ is a consistent system.

In the sequel, instead of $\Phi \in \mathrm{Cn}\, X$, we often write $X \vdash \Phi$ or $\vdash_X \Phi$. If $X = O$ then we use the notation $\vdash \Phi$, which simply means that Φ is provable from the axioms of logic.

11.33. Prove that if X is a consistent system then X is a filter of the Lindenbaum algebra (cf. 11.25).

11.34. Prove that if there is a finite Y such that $Y \in \mathrm{Ax}_S$ then S is a principal filter of a Lindenbaum algebra.

11.35. Prove that if X is a set of axioms of an inconsistent system then there is a finite $Y \subset X$ such that $\mathrm{Cn}\, Y$ is inconsistent.

11.36. Prove that if X is a set of axioms of a complete system and $\Phi \notin \mathrm{Cn}\, X$ then $X \cup \{\Phi\}$ is inconsistent.

Let a language $\mathbf{J} = \langle T_1, T_2, T_3, \vartheta \rangle$ be given and let $S \in S_{\mathbf{J}}$. We say that $\Phi(x_0, \ldots, x_n)$ is a definition of a function symbol in S iff

(a) $\quad \vdash_S \bigwedge_{x_1} \cdots \bigwedge_{x_n} \bigvee_{x_0} \Phi(x_0, \ldots, x_n)$,

(b) $\quad \vdash_S \bigwedge_{x_1} \cdots \bigwedge_{x_n} \bigwedge_{x_0} \bigwedge_{x_0'} [\Phi(x_0, \ldots, x_n) \wedge \Phi(x_0', \ldots, x_n) \Rightarrow x_0 = x_0']$.

Let $\mathscr{F}_{\mathbf{J}}^n$ be the set of formulas of a language $\mathbf{J}$ with n variables. If $\Phi \in \mathscr{F}_{\mathbf{J}}^1$ and Φ satisfies a) and b) then we say that Φ is a *definition of a constant in S.*

11.37. Consider a system S' arising from S by extending the language J by a function symbol f and a new axiom

$$\Phi' = \bigwedge_{x_1} \cdots \bigwedge_{x_n} \Phi[f(x_1, \ldots, x_n), x_1, \ldots, x_n],$$

where $\Phi \in \mathscr{F}_{\mathbf{J}}^{n+1}$ and Φ is a definition of a function symbol in S.

Thus S' is a system in the language:

$$\langle T_1, T_2 \cup \{f\}, T_3, \vartheta' \rangle, \quad \text{where} \quad \vartheta = \vartheta' \restriction T_1 \cup T_2, \vartheta' f = n.$$

Prove that $S = S' \cap \mathscr{F}_{\mathbf{J}}$.

11.38. Given a system S and a formula Φ, we introduce a system S' as follows: The language $\mathbf{J}$ is extended by new relational symbol R and a new axiom:

$$\Phi' = [\bigwedge_{x_1} \cdots \bigwedge_{x_n} R(x_1, \ldots, x_n) \Leftrightarrow \Phi(x_1, \ldots, x_n)].$$

Thus our system is constructed in the language

$$\langle T_1 \cup \{R\}, T_2, T_3, \vartheta' \rangle, \quad \text{where} \quad \vartheta = \vartheta' | T_1 \cup T_2, \ \vartheta' R = n.$$

Prove that $S = S' \cap \mathscr{F}_{\mathbf{J}}$.

Note: Systems arising by a certain number of consecutive application of the operations from 11.37 and 11.38 are called *improper extensions* of S.

11.39. Prove that if S' is an improper extension of S then S is a complete system iff S' is a complete system.

11.40. Let $S_1 \leqslant S_2$ mean that the system S_2 is an improper extension of the system S_1. Is $\leqslant$ and ordering relation?

11.41. Prove that if $S_\alpha (\alpha < \mu)$ is a family of systems such that $\alpha_1 \leqslant \leqslant \alpha_2 \Rightarrow S_{\alpha_1} \leqslant S_{\alpha_2}$ then $\bigcup_{\alpha < \mu} S_\alpha$ is a system and for each $\beta < \mu$ we have $S_\beta \leqslant \bigcup_{\alpha < \mu} S_\alpha$.

11.42. Prove that if $S_0 \leqslant S_1$ and $S_0 \leqslant S_2$ then there exists a system S_3 such that $S_1 \leqslant S_3$ and $S_2 \leqslant S_3$ and the constants added in passing from S_1 to S_3 are different from all the constants added in passing from S_1 to S_3.

11.43. Do there exist elements maximal or minimal in $\leqslant$? Does there exist a least element?

11.44. Given a system S in the language $\mathbf{J}$ and its improper extension S', prove that $S = S' \cap \mathscr{F}_{\mathbf{J}}$.

Consider the following formulas of the language $J=\langle\{\leqslant\}, O, \{x_1, \ldots\}, \vartheta\rangle$, where $\vartheta(\leqslant)=2$

(a) $\bigwedge_{x_1} x_1 \leqslant x_1$,

(b) $\bigwedge_{x_1}\bigwedge_{x_2}\bigwedge_{x_3}[(x_1\leqslant x_2)\wedge(x_2\leqslant x_3)\Rightarrow(x_1\leqslant x_3)]$,

(c) $\bigwedge_{x_1}\bigwedge_{x_2}[(x_1\leqslant x_2)\wedge(x_2\leqslant x_1)\Rightarrow x_1=x_2]$,

(d) $\bigvee_{x_1}\bigwedge_{x_2} x_1\leqslant x_2$,

(e) $\bigvee_{x_1}[\sim\bigvee_{x_2}(x_1\leqslant x_2)\wedge(x_1\neq x_2)]$,

(f) $\sim\bigvee_{x_1}\bigwedge_{x_2} x_2\leqslant x_1$,

(g) $\bigwedge_{x_1}\bigvee_{x_2}\bigwedge_{x_3}[(x_1\leqslant x_3\wedge x_3\leqslant x_2)\Rightarrow((x_1=x_3)\vee(x_3=x_2))]$,

(h) $\bigwedge_{x_1}\bigwedge_{x_2}[x_1\leqslant x_2)\vee(x_2\leqslant x_1)]$,

(i) $\bigvee_{x_1}\bigwedge_{x_2} x_2\leqslant x_1$,

(j) $\sim\bigvee_{x_1}\bigwedge_{x_2} x_1\leqslant x_2$,

(k) $\bigwedge_{x_1}\bigwedge_{x_2}[x_1\neq x_2\wedge x_1\leqslant x_2\Rightarrow\bigvee_{x_3}(x_1\neq x_3\wedge x_1\leqslant x_3)\wedge$
$\wedge(x_3\neq x_2\wedge x_3\leqslant x_2)]$.

11.45. Let S be the set of consequences of formulas (a), (b), (c), (d). Show that the formula $\bigwedge_{x_2} x_1\leqslant x_2$ is a definition of a constant in $\boldsymbol{S}$. Is the formula $\bigvee_{x_2} x_1\leqslant x_2$ a definition of a constant in S?

11.46. Let S be the set of consequences of formulas (a), (b), (c), (d), (f), (g). Prove that the formula

$$\bigwedge_{x_3}(x_2\leqslant x_3)\wedge(x_3\leqslant x_1)\Rightarrow(x_2=x_3)\vee(x_3=x_1)$$

is a definition in S of a (unary) function symbol.

11.47. Prove that in the system of 11.45 there is only one constant definable.

11.48. Prove that the system consisting of the consequences of (a), (b), (c), (d), (f), (h), (k) is complete.

11.49. Prove that the system consisting of the consequences of (a), (b), (c) (f), (j), (h), (k) is complete.

11.50. Prove that the system consisting of the consequences of (a), (b), (c), (d), (h), (i), (k) is complete.

11.51. Prove that the system consisting of the consequences of (a), (b), (c), (f), (h), (j), (k) is complete.

11.52. Prove that there is no constant definable in the system of 11.51.

11.53. Prove that in the system of 11.50 there are exactly two constants definable.

Let systems S_1, S_2 be defined in languages $\mathbf{J}^1$ and $\mathbf{J}^2$, respectively, $\mathbf{J}^1 = \langle T_1^1, T_2^1, T_3^1, \vartheta^1 \rangle$, $\mathbf{J}^2 = \langle T_1^1, T_2^2, T_3^2, \vartheta^2 \rangle$. Assume moreover that $T_3^1 = T_3^2$.

Now let φ_1 be a mapping of T_1^1 into T_1^2 and let φ_2 be a mapping of T_2^1 into T_2^2 such that $\vartheta^2(\varphi_i(x)) = \vartheta^1(x)$. Moreover, let Θ be a formula (with one free variable) of $\mathbf{J}^2$. We define the mappings φ_1' and φ_2' of terms and formulas of the language $\mathbf{J}_1$ into the corresponding sets of the language $\mathbf{J}_2$ as follows.

We start with φ_2', the mapping on the set of terms of $\mathbf{J}_1$.

(a) If $F \in T_2^1$, $\vartheta^1(F) = n$, then

$$\varphi_2' F(x_1, \ldots, x_n) = (\varphi_2 F)(x_1, \ldots, x_n).$$

(b) If $F \in T_2^1$, $\vartheta^1(F) = n$, $t_1, \ldots, t_n \in T_J^1$, then

$$\varphi_2' F(t_1, \ldots, t_n) = \varphi_2(F)(\varphi_2' t_1, \ldots, \varphi_2' t_n).$$

Having defined φ_2', we are able to construct φ_1' as follows

(a) If $R \in T_1^1$, $\vartheta^1(R) = n$, $t_1, \ldots, t_n \in \mathscr{F}_{\mathbf{J}^1}$, then

$$\varphi_1' [R(t_1, \ldots, t_n)] = (\varphi_1 R)(\varphi_2' t_1, \ldots, \varphi_2' t_n).$$

(b) If $t_1, t_2 \in T_1^1$, then $\varphi'(t_1 = t_2) = (\varphi_2' t_1 = \varphi_2' t_2)$.

(c) If $\Phi, \Psi \in \mathscr{F}_{\mathbf{J}^1}$, then $\varphi_1'(\Phi \vee \Psi) = (\varphi_1' \Phi) \vee (\varphi_1' \Psi)$,

$$\varphi_1'(\Phi \wedge \Psi) = (\varphi_1' \Phi) \wedge (\varphi_1' \Psi),$$

$$\varphi_1'(\Phi \Rightarrow \Psi) = (\varphi_1' \Phi \Rightarrow \varphi_1' \Psi),$$

$$\varphi_1'(\sim \Phi) = \sim \varphi_1' \Phi,$$

$$\varphi_1' = (\bigwedge_x \Phi(x)) = \bigwedge_{\theta(x)} \varphi_1' \Phi(x),$$

$$\varphi_1' = (\bigvee_x \Phi(x)) = \bigvee_{\theta(x)} \varphi_1' \Phi(x).$$

A function φ_1' is called an *interpretation* of the language $\mathbf{J}^1$ in the language $\mathbf{J}^2$. Thus the interpretation is determined by functions φ_1, φ_2 and formula Θ.

We say that a system $X_1 \in S_{\mathbf{J}^1}$ is *interpretable* in a system X_2 if there exists an improper extension X_2' of X_2 and an interpretation

φ of the language $\mathbf{J}^1$ in the language $\mathbf{J}^2$ of the system X_2' such that for an arbitrary Φ we have

$$\Phi \in X_1 \Rightarrow \varphi\Phi \in X_2'.$$

We denote this situation by X_1 Int X_2.

11.55. Prove that the relation Int is reflexive and transitive. Is it antisymmetric? Is it symmetric?

11.56. Let $\{X_i : i \in I\}$ be a family of systems in the languages $\mathbf{J}^i = \langle T_1^i, T_2^i, T_3^i, \vartheta^i \rangle$, respectively, such that $T_3^i = T_3^j$ for all $i, j \in I$, $\bigcap_{i \in I} T_1^i \neq 0$. Define the relation $\sim$ as follows:

$$X_i \sim X_j \Leftrightarrow X_i \operatorname{Int} X_j \wedge X_j \operatorname{Int} X_i.$$

Prove that the relation $\sim$ is reflexive, symmetric, and transitive. Prove that the relation Int/$\sim$ is an ordering. Are there minimal elements in Int/$\sim$? What about a least one?

11.57. Let $S_1 \in S_{\mathbf{J}^1}$, $S_2 \in S_{\mathbf{J}^2}$. Prove that S_1 Int S_2 if and only if there exist an interpretation φ of $\mathbf{J}^1$ in $\mathbf{J}^2$ and $X \in \mathrm{Ax}_{S_1}$ such that $\varphi X \subset S_2$.

11.58. Prove that if X_1 Int X_2 and X_2 is a consistent system then X_1 is also consistent.

11.59. Let S_1 be a system consisting of the consequences of the following (a_1), (b_1), (c_1), and (d_1):

(a_1) $\bigwedge_{x_1} x_1 \leqslant x_1$,

(b_1) $\bigwedge_{x_1} \bigwedge_{x_2} (x_1 \leqslant x_2 \wedge x_2 \leqslant x_1 \Rightarrow x_2 = x_1)$,

(c_1) $\bigwedge_{x_1} \bigwedge_{x_2} \bigwedge_{x_3} [(x_1 \leqslant x_2) \wedge (x_2 \leqslant x_3) \Rightarrow (x_1 \leqslant x_3)]$,

(d_1) $\bigwedge_{x_1} \bigwedge_{x_2} [(x_1 \leqslant x_2) \vee (x_2 \leqslant x_1)]$.

Moreover, let S_2 be a system consisting of the consequences of the following (a_2), (b_2), (c_2):

(a_2) $\bigwedge_{x_1} \sim x_1 R x_1$,

(b) $\bigwedge_{x_1} \bigwedge_{x_2} \bigwedge_{x_3} (x_1 R x_2 \wedge x_2 R x_3 \Rightarrow x_1 R x_3)$,

(c_2) $\bigwedge_{x_1} \bigwedge_{x_2} (x_1 R x_2 \vee x_2 R x_1 \vee x_1 = x_2)$.

Prove that S_1 Int S_2 and S_2 Int S_1.

11.60. Let S_1 be a system consisting of the consequences of the following (a)–(h):

(a) $\bigvee_{x_1}\bigwedge_{x_2} x_1 \leqslant x_2,$

(b) $\bigwedge_{x_1} x_1 \leqslant x_1,$

(c) $\bigwedge_{x_1}\bigwedge_{x_2}\bigwedge_{x_3}[(x_1 \leqslant x_2)\wedge(x_2 \leqslant x_3) \Rightarrow (x_1 \leqslant x_3)],$

(d) $\bigwedge_{x_1}\bigwedge_{x_2}[(x_1 \leqslant x_2)\wedge(x_2 \leqslant x_1) \Rightarrow (x_1 = x_2)],$

(e) $\bigwedge_{x_1}\bigwedge_{x_2}[(x_1 \leqslant x_2)\vee(x_2 \leqslant x_1)],$

(f) $\bigwedge_{x_1}\bigvee_{x_2}\bigwedge_{x_3}(x_1 \neq x_2)\wedge[(x_1 \leqslant x_3 \wedge x_3 \leqslant x_2) \Rightarrow (x_1 = x_3)\vee(x_3 = x_2)],$

(g) $\bigwedge_{x_1}[\bigvee_{x_2} x_2 \neq x_1 \wedge x_2 \leqslant x_1 \Rightarrow \bigvee_{x_2}\bigwedge_{x_3}((x_1 \neq x_2 \wedge x_1 \leqslant x_3 \wedge x_3 \leqslant x_2)$
$\Rightarrow (x_1 = x_3)\vee(x_3 = x_2))],$

(h) $\sim\bigvee_{x_1}\bigwedge_{x_2} x_2 \leqslant x_1 .$

Prove that S_1 is complete.

11.61. Let S_2 be a system consisting of the consequences of (b)–(d) of 11.60 and the following (i) and (j):

(i) $\sim\bigvee_{x_1}\bigwedge_{x_2} x_1 \leqslant x_2,$

(j) $\bigwedge_{x_1} x_1 \leqslant 0 \vee 0 \leqslant x_1 .$

Prove that S_2 is complete.

11.62. Prove that S_1 Int S_2.

The *Zermelo–Fraenkel set theory* (usually denoted by ZFC) is a system with the following axioms:

(a) Extensionality axiom:

$$\bigwedge_{y}\bigwedge_{z}[(\bigwedge_{x} x \in y \Leftrightarrow x \in z) \Rightarrow y = z].$$

(b) Empty set axiom:

$$\bigvee_{y}\bigwedge_{x} x \notin y .$$

Extensionality implies that the empty set is unique.

(c) Sum set axiom:

$$\bigwedge_{x}\bigvee_{u}\bigwedge_{w}[w \in u \Leftrightarrow \bigvee_{z}(z \in x \wedge w \in z)].$$

(d) Infinity axiom:

$$\bigvee_x x\neq O \wedge \bigwedge_y (y\in x\Rightarrow \bigvee_z (z\in x\wedge y\neq z\wedge y\subset z)),$$

where $y\subset z$ is an abbreviation of the formula $\bigwedge_u (u\in y\Rightarrow u\in z)$.

(e) Axiom of choice:

$$\bigwedge_x [x\neq O\wedge \sim O\in x\wedge \bigwedge_{y_1}\bigwedge_{y_2}[y_1\in x\wedge y_2\in x$$

$$\Rightarrow (\bigvee_z (z\in y_1\wedge z\in y_2)\Leftrightarrow y_1=y_2)]\Rightarrow \bigvee_w (\bigwedge_z z\in x\Rightarrow \dot{\bigvee_y} y\in z\wedge y\in w)],$$

where $\dot{\bigvee_y}\Phi(y)$ is an abbreviation of the formula $\bigvee_y [\Phi(y)\wedge \bigwedge_z (\Phi(z) \Rightarrow z=y)]$.

(f) Power-set axiom:

$$\bigwedge_x\bigvee_y\bigwedge_z (z\in y\Leftrightarrow z\subset x),$$

(g) Substitution:
Given the formula Φ, the formula

$$\bigwedge_x \dot{\bigvee_y}\Phi(x, y, ...)\rightarrow \bigwedge_u\bigvee_w\bigwedge_z [z\in w\Leftrightarrow \bigvee_x x\in u\wedge \Phi(x, z, ...)]$$

is accepted as an axiom.

Thus the system ZFC is based on infinite axiomatization.

The system ZF′ arises from ZFC by dropping axiom (d).

The *Peano Arithmetic*, Ar, is a system in a language with two binary function symbols $+$and $\cdot$, one unary function symbol S and a constant O with the following axioms:

(h) $\sim \bigvee_x S(x)=0$,

(i) $S(x)=S(y)\Rightarrow x=y$,

(j) $x+0=x$,

(k) $x+S(y)=S(x+y)$,

(l) $x\cdot 0=0$,

(m) $x\cdot S(y)=x+(x\cdot y)$,

(n) For every formula Φ the following is an axiom:

$$\Phi(0) \wedge \bigwedge_x \{\Phi(x) \Rightarrow \Phi[S(x)]\} \to \bigwedge_x \Phi(x).$$

11.63. Prove that S_1 Int Ar where S_1 is the system from 11.59.

11.64. Prove that ZF′ Int Ar.

11.65. Prove that Ar Int ZF.

CHAPTER 12

MODEL THEORY

A *relational system* (*relational structure*) is a pair $\mathfrak{a}=\langle A, \mathfrak{v}\rangle$ where $\mathfrak{v}$ is a function defined on the union of disjoint sets T_1 and T_2 (we assume $T_1 \cup T_2 \neq O$). If $t \in T_1$ then $\mathfrak{v}(t)$ is an n_t-argument relation in A (i.e. a subset of A^{n_t}) and if $t \in T_2$ then $\mathfrak{v}(t)$ is an m_t-argument function on A (i.e. an element of ${}^{A^{n_t}}A$). If $m_t = O$ then the function is a constant.

It may happen that one of the sets T_1 and T_2 is empty. In the first of these cases the relational system is called an *algebra*.

The relational system $\mathfrak{a}$ determines (cf. chapter XI) the language $\mathbf{J}_{\mathfrak{a}}$, namely a formalized language which contains, for every relation R_t of a relational symbol P_t (an n_t-ary symbol) and for every function f_t, a function symbol φ_s (an m_s-ary symbol).

In addition to the above symbols, $\mathbf{J}_{\mathfrak{a}}$ contains symbols for (countably many) variables and logical symbols and the symbol $=$ of identity.

Conversely, given a language $\mathbf{J}$ with predicate symbols P_t (respectively n_t-ary) and function symbols φ_s (m_s-ary), we consider relational systems $\mathfrak{a}=\langle A, \ldots, R_t, \ldots, \ldots, f_s, \ldots\rangle$ where $R_t \subset A^{n_t}$ and $f_s\colon A^{m_s} \to A$ and we assume that the symbols P_t and φ_s are the names of R_t and f_s, respectively.

Given the relationship between the symbols of the language and the relations and the functions of $\mathfrak{a}$, we introduce the satisfaction *relation* $\models$:

$$\mathfrak{a} \models \Phi[X], \quad \text{where} \quad X \in {}^{\mathcal{N}}A.$$

(Read: The sequence X satisfies in $\mathfrak{a}$ the formula Φ).

We first define the value $t^{\mathfrak{a}}[X]$ of the term t of the language $\mathbf{J}_{\mathfrak{a}}$ in $\mathfrak{a}$ (under the sequence X) inductively, as follows:

(a) $v_i^{\mathfrak{a}}[X] = x_i$,

(b) $(\varphi_s[t_1, \ldots, t_n])^{\mathfrak{a}}[X] = f_s(t_1^{\mathfrak{a}}[X], \ldots, t_n^{\mathfrak{a}}[X])$.

Now we define the relation $\models$ (remember that $\mathfrak{a}$ is fixed) inductively as follows:

for atomic formulas –

$$\mathfrak{a} \models P_t(t_1, \ldots, t_n)[X] \equiv R_t(t_1^{\mathfrak{a}}[X], \ldots, t_n^{\mathfrak{a}}[X]),$$

$$\mathfrak{a} \models t_1 = t_2[X] \equiv t_1^{\mathfrak{a}}[X] = t_2^{\mathfrak{a}}[X];$$

for more complicated formulas –

$$\mathfrak{a} \models \Phi[X] \equiv \text{non } \mathfrak{a} \models \Phi[X],$$

$$\mathfrak{a} \models \Phi \wedge \Psi[X] \equiv (\mathfrak{a} \models \Phi[X] \quad \text{and} \quad \mathfrak{a} \models \Psi[X]);$$

for formulas with the existential quantifier –

$$\mathfrak{a} \models \bigvee_{v_i} \Phi[X] \Leftrightarrow \text{The exists an } a \in A \text{ such that } \mathfrak{a} \models \Phi\left[X\binom{i}{a}\right],$$

where $X\binom{i}{a}$ arises from X by replacing x_i by a. (Thus formally $X\binom{i}{a} = (X - \{i\} \times A) \cup \{\langle i, a\rangle\}$).

12.1. The authors have simplified their task by defining the satisfaction relation for a formula containing only $\sim$, $\wedge$ and the existential quantifier $\bigvee_{x_i}$. Why is it enough?

12.2. Using the tautology $p \vee q \Leftrightarrow \sim(\sim p \wedge \sim q)$, extend the definition of satisfaction for formulas containing the connective $\vee$.

12.3. Why have the authors written $\equiv$, "non", "and" instead of $\Leftrightarrow$, $\sim$, $\wedge$?

12.4. Extend the definition of the satisfaction relation to formulas containing the connectives $\Rightarrow$, $\Leftrightarrow$ and the universal quantifier $\bigwedge_{x_i}$. What tautologies are you going to use?

12.5. Consider the relational system $\mathfrak{a} = \langle \mathcal{N}, +, \cdot, 0\rangle$ and the formula Φ of the language $\mathbf{J}_{\mathfrak{a}}$, $\Phi = \bigvee_{v_3} (v_1 \cdot v_3 = v_2)$.

(a) Does the sequence $\{x_n\}_{n \in \mathcal{N}}$ defined by: $z_n = n + 2$ satisfy the formula Φ?

(b) Does the sequence $\{z_n\}_{n \in \mathcal{N}}$ defined by: $z_n = n$ satisfy the formula Φ?

12.6. Consider the formula Φ of the language of $\mathfrak{a}$ (c.f 12.5)

$$\Phi = \bigvee_{v_4} \bigvee_{v_3} v_4 \cdot v_3 = v_4 + v_1:$$

(a) Does the sequence $\{x_n\}_{n\in\mathcal{N}}$ defined by: $x_n=1$ satisfy the formula Φ?

(b) Does the sequence $\{x_n\}_{n\in\mathcal{N}}$ defined by $x_0=0$, $x_n=n-1$ (for $n\geqslant 1$) satisfy the formula Φ?

12.7. Given the formulas

$$\Phi_1=\bigwedge_{v_2}\sim[v_1+v_2=v_3], \qquad \Phi_2=\bigvee_{v_2}[v_1+v_2=v_3\cdot v_2]:$$

of the language $\mathbf{J}_{\mathfrak{a}}$ (Problem 12.5).

(a) Does the sequence $\{x_n\}_{n\in\mathcal{N}}$, $x_n=n+4$ satisfy Φ_1, Φ_2, $\Phi_1\Rightarrow\Phi_2$?

(b) Does the sequence $\{x_n\}_{n\in\mathcal{N}}$, $x_n=n^2+3$ satisfy Φ_1, Φ_2, $\Phi_1\vee\Phi_2$?

(c) Does the sequence $\{x_n\}_{n\in\mathcal{N}}$, $x_n=2$ satisfy Φ_1,Φ_2, $\bigvee_{v_1}(\Phi_1\wedge\Phi_2)$?

12.8. Prove that if the free variables of Φ have the numbers $i_0, \ldots, i_n$ and $x_{i_0}=y_{i_0}, \ldots, x_{i_n}=y_{i_n}$ then

$$\mathfrak{a}\models\Phi[X]\equiv\mathfrak{a}\models\Phi[Y].$$

12.9. Using the result of 12.8, prove that if Φ is a sentence (i.e. has no free variables), $\mathfrak{a}=\langle A, R_i, \ldots, f_j, \ldots\rangle$, $A\neq 0$ and $X, Y\in{}^{\mathcal{N}}\!A$, then

$$\mathfrak{a}\models\Phi[X]\equiv\mathfrak{a}\models\Phi[Y].$$

12.10. Prove under the assumptions of 12.9,

$$\bigwedge_{X\in\mathcal{N}_A}\mathfrak{a}\models\Phi[X]\equiv\bigvee_{X\in\mathcal{N}_A}\mathfrak{a}\models\Phi[X].$$

We say that Φ is *true* in $\mathfrak{a}$ (under a fixed interpretation of relational and function symbols) iff, for every sequence $X\in{}^{\mathcal{N}}\!A$, $\mathfrak{a}\models\Phi[X]$. We write in this case $\mathfrak{a}\models\Phi$.

12.11. Prove that if Φ is a sentence, $A\neq 0$, then

$$\mathfrak{a}\models\Phi\equiv\bigvee_{X\in\mathcal{N}_A}\mathfrak{a}\models\Phi[X].$$

12.12. Let the free variables of Φ be $v_{i_0}, \ldots, v_{i_n}$. Prove that

$$\mathfrak{a}\models\Phi\equiv\mathfrak{a}\models\bigwedge_{v_{i_0}}\bigwedge_{v_{i_1}}\ldots\bigwedge_{v_{i_n}}\Phi.$$

12.13. Prove that $\mathfrak{a}\models\bigwedge_{v_0}\bigvee_{v_1}\Phi(v_0, v_1)$ iff there exists a function f: $A\to A$ such that

$$\mathfrak{a}\models\Phi(v_1, v_2)[x_0, f(x_0), \ldots].$$

12.14. Prove that for every formula Φ and sequence $X \in {}^{\mathcal{N}}A$

$$\mathfrak{a} \models \Phi[X] \quad \text{or} \quad \mathfrak{a} \models \sim\Phi[X].$$

12.15. Given a relational system $\mathfrak{a}$, let $\mathrm{Th}\,\mathfrak{a}$ be the set of those sentences Φ of $\mathbf{J}_{\mathfrak{a}}$ for which $\mathfrak{a} \models \Phi$. Prove that $\mathrm{Th}\,\mathfrak{a}$ is a complete and consistent set of sentences.

12.16. Find algebras $\mathfrak{a} = \langle A, +, 0\rangle$ and $\mathfrak{b} = \langle B, +', 0'\rangle$ such that

$$\mathfrak{a} \models \bigvee_{x}(x+x=0) \vee \bigvee_{x,y}(x \neq y) \wedge \bigwedge_{x,y}(x+y=y+x) \wedge$$
$$\wedge \bigwedge_{x,y,z}[x+(y+z)=(x+y)+z]$$

and

$$\mathfrak{b} \models \bigwedge_{x}(x+x \neq 0) \wedge \bigvee_{x,y}(x \neq y) \wedge \bigwedge_{x,y}(x+y=y+x) \wedge$$
$$\wedge \bigwedge_{x,y,z}[x+(y+z)=(x+y)+z].$$

12.17. Prove that if $\Phi \in \mathrm{Cn}S$ and $\bigwedge_{\Psi}(\Psi \in S \Rightarrow \mathfrak{a} \models \Psi)$ then $\mathfrak{a} \models \Phi$.

12.18. Using 12.16 and 12.17 prove that neither the sentence $\bigwedge_{x}(x+x=0)$ nor its negation can be deduced from the set of sentences

$$\bigvee_{x,y}(x \neq y), \quad \bigwedge_{x,y}(x+y=y+x), \quad \bigvee_{x,y,z}[x+(y+z)=(x+y)+z].$$

A sentence Φ is *consistent* with the set of sentences S if there exists a relational system $\mathfrak{a}$ such that $S \cup \{\Phi\} \subset \mathrm{Th}\,\mathfrak{a}$ (see 12.15).

A sentence Φ is *independent* of the set of sentences S if both Φ and $\sim\Phi$ are consistent with S.

12.19. Justify the following method of independence proofs: If there exist relational systems $\mathfrak{a}$ and $\mathfrak{b}$ such that $S \subset \mathrm{Th}\,\mathfrak{a}$, $S \subset \mathrm{Th}\,\mathfrak{b}$, $\mathfrak{a} \models \Phi$ and $\mathfrak{b} \models \sim\Phi$ then the sentence Φ is independent of S.

12.20. Prove that the sentence

$$A_3 = \bigvee_{v_0,v_1,v_2} \bigwedge_{v_3} \{[(v_3=v_0) \vee (v_3=v_1) \vee (v_3=v_2)] \wedge$$
$$\wedge (v_0 \neq v_1) \wedge (v_1 \neq v_2) \wedge (v_0 \neq v_2)\}$$

is independent of the axiomatics of fields.

12.21. Check if the sentence

$$A_6 = \bigvee_{v_0, v_1, \ldots, v_5} \bigwedge_{v_6} \{[(v_6 = v_0) \vee \ldots \vee (v_6 = v_5)] \wedge$$
$$\wedge (v_0 \neq v_1) \wedge \ldots \wedge (v_0 \neq v_5) \wedge (v_1 \neq v_2) \wedge \ldots \wedge (v_4 \neq v_5)\}$$

is independent of the axiomatics of fields.

12.22. Check if the sentence

$$A_4 = \bigvee_{v_0, v_1, v_2, v_3} \bigwedge_{v_4} [(v_4 = v_0) \vee \ldots \vee (v_4 = v_3)]$$

is independent of the axiomatics of fields.

12.23. Prove that the sentence

$$\bigwedge_{v_0, v_1} (v_0 \cdot v_1 = v_1 \cdot v_0) \quad \text{(commutativity)}$$

is independent of the axiomatics of groups.

12.24. Prove that the sentence

$$\bigwedge_{v_0, v_1} \{(v_0 \neq v_1 \wedge v_0 \leqslant v_1) \Rightarrow \bigvee_{v_2} [(v_2 \neq v_0 \wedge v_2 \neq v_1 \wedge v_0 \leqslant v_2 \wedge v_2 \leqslant v_1)]\}$$

is independent of the axiomatics of linear orderings.

12.25. Check if the statement asserting the existence of a least element is independent of the axiomatics of ordering.

A relational system $\mathfrak{b} = \langle B, S_i, \ldots, g_j, \ldots \rangle$ is called a *subsystem* of a system $\mathfrak{a} = \langle A, R_i, \ldots, f_j, \ldots \rangle$ if $R_i \cap B^{n_i} = S_i$, $g_j = f_j \restriction B$ (in particular the functions f_j on the arguments from B take the values in B and all the constants of $\mathfrak{a}$ belong to B). In symbols $\mathfrak{b} \subset \mathfrak{a}$, and we also say in this situation that $\mathfrak{a}$ is an *extension* of $\mathfrak{b}$.

12.26. In this problem we consider formulas in prenex normal form (cf. Problem 3.197, 3.198).

A formula Φ is said to be *open* if it does not contain quantifiers, *universal*—if all its quantifiers are universal ones, and *existential*—if all its quantifiers are existential.

(a) Prove that, if $\mathfrak{b} \subset \mathfrak{a}$, Φ is open or universal, $\mathfrak{a} \models \Phi$, then $\mathfrak{b} \models \Phi$.

(b) Prove that, if $\mathfrak{b} \subset \mathfrak{a}$, Φ is existential and $\mathfrak{b} \models \Phi$, then $\mathfrak{a} \models \Phi$.

12.27. Prove that if $\mathfrak{a}$ is a ring and $\mathfrak{b} \subset \mathfrak{a}$, then $\mathfrak{b}$ is also a ring.

12.28. Prove that if $\mathfrak{a}$ is a group and $\mathfrak{b} \subset \mathfrak{a}$, then $\mathfrak{b}$ is also a group.

12.29. Prove that if a formula Φ is equivalent to an open formula or to a universal formula, then Φ has the property of 12.26 (a). Prove that statement A_4 (Problem 12.22) is not equivalent to a universal formula.

Systems $\mathfrak{a}$ and $\mathfrak{b}$ are called *elementarily equivalent* (in symbols $\mathfrak{a} \equiv \mathfrak{b}$) if $\mathrm{Th}\mathfrak{a} = \mathrm{Th}\mathfrak{b}$ (cf. Problem 12.15), i.e. for every proposition Φ, $\mathfrak{a} \models \Phi \equiv \mathfrak{b} \models \Phi$ (in particular $\mathbf{J}_\mathfrak{a} = \mathbf{J}_\mathfrak{b}$).

12.30. Prove that the relation $\equiv$ is an equivalence, i.e.

(a) $\mathfrak{a} \equiv \mathfrak{a}$,

(b) $\mathfrak{a} \equiv \mathfrak{b} \Rightarrow \mathfrak{b} \equiv \mathfrak{a}$,

(c) $\mathfrak{a} \equiv \mathfrak{b} \wedge \mathfrak{b} \equiv \mathfrak{c} \Rightarrow \mathfrak{a} \equiv \mathfrak{c}$.

12.31. Prove that the system $\langle \mathscr{Q}, \leqslant \rangle$ is elementarily equivalent to $\langle \mathscr{R}, \leqslant \rangle$ (where $\mathscr{R}$ is the set of all real numbers).

12.32. Prove that if a relation R orders a set T in type $\omega + (\omega^* + \omega)$ then $\langle \mathscr{N}, \leqslant \rangle \equiv \langle T, R \rangle$.

A subsystem $\mathfrak{b} \subset \mathfrak{a}$ is an *elementary subsystem* of the system $\mathfrak{a}$ (in symbols: $\mathfrak{b} \prec \mathfrak{a}$) if

$$\bigwedge_{X \in {}^{\mathscr{N}}B} \bigwedge_{\Phi} (\mathfrak{a} \models \Phi[X] \equiv \mathfrak{b} \models \Phi[X]).$$

In this case we say that $\mathfrak{a}$ is an *elementary extension* of $\mathfrak{b}$.

12.33. Prove that if $\mathfrak{a} \prec \mathfrak{b}$ and $\mathfrak{a} \neq \mathfrak{b}$ then $\mathfrak{a} \equiv \mathfrak{b}$.

12.34. Find $\mathfrak{a}$ and $\mathfrak{b}$ such that $\mathfrak{a} \subset \mathfrak{b}$, $\mathfrak{a} \equiv \mathfrak{b}$ but not $\mathfrak{a} \prec \mathfrak{b}$.

12.35. Prove that the relation $\prec$ is an ordering, i.e.

(a) $\mathfrak{a} \prec \mathfrak{a}$,

(b) $\mathfrak{a} \prec \mathfrak{b} \wedge \mathfrak{b} \prec \mathfrak{a} \Rightarrow \mathfrak{a} = \mathfrak{b}$,

(c) $\mathfrak{a} \prec \mathfrak{b} \wedge \mathfrak{b} \prec \mathfrak{c} \Rightarrow \mathfrak{a} \prec \mathfrak{c}$.

12.36. Prove that if $\mathfrak{a} \subset \mathfrak{b} \subset \mathfrak{c}$, $\mathfrak{a} \prec \mathfrak{c}$ and $\mathfrak{b} \prec \mathfrak{c}$ then $\mathfrak{a} \prec \mathfrak{b}$.

12.37. Given a sequence $Y \in {}^{\mathscr{N}}B$, let $X\binom{k}{t}$ be a sequence arising from X by replacing the term x_k by t (i.e. $X\binom{k}{t} = (X - (\{k\} \times B)) \cup \cup \{\langle k, t \rangle\}$).

Prove the following test of Tarski:

Let $\mathfrak{a} \subset \mathfrak{b}$. Then $\mathfrak{a} \prec \mathfrak{b} \Leftrightarrow$ For every formula Φ and sequence $X \in {}^{\mathscr{N}}A$: If $\mathfrak{b} \models \bigvee_{x_k} \Phi[X]$ then for a certain $a \in A$, $\mathfrak{b} \models \Phi\left[X\binom{k}{a}\right]$.

12.38. Let $\{\mathfrak{a}_\xi\}_{\xi<\alpha}$ be a family of relational structures such that $\eta \leqslant \xi \Rightarrow \mathfrak{a}_\eta \prec \mathfrak{a}_\xi$. Let $\mathfrak{b}$ be the union of the family $\{\mathfrak{a}_\xi\}_{\xi<\alpha}$ (in particular $B = \bigcup_{\xi<\alpha} A_\xi$, $S_i = \bigcup_{\xi<\alpha} R_i^\xi$, $g_j = \bigcup_{\xi<\alpha} f_j^\xi$). Prove that $\bigwedge_{\xi<\alpha} \mathfrak{a}_\xi \prec \mathfrak{b}$.

12.39. Generalize 12.38 to directed families of systems.

Let $\mathfrak{a} = \langle A, R_i, \ldots, f_j, \ldots \rangle$, $\mathfrak{b} = \langle B, S_i, \ldots, g_j, \ldots \rangle$ be two relational systems. Let φ be an injection of A into B.

We say that φ is an *imbedding* of $\mathfrak{a}$ into $\mathfrak{b}$ (in symbols φ: $\mathfrak{a} \to \mathfrak{b}$) if the following conditions are satisfied:

$$R_i(a_i, \ldots, a_n) \Leftrightarrow S_i(\varphi a_1, \ldots, \varphi a_n),$$

$$\varphi[f_j(a_1, \ldots, a_n)] = g_j(\varphi a_1, \ldots, \varphi a_n).$$

If moreover φ is a bijcction, then we say that φ is an *isomorphism* of $\mathfrak{a}$ and $\mathfrak{b}$.

12.40. Prove that if $\mathfrak{a} \subset \mathfrak{b}$ then Id_A is an imbedding of $\mathfrak{a}$ into $\mathfrak{b}$.

Let $\mathfrak{a}_t = \langle A_t, R_{ti}, \ldots, g_{tj}, \ldots \rangle$ be a family of relational systems indexed by the elements of set I, and F an ultrafilter on I. In the set $\prod_{t \in I} A_t$ introduce the relation $\sim_F$ (remember that $\prod_{t \in I} A_t$ consists of functions with the domain equal to I such that $f(t) \in A_t$).

$$f \sim_F g \Leftrightarrow \{t : f(t) = g(t)\} \in F.$$

The relation $\sim_F$ is an equivalence (cf. 4.81, D2.61). Let $B = \prod_{t \in I} A_t / F$. We define in B relations S_i $(i \in T_1)$ and functions $g_j (j \in T_2)$ as follows:

$$S_i([f_1], \ldots, [f_{ni}]) \Leftrightarrow \{t : R_{t_i}(f_1(t), \ldots, f_{n_i}(t))\} \in F,$$

$$g_j([f_1], \ldots, [f_{n_j}]) = [h] \Leftrightarrow \{t : g_{t_j}(f_1(t), \ldots, f_{n_j}(t)) = h(t)\} \in F.$$

The relational system $\mathfrak{b} = \langle B, S_i, \ldots, g_j, \ldots \rangle$ is called the *reduced product* of the family $\mathfrak{a}_t$ modulo F or simply the ultraproduct. In symbols: $\mathfrak{b} = \prod_{t \in I} \mathfrak{a}_t / F$.

12.41. Prove that the definition of functions and relations of the ultraproduct is sound.

12.42. Prove that if the ultrafilter F is principal, i.e. $\cap F=\{i_0\}$, then the product $\prod_{i\in I} \mathfrak{a}_i/F$ is isomorphic to $\mathfrak{a}_{i_0}$.

12.43. Prove the following fundamental theorem on ultraproducts: (Łoś's theorem): Let $\Phi(v_0, \ldots, v_n)$ be a formula and $[f_0], [f_1], \ldots, [f_n], \ldots$ an arbitrary sequence of elements of $\prod_{i\in I} \mathfrak{a}_i/F$. Then

$$\prod_{i\in I} \mathfrak{a}_i/F \models \Phi(v_0, \ldots, v_n)[[f_0], [f_1], \ldots]$$

if and only if

$$\{t : \mathfrak{a}_t \models \Phi(v_0, \ldots, v_n)[f_0(t), f_1(t), \ldots]\} \in F.$$

12.44. Let $X\subset I$. The *trace* of the filter F on X is the family $\{X\cap U: U\in F\}$. It is denoted by $F\restriction X$. If $X\in F$ then $F\restriction X$ is an ultrafilter on X. Prove that if $X\in F$ then $\prod_{i\in I} \mathfrak{a}_i/F$ is isomorphic to the system $\prod_{j\in X} \mathfrak{a}_j/F\restriction X$. What is the connection of our problem with 12.42?

12.45. For every $i\in I$ let $\mathfrak{a}_i=\mathfrak{a}$. We then write $\mathfrak{a}^I/F$ instead of $\prod_{i\in I} \mathfrak{a}_i/F$. Prove that the mapping φ: $\mathfrak{a}\to\mathfrak{a}^I/F$ defined by: $\varphi(a)=[\underline{a}]$ where a is a constant function taking the value a is an imbedding of $\mathfrak{a}$ into $\mathfrak{a}^I/F$. Prove that $\varphi * \mathfrak{a}\prec\mathfrak{a}^I/F$.

12.46. Suppose that φ is an imbedding of $\mathfrak{a}$ into $\mathfrak{b}$ such that $\varphi * \mathfrak{a}\prec\mathfrak{b}$. Prove that $\mathfrak{a}\equiv\mathfrak{b}$.

12.47. Prove that if $\mathfrak{a}$ is a finite relational system then, for all I and F, $\mathfrak{a}^I/F$ is finite and iscmorphic to $\mathfrak{a}$.

12.48. Prove that if all systems $\mathfrak{a}_i$ are finite and for a certain n, $\{i : \bar{\bar{A}}_i=n\}\in F$ then the system $\prod\mathfrak{a}_i/F$ is finite.

12.49. Find a family $\{\mathfrak{a}_i\}_{i\in I}$ of finite relational systems, a set I and an ultrafilter F on I such that $\prod_{i\in I} \mathfrak{a}_i/F$ is not finite.

In Problems 12.50–12.56 we investigate the properties of ultrafilters.

12.50. Let S be a subfamily of the set $\mathscr{P}(I)$ with the so-called *finite intersection property*, i.e. $\bigwedge_{a_1,\ldots,a_n\in S} a_1\cap, \ldots, \cap a_n\neq O$ (for all $n\in\mathscr{N}$). Prove that there exists an ultrafilter F on I such that $S\subset F$.

12.51. An ultrafilter F on I is called *principal* if $A_F=\cap F\neq O$. Prove hat if F principal, then there exists an $i_F\in I$ such that $A_F=\{i_F\}$.

12.52. Prove that the family $\mathcal{N}_\infty=\{X\subset\mathcal{N}: \overline{\overline{\mathcal{N}-X}}<\aleph_0\}$ can be extended to a nonprincipal ultrafilter.

12.53. Let $\overline{\overline{A}}=\aleph_0$, $F\subset\mathcal{P}(\mathcal{N})$, be a nonprincipal ultrafilter. Prove that $\overline{\overline{A^{\mathcal{N}}/F}}=2^{\aleph_0}$.

12.54. Let $\{A_i\}_{i\in I}$ be a family of pairwise disjoint, nonempty sets. For each $i\in I$, let F_i be an ultrafilter on A_i, and finally let G be an ultrafilter on I. Prove that the family

$$\bigcup_{i\in I} F_i/G=\{X\subset \bigcup_{i\in I} A_i : \{i : X\cap A_i\in F_i\}\in G\}$$

is an ultrafilter on $\bigcup_{i\in I} A_i$.

12.55. Let $\overline{\overline{I}}=\mathfrak{m}\geqslant\aleph_0$. Prove that the cardinality of the set of all ultrafilters on I is $2^{2^{\mathfrak{m}}}$.

12.56. Let F and G be ultrafilters on I. We say that F is isomorphic to G iff there exists a permutation φ of I (i.e. a bijection of I into I) such that $G=\{\varphi * X: X\in F\}$.

(a) Prove that if F and G are principal then F is isomorphic to G.

(b) Prove that for every infinite set X there exist nonisomorphic ultrafilters on X.

12.57. Assume $\bigwedge_{i\in I} \mathfrak{a}_i\equiv\mathfrak{b}_i$. Prove that $\prod_{i\in I} \mathfrak{a}_i/F\equiv \prod_{i\in I} \mathfrak{b}_i/F$. How can one weaken the assumption without weakening the assertion of our theorem?

The systems $\mathfrak{a}$ and $\mathfrak{b}$ are *similar* if $\mathbf{J}_\mathfrak{a}=\mathbf{J}_\mathfrak{b}$ (notice that the product $\prod_{i\in I} \mathfrak{a}_i/F$ exists only if for all $i,j\in I$ $\mathfrak{a}_i$ is similar to $\mathfrak{a}_j$). Let K be a class of similar systems. The set Th K of sentences is defined as follows: Th $K=\bigcap_{\mathfrak{a}\in K}$ Th $\mathfrak{a}$ (cf. Problem 12.15). Thus $\Phi\in\text{Th}\,K\Leftrightarrow \bigwedge_{\mathfrak{a}\in K} \mathfrak{a}\models\Phi$.

Similarly, the set of sentences $S\subset\mathbf{J}$ determines a class Mod S of a similar relational system, namely: $\mathfrak{a}\in\text{Mod}\,S\Leftrightarrow S\subset\text{Th}\,\mathfrak{a}$. Thus $\mathfrak{a}\in$ $\in$ Mod S iff all the sentences from S are true in $\mathfrak{a}$. In that case we say that $\mathfrak{a}$ is a *model* of S.

12.58. Prove that, for all K, $K\subset\text{Mod}\,(\text{Th}\,K)$.

12.59. Prove that if S has a model then $S\subset\text{Th}\,(\text{Mod}\,S)$.

A class K is called a *basic elementary class* (in symbols $K\in EC$), if there exists a sentence Φ such that $K=\text{Mod}\,\Phi$. A class K is called an *elementary class* ($K\in EC_\Delta$) if there exists an S such that $K=\text{Mod}\,S$.

COMPLETENESS THEOREM (GÖDEL'S THEOREM):

$$S \vdash \Phi \Leftrightarrow \bigwedge_{\mathfrak{a}} (\mathfrak{a} \in \operatorname{Mod} S \Rightarrow \mathfrak{a} \models \Phi).$$

Thus this theorem tells us that Φ can be derived from S if and only if Φ is true in all models of S.

12.60. Prove the following consistency theorem: Let S be a set of sentences; then S is consistent iff $\operatorname{Mod} S \neq 0$.

12.61. Prove the following compactness theorem: Let S be a set of sentences. If for every finite $S_0 \subset S$, $\operatorname{Mod} S_0 \neq O$ then $\operatorname{Mod} S \neq O$. In other words: if every finite subset S_0 of S has a model then S itself has a model.

12.62. Let A be the set of all finite subsets of a set T. Given $U \in A$, let $B_U = \{W\colon\ U \subset W\}$.

(a) Prove that $B_{U_1} \cap B_{U_2} = B_{U_1 \cup U_2}$.

(b) Prove that the family $\{B_U\colon U \text{ is a finite subset of } T\}$ can be extended to an ultrafilter on A.

(c) Use (b) to prove the compactness theorem (without the use of the completeness theorem).

12.63. Prove the following result due to Frayne:

Let $\mathfrak{a} \equiv \mathfrak{b}$. Then there exist a set I, an ultrafilter F on I and an imbedding φ of $\mathfrak{a}$ into $\mathfrak{b}^I/F$ such that $\varphi * \mathfrak{a} \prec \mathfrak{b}^I/F$.

12.64. Prove that if $\mathfrak{a}$ and $\mathfrak{b}$ are finite relational systems, $\mathfrak{a} \equiv \mathfrak{b}$, then $\mathfrak{a}$ is isomorphic to $\mathfrak{b}$.

12.65. Let $\mathfrak{a} = \langle A, R_t \rangle_{t \in T}$ be a relational system in which all relations R_t are unary and run through all subsets of A. Let $\mathfrak{a}' = \langle A', R_t' \rangle_{t \in T}$ be a proper elementary extension of $\mathfrak{a}$. Finally, let $x \in A' - A$. Prove that the set $F_x = \{R_t\colon\ x \in R_t'\}$ is a nonprincipal ultrafilter in A.

12.66. Let X be the class of all systems of some similarity type. Prove that if $K \subset X$ then $K \in EC \Leftrightarrow X - K \in EC$.

12.67. Let X be as in 12.66. Prove that if $K \in EC_\Delta$ and $X - K \in EC_\Delta$ then $K \in EC$.

12.68. Prove that if $K \in EC_\Delta$ then there exist classes $K_1, \ldots, K_t, \ldots$ $\ldots \in EC$ such that $K = \bigcap_{t \in T} K_t$.

12.69. Prove that every elementary class is closed with respect to:

(a) Ultraproducts.

(b) Ultrapowers (i.e. ultraproducts of identical terms).

(c) Elementary equivalence.

12.70. Prove that if the class K is closed with respect to the ultraproducts and the elementary equivalence then $K \in EC_\Delta$.

12.71. Prove that if the classes K and $X-K$ are both closed with respect to ultraproducts and isomorphisms and K is closed with respect to elementary equivalence then $K \in EC$.

A class K is said to be *universal* ($K \in UC_\Delta$) if $K = \text{Mod}\, S$, where S consists of universal formulas only (cf. 12.26).

12.72 Prove that if $K \in UC_\Delta$, $\mathfrak{a} \in K$, $\mathfrak{b} \subset \mathfrak{a}$, then $\mathfrak{b} \in K$.

12.73. Prove that $K \in UC_\Delta$ iff the conjuction of the following a, b and c holds:

(a) K is closed with respect to isomorphisms.

(b) K is closed with respect to subsystems.

(c) K is closed with respect to ultraproducts.

12.74. Are the following statements true?

(a) The class of groups is a UC_Δ class.

(b) The class of rings is a UC_Δ class.

(c) The class of fields is a UC_Δ class.

12.75. (a) Is the class of all systems $\langle A, A^2 \rangle$ where A is finite an EC_Δ class?

(b) Is the class of all systems $\langle A, A^2 \rangle$ where A is infinite an EC_Δ class?

(c) Is the class of all systems $\langle A, R \rangle$ where R is a linear ordering of A an EC_Δ class?

(d) Is the class of all systems $\langle A, R \rangle$ where R is a well-ordering of A an EC_Δ class?

LÖWENHEIM–SKOLEM–TARSKI THEOREM:

(a) Upper: Let $\mathfrak{a}$ be an infinite relational system, $\mathfrak{m}$ the cardinality of $\mathbf{J}_\mathfrak{a}$, $\mathfrak{n} \geqslant \mathfrak{m} + \bar{\bar{\mathfrak{a}}}$. Then there exists a $\mathfrak{b}$ such that $\bar{\bar{\mathfrak{b}}} = \mathfrak{n}$ and $\mathfrak{a} \prec \mathfrak{b}$.

(b) Lower: Let $\mathfrak{a}$ be a relational system, $\mathfrak{m}$ the cardinality of $\mathbf{J}_\mathfrak{a}$, $\bar{\bar{\mathfrak{a}}} \geqslant \mathfrak{m}$. Then for every $\mathfrak{n}$ such that $\mathfrak{m} \leqslant \mathfrak{n} \leqslant \bar{\bar{\mathfrak{a}}}$ there exists a $\mathfrak{b}$ such that $\bar{\bar{\mathfrak{b}}} = \mathfrak{n}$ and $\mathfrak{b} \prec \mathfrak{a}$. One can stipulate in addition that $\mathfrak{b}$ contains the required subset $A_0 \subset A$ (provided $\bar{\bar{A}}_0 \leqslant \mathfrak{n}$).

12.76. Prove the upper Löwenheim–Skolem–Tarski theorem using the lower Löwenheim–Skolem–Tarski theorem and a suitable ultrapower.

12.77. A set of sentences X is *categorical* if every two of its models are isomorphic. Prove that if X has an infinite model then X is not categorical.

12.78. Prove the following Łoś–Vaught completeness criterion: If for some cardinal $\mathfrak{m} \geqslant \overline{\overline{\mathbf{J}}}$ every two models of cardinality $\mathfrak{m}$ of the set of sentences S are isomorphic and S possesses infinite models then S is complete.

12.79. Prove that the theory of dense linear ordering without extremes is complete.

12.80. Prove that the theory of algebraically closed fields of fixed characteristics is complete.

12.81 Prove that the theory of linear spaces of fixed dimensions n over the algebraically closed field of given characteristics p is complete.

12.82. Prove that the set of sentences S is complete iff

$$\bigwedge_{\mathfrak{a} \in \operatorname{Mod} S} \bigwedge_{\mathfrak{b} \in \operatorname{Mod} S} \mathfrak{a} \equiv \mathfrak{b} .$$

12.83. A set of sentences $\{\Phi_n\}_{n \in \mathcal{N}}$ is called *increasing* if

(a) if $n \leqslant m$ then $\Phi_m \Rightarrow \Phi_n$ is true,

(b) if $n \leqslant m$ then $\sim(\Phi_n \Rightarrow \Phi_m)$ is true.

Prove that if $\{\Phi_n\}_{n \in \mathcal{N}}$ is increasing then there is no sentence Ψ such that $\operatorname{Mod}\{\Phi_n \colon n \in \mathcal{N}\} = \operatorname{Mod} \Psi$.

12.84. Prove that the theory of fields of characteristic 0 is not finitely axiomatizable.

12.85. Use 12.83 to show that there exist EC_Δ classes which are not EC classes.

12.86. Prove that the class of all systems $\langle A, A^2 \rangle$ where $\overline{\overline{A}} \geqslant \aleph_0$ is not finitely axiomatizable.

12.87. Let $\mathfrak{a} = \langle A, R_i, \ldots, f_j, \ldots \rangle$ be a relational system. For every $a \in A$ introduce a new constant c_a (thus the language is extended) and interpret this constant as a. The set T_2 is thus enlarged and we get a system $\mathfrak{a}'$, which differs from $\mathfrak{a}$ by new constants. Prove that $\operatorname{Th} \mathfrak{a} \subset \operatorname{Th} \mathfrak{a}'$.

The set $D(\mathfrak{a})$ – the *diagram* of the system $\mathfrak{a}$ – is the set of all true atomic sentences of $\mathbf{J}_\mathfrak{a}$. Thus $P_i(c_{a_1}, \ldots, c_{a_n}) \in D(\mathfrak{a})$ if $R_i(a_1, \ldots, a_n)$ and $\sim P_i(c_{a_1}, \ldots, c_{a_n}) \in D(\mathfrak{a})$ if non $R_1(a_1, \ldots, a_n)$. Also $t(c_{a_1}, \ldots, c_{a_n}) = t'(c_{b_1}, \ldots, c_{b_n}) \in D(\mathfrak{a})$ if it is true in $\mathfrak{a}'$ and the same applies to inequalities of terms.

A set X is called *model-complete* iff, for every $\mathfrak{a} \in \operatorname{Mod} X$, the set $X \cup D(\mathfrak{a})$ is a complete set of sentences.

12.88. Find X which is model-complete but not complete.

12.89. Find X which is complete but not model-complete.

12.90. Prove that X is model-complete iff

$$\bigwedge_{\mathfrak{a},\mathfrak{b} \in \mathrm{Mod}\, X} (\mathfrak{a} \subset \mathfrak{b} \Leftrightarrow \mathfrak{a} \prec \mathfrak{b}).$$

12.91. A relational system $\mathfrak{a}$ is called a *weakly-prime model* for a set of sentences S if for every $\mathfrak{b} \in \mathrm{Mod}\, S$ there exists an imbedding φ of $\mathfrak{a}$ into $\mathfrak{b}$. Prove that $\langle \mathcal{N}, +, \cdot, S, 0 \rangle$ is a weakly prime model of Peano arithmetic.

12.92. Prove that $\langle Q, +, \cdot, 0, 1 \rangle$ is a weakly-prime field of characteristic 0.

12.93. Prove that $\langle \mathscr{Z}_p, +, \cdot, 0, 1 \rangle$ is a weakly-prime field of characteristic p.

12.94. Prove that $\langle \mathscr{Z}, +, \cdot, 0, 1 \leqslant \rangle$ is a weakly-prime ordered ring with unity.

12.95. Prove that if the set X is model complete and possesses a weakly-prime model $\mathfrak{a}_0$ then

(a) X is complete.

(b) $\mathrm{Cn}\, X = \mathrm{Th}\, (\mathfrak{a}_0)$.

12.96. Give an example of a theory without a weakly-prime model.

12.97. An isomorphism of **a** into itself is called an *automorphism*. Assume $\mathfrak{a} \subset \mathfrak{b}$ and that for every finite $A' \subset A$ and $b \in B$ there exists an automorphism $f_{A',b}$ of $\mathfrak{b}$ such that $f_{A',b} \restriction A = \mathrm{Id}$, $f_{A',b}(b) \in A$. Prove that $\mathfrak{a} \prec \mathfrak{b}$.

12.98. Let X be a model-complete set of sentences and, for every $\mathfrak{a}, \mathfrak{b} \in \mathrm{Mod}\, X$, let there exist $\mathfrak{c} \in \mathrm{Mod}\, X$ and imbeddings $\varphi_1\colon \mathfrak{a} \to \mathfrak{c}$ and $\varphi_2\colon \mathfrak{b} \to \mathfrak{c}$. Prove that X is complete.

12.99. Given $\Phi \in \mathbf{J}$, let U_Φ be the family of those complete consistent sets of sentences S of $\mathbf{J}$ for which $\Phi \in S$.

In the set of all complete consistent sets of sentences of $\mathbf{J}$ we define a topology whose basis is $\{U_\Phi\colon \Phi \in \mathbf{J}\}$.

(a) Show that our space is 0-dimensional.

(b) Show that our space is compact, Hausdorff.

(c) Which sets in our space are closed?

12.100. Show that the Löwenheim–Skolem–Tarski theorem is equivalent to the axiom of choice (see Kuratowski and Mostowski "Set Theory").

12.101. Prove that if $K_1 \in EC$, $K_2 \in EC$, then $K_1 \cap K_2 \in EC$, $K_1 \cup K_2 \in EC$.

12.102. Prove that if $K_1 \in EC_\Delta$, $K_2 \in EC_\Delta$ then $K_1 \cap K_2 \in EC_\Delta$, $K_1 \cup K_2 \in EC_\Delta$.

12.103. Find the topological interpretation of 12.101, 12.102 (cf. 12.99).

12.104. Prove – using model theoretic methods – that every ordering can be extended to a linear ordering.

12.105. Let φ_1, φ_2 be statements of language **J**, $\varphi_1 \approx \varphi_2 \Leftrightarrow \vdash \varphi_1 \Leftrightarrow \varphi_2$. Prove that $\approx$ is an equivalence (cf. 9.II.1).

12.106. In the set of equivalence classes of **J** (cf. 12.105) define the operations $\cap$, $\cup$, $-$, 1, 0 as follows:

$$[\varphi_1] \cap [\varphi_2] = [\varphi_1 \wedge \varphi_2], \quad 1 = [\varphi \vee \sim \varphi],$$

$$[\varphi_1] \cup [\varphi_2] = [\varphi_1 \vee \varphi_2], \quad 0 = [\varphi \wedge \sim \varphi],$$

$$-[\varphi_1] = [\sim \varphi_1].$$

Prove that the set of equivalence classes with the above operations is a Boolean algebra (called a *Lindenbaum algebra* of the language **J**).

CHAPTER 14

RECURSIVE FUNCTIONS

Let f be a function with the values in the set $\mathcal{N}$ of natural numbers. If, for a certain k, $Df \subset \mathcal{N}^k$ then we say that f is *defined in the set of natural numbers*. If $Df = \mathcal{N}^k$ then we say that f is a *total function*.

In this chapter we deal with functions defined in the set of natural numbers.

We say that a set of functions F is *closed with respect to an n-argument operation* ϑ if, whenever $f_i \in F$ $(i \leqslant n)$ the function $\vartheta(f_1, \ldots, f_n)$ belongs to the family F as well.

Let f be an $(n+1)$-argument function. A function $g(x_1, \ldots, x_n)$ defined as

$$g(x_1, \ldots, x_n) = \text{the least } x \text{ such that } f(x, x_1, \ldots, x_n) = 0$$

is denoted by $(\mu x)(f(x, x_1, \ldots, x_n) = 0)$.

We say that the function g is defined from f by means of *minimum operation*.

If the function f possesses the property $\bigwedge_{x_1} \ldots \bigwedge_{x_n} \bigvee_{x} f(x, x_1, \ldots, x_n) = 0$ then we say that g — defined as above — is defined from f by means of the *effective minimum operation*.

We introduce an operation of *bounded minimum* as follows: given $f(x, x_1, \ldots, x_n)$ define $g(y, x_1, \ldots, x_n)$ by the conditions:

$$g(y, x_1, \ldots, x_n) =$$

$$= \begin{cases} (\mu x)\,[f(x, x_1, x_2, \ldots, x_n) = 0 \wedge x \leqslant y] & \text{if such an } x \text{ exists,} \\ 0 & \text{otherwise.} \end{cases}$$

Another operation on functions is the so-called *recursive definition* (or just: *recursion*):

Given functions $f(x_1, \ldots, x_n)$ and $g(x_1, \ldots, x_n, y, z)$, define a function h as follows:

$$h(0, x_1, \ldots, x_n) = f(x_1, \ldots, x_n),$$

$$h(n+1, x_1, \ldots, x_n) = g[x_1, \ldots, x_n, n, h(n, x_1, \ldots, x_n)].$$

Iteration (which is a special case of recursion) is the following operation on unary functions. Given a function $g(x)$ define a function f as follows:

$$f(0) = 0,$$

$$f(n+1) = g[f(n)].$$

The *class of primitive recursive functions* is the least family of functions which contains the functions

$$S(x) = x+1, \quad I(x) = x, \quad I_2(x, y) = y, \quad 0(x) = 0$$

and is closed with respect to superposition and recursion operations.

The *class of recursive functions* is the smallest family containing the same initial functions, $S(x)$, $I(x)$, $I_2(x, y)$ and $0(x)$ and closed under superposition, recursion and effective minimum operations.

The *class of partial recursive functions* is the least family of functions containing the functions $S(x)$, $I(x)$, $I_2(x, y)$ and $0(x)$ and closed with respect to the operations of superposition, recursion, and minimum.

A relation R is called *primitive recursive* (*recursive*) iff there exists a primitive recursive (recursive) function f such that:

$$R(m_1, \ldots, m_n) \Leftrightarrow f(m_1, \ldots, m_n) = 0.$$

In particular a set $A \subset \mathcal{N}$ is called *primitive recursive* (*recursive*) iff there exists a primitive recursive (recursive) function f such that:

$$m \in A \Leftrightarrow f(m) = 0.$$

A set $A \subset \mathcal{N}$ is called *recursively enumerable* iff there exists a recursive function f such that $f * \mathcal{N} = A$.

13.1. Prove that every primitive recursive function is total.

13.2. Prove that every recursive function is total.

13.3. Prove that every primitive recursive function is recursive.

13.4. Prove that every recursive function is partial recursive.

13.5. Prove that:

(a) The class of primitive recursive functions has cardinality $\aleph_0$.

(b) The class of partial recursive functions has cardinality $\aleph_0$.

Prove that the following functions are primitive recursive (Problems 13.6–13.11):

13,6. $f(x, y) = x + y$.

13.7. $f(x, y) = x \cdot y$.

13.8. $f_n(x) = n$.

13.9. $f(x, y) = x^y$.

13.10. $f(x) = (x+1)!$

13.11. $f(x, y) = x \dot{-} y = \begin{cases} x - y, & \text{if} \quad x \geqslant y. \\ 0, & \text{if} \quad x < y. \end{cases}$

Prove that the following relations are primitive recursive (Problems 13.12–13.14):

13.12. $x \leqslant y$.

13.13. $x = y$.

13.14. $x < y$.

13.15. Prove that if $R_1(x, y)$ and $R_2(x, y)$ are primitive recursive (recursive) relations then also $R_1(x, y) \vee R_2(x, y)$, $R_1(x, y) \wedge R_2(x, y)$ and $\sim R_1(x, y)$ are primitive recursive (recursive) relations.

13.16. Prove that if $R(x, y)$ is a recursive relation then the relations $\bigwedge_{x<n} R(x, y)$ and $\bigvee_{x<n} R(x, y)$ are also recursive.

13.17. Prove that $A \subset \mathcal{N}$ is primitive recursive iff there exists a primitive recursive function f, $f: \mathcal{N} \to \{0, 1\}$ such that $a \in A \Leftrightarrow f(n) = 1$.

13.18. Prove that if A and B are recursive sets then the sets $A \cup B$, $A \cap B$ and $\mathcal{N} - A$ are also recursive.

13.19. Prove that if $f(x)$ and $g(x)$ are primitive recursive and $R(x)$ is a primitive recursive relation then the function $h(x)$ defined as follows:

$$h(x) = \begin{cases} f(x) & \text{if} \quad R(x), \\ g(x) & \text{if} \quad \sim R(x), \end{cases}$$

is primitive recursive.

13.20. Prove that every finite set is primitive recursive.

13.21. Prove that every recursive set is recursively enumerable.

13.22. Let $n_0 \in \mathcal{N}$. Prove that if $f(x)$ is almost equal to n_0 (i.e. $f(x) \neq n_0$ for finitely many x) then f is a primitive recursive function.

13.23. Prove that every recursively enumerable infinite subset of $\mathcal{N}$ contains an infinite recursive subset.

13.24. Prove that an infinite subset $A \subset \mathcal{N}$ is recursive iff there exists an increasing recursive function f such that $A = f * \mathcal{N}$.

13.25. Prove that if f is a recursive permutation of $\mathcal{N}$ then the function f^{-1} is recursive.

13.26. Prove that if A is a recursively enumerable set and f is a recursive function then the set $f^{-1} * A$ is recursively enumerable.

13.27. Prove that if $R(x, y)$ is a recursive relation such that $\bigwedge_x \bigvee_y R(x, y)$ then the function $f(x) = \mu y R(x, y)$ is recursive.

13.28. Prove that if $R(x, y)$ is a recursive relation such that $\bigwedge_x \bigvee_y R(x, y)$ and $g(x)$ is a recursive function then the function $f(x) = \mu y R(g(x), y)$ is recursive and the frnction $f(x) = \mu y R(x, g(y))$ is partial recursive.

Prove that the following functions and relations are primitive recursive (Problems 13.29–13.32):

13.29. $E\left[\frac{x}{y}\right]$.

13.30. $x \mid y$.

13.31. $E[\sqrt{x}]$.

13.32. $x - E[\sqrt{x}]^2$.

13.33. Prove that the class of primitive recursive functions is closed with respect to the bounded minimum operation.

13.34. Prove that if $R(x, y)$ is a primitive recursive relation then the relations

$$\bigwedge_{x \leqslant n} R(x, y) \quad \text{and} \quad \bigvee_{x \leqslant n} R(x, y)$$

are also primitive recursive.

13.35. Prove that for every $n \geqslant 2$ there exist primitive recursive functions $J_n, J_n^1, \ldots, J_n^n$ such that

(a) J_n is a bijection of $\mathcal{N}^n$ and $\mathcal{N}$.

(b) $J_n^i : \mathcal{N} \to \mathcal{N}$.

(c) $J_n(J_n^1(x), \ldots, J_n^n(x)) = x$.

13.36. Prove that if the relation $R(x, x_1, \ldots, x_n)$ is primitive recursive then there exists a primitive recursive relation $R'(x, y)$ such that

$$\bigvee_{x_1} \cdots \bigvee_{x_n} R(x, x_1, \ldots, x_n) \Leftrightarrow \bigvee_{y} R'(x, y),$$

$$\bigwedge_{x_1} \cdots \bigwedge_{x_n} R(x, x_1, \ldots, x_n) \Leftrightarrow \bigwedge_{y} R'(x, y).$$

13.37. Prove that the class of recursive functions can be defined as the least family of functions containing the initial functions $O(x)$, $S(x)$, $I(x)$, $I_2(x, y)$ and J_n, J_n^i ($n \geqslant 2$, $i \in \mathcal{N}$) and closed with respect to the superposition and the following simplified recursion scheme: Given a unary function $g(x)$ define $f(x, y)$ as follows:

$$f(x, 0) = x,$$

$$f(x, n+1) = g[f(x, y)].$$

Let S be the least family of functions closed with respect to the operations of addition, superposition, and iteration. Prove that the following functions belong to S (Problems 13.38–13.43):

13.38. $sq(x) = \begin{cases} 1, & \text{if } x \text{ is a square,} \\ 0, & \text{otherwise.} \end{cases}$

13.39. x^2.

13.40. $E\left[\dfrac{x}{2}\right]$.

13.41. $E[\sqrt{x}]$.

13.42. $x \cdot y$.

13.43. $x - y$.

13.44. Prove that the class of primitive recursive functions of one variable can be defined as the least family of functions containing the functions $S(x) = x + 1$ and $Q(x) = x - [\sqrt{x}]^2$ and closed with respect to addition, superposition, and iteration.

13.45. A function $f(x, x_1, \ldots, x_n)$ is called a *universal function* for a family of n-ary functions F iff, for every $g \in F$, there exists a u such that $g(x_1, \ldots, x_n) = f(u, x_1, \ldots, x_n)$ and, for every u, $f(u, x_1, \ldots, x_n) \in F$. Prove that there exists a recursive universal function for the family of unary primitive recursive functions.

13.46. Prove that if a family of functions F is closed with respect to superposition and contains the function $S(x)$ then it does not contain its own universal function.

13.47. Prove that there exist recursive functions which are not primitive recursive.

13.48. Given a recursive function f, prove that there exist primitive recursive relations R_1 and R_2 such that:

$$f(m)=n \Leftrightarrow \bigvee_x R_1(m,n,x) \Leftrightarrow \bigwedge_x R_2(m,n,x).$$

13.49. Given a recursive function f, prove that there exists a primitive recursive relation R such that:

$$f(n)=J_2^1[(\mu x)R(x,n)].$$

SUPPLEMENT 1

INDUCTION

The *principle of induction* is formulated as follows:

Let Z be a set such that:

(1) $0 \in Z$.

(2) For every natural number n, $n \in Z$ then $n+1 \in Z$.

Then $\mathcal{N} \subset Z$.

Thus, if Z consists of natural numbers only then $Z = \mathcal{N}$.

The sum $a_1 + \ldots + a_n$ is denoted by $\sum_{i=1}^{n} a_i$, the product $a_1 \ldots a_n$ by $\prod_{i=1}^{n} a_i$.

Some useful definitions:

$$n! = 1 \cdot 2 \cdot \ldots \cdot n, \quad 0! = 1,$$

$$\binom{n}{k} = \frac{n!}{k!(n-k)!}.$$

S1.1. Prove that every nonempty set of natural numbers contains a least element (this is the so-called *minimum principle*).

S1.2. Prove that every nonempty bounded set of natural numbers contains a largest element (this is the so-called *maximum principle*).

S1.3. Prove the following *gneralized induction principle*:

Let Z be a set such that:

(1) $k \in Z$.

(2) For all natural $n \geqslant k$, if $n \in Z$ then $n+1 \in Z$.

Then $Z \supset \mathcal{N} - \{0, \ldots, k-1\}$.

S1.4. Prove the following *order-induction principle*:

If Z is a set such that, for all $m \geqslant k$ and $m < n$, $m \in Z$ implies $n \in Z$ then $Z \supset \mathcal{N} - \{0, \ldots, k-1\}$.

S1.5. Prove the induction principle using the minimum principle.

S1.6. Prove the maximum principle using the minimum principle.

S1.7. Prove the generalized induction principle using the minimum principle.

S1.8. Prove that every nonempty set of integers bounded from below contains a least element.

S1.9. Prove that every nonempty set of integers bounded from above contains a largest element.

S1.10. Prove that if Z is a set of integers satisfying the conditions

(1) $0 \in Z$,

(2) for all integer n, if $n \in Z$ then both $n-1$ and $n+1$ belong to Z,

then Z is the set of all integers.

Prove that (Problems S1.11–S1.15):

S1.11. $1+2+3+\ldots+n=\dfrac{n(n+1)}{2}$.

S1.12. $1^2+2^2+3^2+\ldots+n^2=\dfrac{n(n+1)(2n+1)}{6}$.

S1.13. $1^3+2^3+3^3+\ldots+n^3=(1+2+\ldots+n)^2=\dfrac{n^2(n+1)^2}{4}$.

S1.14. $1^2-2^2+3^2-4^2+\ldots+(-1)^{n-1}n^2=(-1)^{n-1}\cdot\dfrac{n(n+1)}{2}$.

S1.15. $1\cdot 2+2\cdot 3+\ldots+n\cdot(n+1)=\dfrac{n(n+1)(n+2)}{3}$.

S1.16. Find a natural number k such that for all n

$$1\cdot 4+2\cdot 7+\ldots+n(3n+1)=n(n+1)(n+k)$$

holds.

S1.17. Find a natural number k such that for all n

$$2\cdot 1^2+3\cdot 2^2+\ldots+n(n-1)^2+(n+1)\cdot n^2=$$

$$=\frac{n(n+1)(n+2)(3n+k)}{12}$$

holds.

Prove that (Problems S1.18–S1.33):

S1.18. $\sum_{i=1}^{n}\dfrac{1}{i(i+1)}=\dfrac{n}{n+1}$.

S1.19. $\sum_{i=1}^{n} \frac{1}{(2i-1)(2i+1)} = \frac{n}{2n+1}$.

S1.20. $\sum_{i=1}^{n} \frac{1}{(3i-2)(3i+1)} = \frac{n}{3n+1}$.

S1.21. $\sum_{i=1}^{n} \frac{1}{(a+i-1)(a+i)} = \frac{n}{a(a+n)}, \quad a>0$.

S1.22. $\sum_{i=1}^{n} \frac{1}{i(i+1)(i+2)} = \frac{1}{2}\left[\frac{1}{2} - \frac{1}{(n+1)(n+2)}\right]$,

S1.23. $\sum_{i=1}^{n} \frac{1}{(2i-1)(2i+1)(2i+3)} = \frac{n(n+1)}{2(2n+1)(2n+3)}$.

S1.24. $\prod_{j=1}^{k} \sum_{i=1}^{n} (i+j-1) = \frac{n^2(n+k+1)}{2}$.

S1.25. $\sum_{i=1}^{n} (2i-1) = n^2$.

S1.26. $\sum_{i=1}^{n} (2i-1)^3 = n^2(2n^2-1)$.

S1.27. $\sum_{i=1}^{n} (4i-3) = n(2n-1)$.

S1.28. $\sum_{i=1}^{n} i \cdot 2^i = 2+(n-1)\cdot 2^{n+1}$.

S1.29. $\sum_{i=1}^{n} 2\cdot 3^{i-1} = 3^n - 1$.

S1.30. $(x+y)^n = \sum_{i=0}^{n} \binom{n}{i} x^{n-i}\cdot y^i$.

S1.31. $\binom{k}{l} + \binom{k}{l+1} = \binom{k+1}{l+1}$.

S1.32. $\sum_{i=0}^{n} \binom{n}{i}^2 = \binom{2n}{n}$.

S1.33. $\sum_{i=0}^{k} \binom{m}{i}\cdot\binom{n}{k-i} = \binom{m+n}{k}$.

S1.34. Prove that if $f\colon Z\to Z$ is an increasing function and the set Z satisfies the conditions

(1) $f(n)\in Z$ for all $n\in\mathcal{N}$,

(2) for all $k\in Z$, if $k\in Z$ then $k-1\in Z$,

then Z contains all the integers.

Prove that (Problems S1.35–S1.45):

S1.35. $\sum_{i=1}^{n} i\cdot i!=(n+1)!-1$.

S1.36. $\sum_{i=0}^{n}\cos i\theta=\frac{1}{2}+\frac{\sin(n+\frac{1}{2})\theta}{2\sin\frac{1}{2}\theta}$

S1.37. $\sum_{i=1}^{n}\cos(2i-1)\theta=\frac{\sin 2n\theta}{2\sin\theta}$.

S1.38. $x^n-1=(x-1)\sum_{i=0}^{n-1}x^i$.

S1.39. $\sum_{i=0}^{n}\frac{2^i}{1+x^{2^i}}=\frac{1}{x-1}+\frac{2^{n+1}}{1-x^{2^{n+1}}}$ for $x\neq 1$.

S1.40. $\sum_{i=1}^{n}\frac{x^{2^{i-1}}}{1-x^{2^i}}=\frac{1}{1-x}\cdot\frac{x-x^{2^n}}{1-x^{2^n}}$ for $|x|\neq 1$.

S1.41. $\sum_{i=1}^{n} ix^i=\frac{x-(n+1)x^{n+1}+nx^{n+2}}{(1-x)^2}$ for $x\neq 1$.

S1.42. $\prod_{i=1}^{n}(n+i)=2^n\cdot\prod_{i=1}^{n}(2i-1)$.

S1.43. $\prod_{i=1}^{n}(1+x^{2^{i-1}})=1+\sum_{i=1}^{2^n-1}x^i$.

S1.44. $\prod_{i=0}^{n}\cos(2^i\cdot\theta)=\frac{\sin(2^{n+1}\cdot\theta)}{2^{n+1}\cdot\sin\theta}$.

S1.45. $1+\frac{1}{a_1}+\sum_{j=1}^{n}\frac{\prod_{i=1}^{j}(a_i+1)}{\prod_{i=1}^{j+1}a_i}=\frac{\prod_{i=1}^{n}(a_i+1)}{\prod_{i=1}^{n+1}a_i}$.

Prove that for every natural number n (Problems S1.46–S1.53):

S1.46. $2^n > n$.

S1.47. $n! > n,\ n > 2$.

S1.48. $2 \mid n^2 + n$.

S1.49. $6 \mid n^3 - n$.

S1.50. $30 \mid n^5 - n$.

S1.51. $42 \mid n^7 - n$.

S1.52. $43 \mid 6^{n+2} + 7^{2n+1}$.

S1.53. $k^2 + k + 1 \mid k^{n+2} + (k+1)^{2n+1}$.

Prove that for every natural number n and every prime number p (Problems S1.54–S1.60.):

S1.54. $p \mid n^p - n$.

S1.55. $6 \mid n^3 + 5n$.

S1.56. $120 \mid n^5 - 5n^3 + 4n$.

S1.57. $24 \mid n^6 - 3n^5 + 6n^4 - 7n^3 + 5n^2 - 2n$.

S1.58. $133 \mid 11^{n+2} + 12^{2n+1}$.

S1.59. $25 \mid 2^{n+2} \cdot 3^n + 5n - 4$.

S1.60. $64 \mid 3^{2n+1} + 40n - 67$.

For what natural number n do the following inequalities hold? (Problems S1.61–S1.68):

S1.61. $2n + 1 < 2^n$.

S1.62. $n^2 < 3^{n-1}$.

S1.63. $n^3 < 2^n$.

S1.64. $n^3 < n!$.

S1.65. $(n+1)^n < n^{n+1}$.

S1.66. $n^2 \cdot 2^n < n^2 + n - 2$.

S1.67. $3^n < n^2 + 2n - 4$.

S1.68. $n^2 + 1 < 2^{n-1}$.

S1.69. Prove that for all natural $n > 2$

$$2^{\frac{1}{2}n(n-1)} > n!.$$

Prove that for all natural $n > 1$ (Problems S1.70–S1.71):

S1.70. $n! < \left(\dfrac{n+1}{2}\right)^n$.

S1.71. $\dfrac{4^n}{n+1} < \dfrac{(2n)!}{(n!)^2}$.

Prove that for all $x \geqslant 0$, natural n and $k \leqslant n$ (Problems S1.72–S1.73):

S1.72. $(1+x)^n \geqslant 1+nx$.

S1.73 $1+\frac{k}{n} \leqslant \left(1+\frac{1}{n}\right)^k < 1+\frac{k}{n}+\left(\frac{k}{n}\right)^2$.

Prove that (Problems S1.74–S1.76):

S1.74. $\sum_{i=1}^{n} \frac{1}{\sqrt{i}} > \sqrt{n}$ for $n \geqslant 2$.

S1.75. $\sum_{i=2}^{n} \frac{1}{i^2} < 1$ for $n > 1$.

S1.76. $\sum_{i=1}^{n} \frac{1}{n+1} > \frac{13}{24}$ for $n > 1$.

S1.77. Prove that if $a_1 \geqslant 0, \ldots, a_n \geqslant 0$ then:

$$\frac{\sum_{i=1}^{n} a_i}{n} \geqslant \sqrt[n]{\prod_{i=1}^{n} a_i}.$$

S1.78. Prove that for positive $a_1, \ldots, a_n$:

$$\frac{a_n}{a_1} + \sum_{i=1}^{n-1} \frac{a_i}{a_{i+1}} \geqslant n.$$

S1.79. Prove that for positive $a_1, \ldots, a_n$:

$$\sum_{i=1}^{n} \frac{1}{a_i} \geqslant \frac{n^2}{\sum_{i=1}^{n} a_i}.$$

S1.80. Prove that $\sin 2nx + \cos 2nx \geqslant \frac{1}{2^{n-1}}$.

S1.81. Prove (without the use of S1.77) that for positive $a_1, \ldots$ $\ldots, a_n$

$$\text{if } \prod_{i=1}^{n} a_i = 1, \quad \text{then } \sum_{i=1}^{n} a_i \geqslant n.$$

When does the equality hold?

S1.82. Prove that for non-negative $a_1, \ldots, a_n$

$$\left(\frac{\sum_{i=1}^{n} a_i}{n}\right)^2 \leqslant \frac{\sum_{i=1}^{n} a_i^2}{n}.$$

Prove that the equality holds only when $a_i = a_j$ for all $i, j \leqslant n$.

S1.83. Prove that for a and b such that $a+b>0$ the following inequality holds:

$$\frac{a^n+b^n}{2} \geqslant \left(\frac{a+b}{2}\right)^n.$$

S1.84. Prove that, for every natural number n, $\frac{n^5}{5}+\frac{n^3}{3}+\frac{7n}{15}$ is a natural number.

S1.85. Prove that if the terms of the sequence a_n satisfy the conditions $a_0=2$, $a_1=3$, $a_{n+1}=3a_n-2a_{n-1}$ then, for all n, $a_n=2^n+1$.

S1.86. Prove that if the terms of the sequence a_n satisfy the conditions $a_1=1$, $a_2=2$, $a_n=a_{n-1}+a_{n-2}$ then, for all n, $a_n<(\frac{7}{4})^n$.

S1.87. Prove that if the terms of the sequence a_n satisfy the condition

$$a_n=\frac{a_{n-1}}{2a_{n-1}+1},$$

then

$$a_n=\frac{a_0}{2na_0+1}.$$

S1.88. Given two sequences a_n and b_n such that $a_1=1$, $b_1=1$ and $a_{n+1}=a_n+2b_n$, $b_{n+1}=a_n+b_n$, prove that

$$a_n=\tfrac{1}{2}[(1+\sqrt{2})^n+(1-\sqrt{2})^n], \quad b_n=\tfrac{1}{4}\sqrt{2}[(1+\sqrt{2})^n-(1-\sqrt{2})^n].$$

S1.89. Given two sequences a_n and b_n such that $a_1=3$, $b_1=2$, $a_{n+1}=a_n+2b_n$, $b_{n+1}=a_n+b_n$ prove, without the use of S1.88, that $a_n^2-2b_n^2=1$.

S1.90. Given the sequence a_n such that $a_0=0$, $a_1=1$ and $a_{n+1}=a_n+a_{n+1}$, prove that for all $n>0$:

(a) $\quad a_{n+1}=1+\sum_{i=0}^{n-1} a_i,$

(b) $\quad a_n^2=1+a_{n-1}a_{n+1},$

(c) $\quad a_{2n+1}=a_n^2+a_{n+1}^2.$

(the sequence of S1.90 is called the *Fibonacci sequence*).

S1.91. The terms of the sequence a_n satisfy the conditions

$$a_1=2, \quad a_n=3a_{n+1}+1.$$

Find $\sum_{i=1}^{n} a_i$.

S1.92. The terms of the sequence a_n satisfy the conditions

$$a_{n+1}-2a_n+a_{n-1}=1.$$

Express a_n in terms of a_0, a_1 and n.

S1.93. The terms of the sequences a_n and b_n satisfy the conditions

$$a_0=2, \quad b_0=1, \quad a_{n+1}=2a_n+5b_n, \quad b_{n+1}=a_n+2b_n.$$

Express a_n and b_n in terms of n.

S1.94. Prove that every n-gon can be partitioned into $n-2$ triangles by means of non-intersecting diagonals.

S1.95. Prove that the regions into which the plane is partitioned by n straight lines can be coloured with two colours in such a way that adjacent regions (i.e. regions with a common node) have different colours.

S1.96. Prove that, for all $n \geqslant 1$, n squares can be partitioned with straight lines in such a way that from the resulting pieces a square can be obtained.

SUPPLEMENT 2

LATTICES AND BOOLEAN ALGEBRAS

A relational system $\mathscr{L}=\langle L, \wedge, \vee\rangle$ is called a *lattice* if $\wedge$ and $\vee$ are binary operations and the following formulas are true in $\mathscr{L}$:

(L1) (a) $a \wedge a=a$, (b) $a \vee a=a$.

(L2) (a) $a \wedge b=b \wedge a$, (b) $a \vee b=b \vee a$.

(L3) (a) $a \wedge(b \wedge c)=(a \wedge b) \wedge c$, (b) $a \vee(b \vee c)=(a \vee b) \vee c$.

(L4) (a) $a \wedge(a \vee b)=a$, (b) $a \vee(a \wedge b)=a$.

If – in addition – the following is true:

(L5) (a) $a \wedge(b \vee c)=(a \wedge b) \vee(a \wedge c)$,

(b) $a \vee(b \wedge c)=(a \vee b) \wedge(a \vee c)$,

then we say that $\mathscr{L}$ is a *distributive lattice*.

S2.1. Prove that the following is true in every lattice:

$$(a=a \vee b) \Leftrightarrow(a \wedge b=b).$$

S2.2. Let $\leqslant$ be a binary relation defined by $a \leqslant b \Leftrightarrow a \vee b=b$. Prove that $\leqslant$ is an ordering.

S2.3. Let $\mathscr{L}=\langle L, \vee, \wedge\rangle$ be a lattice and $\leqslant$ an ordering introduced in S2.2. Prove that $\langle L, \leqslant\rangle$ is a directed set, i.e. that

$$\bigwedge_{a,b} \bigvee_{c}(a \leqslant c \wedge b \leqslant c).$$

S2.4. Consider the system $\langle L, \leqslant\rangle$ introduced in S2.3. Show that in L there exist suprema and infima, i.e.

(a) $\bigwedge_{a,b} \bigvee_{c} \bigwedge_{d}[(d \leqslant a \wedge d \leqslant b) \Leftrightarrow d \leqslant c]$,

(b) $\bigwedge_{a,b} \bigvee_{c} \bigwedge_{d}[(a \leqslant d \wedge b \leqslant d) \Leftrightarrow c \leqslant d]$.

S2.5. Prove that if $\mathscr{L}$ is a lattice and $\leqslant$ is the derived ordering relation then:

(a) $a \wedge b \leqslant a$,

(b) $(a \leqslant b \wedge c \leqslant d) \Rightarrow(a \wedge c \leqslant b \wedge d)$,

(c) $a \leqslant a \vee b$,

(d) $(a \leqslant b \wedge c \leqslant d) \Rightarrow(a \vee c \leqslant b \vee d)$.

S2.6. Consider the system $N=\langle\mathcal{N}-\{0\}, \wedge, \vee\rangle$ where $x\wedge y=$ $=GCD(x, y)$ and $x\vee y=LCM(x, y)$. Prove that it is a lattice. Is the distributivity axiom (L5) true in N?

S2.7. Prove that if $\langle X, \leqslant\rangle$ is a linearly ordered set then there exist operations $\wedge$ and $\vee$ in X such that $\langle X, \wedge, \vee\rangle$ is a lattice and the derived ordering is identical with $\leqslant$.

S2.8. Given a lattice $\langle L, \wedge, \vee\rangle$ and a subset $X\subset L$, we say that an element $t\in L$ is an *infimum* of X if the following is true:

$$\bigwedge_x\bigwedge_y[(y\in X\Rightarrow x\leqslant y)\Leftrightarrow x\leqslant t].$$

Prove that every finite subset of a lattice possesses an infimum.

S2.9. Define the notion of the *supremum* of the set X. Prove a theorem analogous to that of S2.8

S2.10. Find a lattice $\langle L, \wedge, \vee\rangle$ and a subset $X\subset L$ without an infimum. Can such an example be found for a finite L?

S2.11. Let $\cap X$, $\cup X$ denote, respectively, the infimum and the supremum of the set $X\subset L$ (providing they exist). Let X and Y be finite subsets of L. Prove that

(a) $\cap X\wedge\cap Y=\cap(X\cup Y)$,

(b) $\cup X\vee\cup Y=\cup(X\cup Y)$.

S2.12. In the set $\mathcal{P}(\mathcal{N})$ introduce a relation $\sim$ as follows: $X\sim Y$ if $X\dot{-}Y$ is a finite set. The relation $\sim$ is an equivalence. In the family of equivalence classes introduce a relation as follows: $[A]\leqslant[B]$ iff $A-B$ is a finite set. Prove that in the set $\mathcal{P}(\mathcal{N})/\sim$ there exist operations $\wedge$ and $\vee$ such that $\langle\mathcal{P}(\mathcal{N})/\sim, \wedge, \vee\rangle$ is a lattice and its derived ordering is identical with $\leqslant$. Show that the definition of the relation $\leqslant$ is sound, i.e. does not depend on the choice of representatives.

S2.13. Let $\mathcal{L}=\langle L, \wedge, \vee\rangle$ be a lattice. Define the operations $\wedge'$ and $\vee'$ as follows: $a\wedge'b=a\vee b$, $a\vee'b=a\wedge b$. Prove that $\mathcal{L}'=\langle L, \wedge', \vee'\rangle$ is a lattice.

S2.14. Let $\langle X, D\rangle$ be a topological space and D its family of closed subsets. In the set D we define the operations $\wedge$ and $\vee$ as follows $A\wedge B=A\cap B$, $A\vee B=A\cup B$. Prove that the resulting system is a distributive lattice.

S2.15. A lattice $\mathcal{L}=\langle L, \wedge, \vee\rangle$ is called a *modular lattice* iff the following is true in $\mathscr{L}$:

(L4 1/2) $a \leqslant c \Rightarrow a \vee (b \wedge c) = (a \vee b) \wedge c.$

Prove that every distributive lattice is a modular one.

S2.16. Prove that a lattice $\mathscr{L}$ is modular iff it does not possesses a five element sublattice $\langle \{a, b, c, d, e\}, \wedge, \vee \rangle$ with a diagram as in Fig. 45 p. 259.

S2.17. Give examples of a distributive lattice and of a lattice which is not distributive. What property of statement (L5) is thus proved?

S2.18. Give an example of:

(a) a non-modular lattice.

(b) a lattice which is modular but not distributive.

S2.19. In the set $\mathscr{P}(X^2)$, i.e. in the family of all relations in X, the relation $\subset$ is an ordering. Prove that there exist operations $\wedge$ and $\vee$ such that $\langle \mathscr{P}(X^2), \wedge, \vee \rangle$ is a lattice and the derived ordering is identical with $\subset$.

S2.20. In the family T of all equivalences in the set X, the relation $\subset$ is an ordering. Prove that there exist operations $\wedge$ and $\vee$ in the set T such that $\langle T, \wedge, \vee \rangle$ is a lattice and the derived ordering is identical with $\subset$.

S2.21. Prove that the set of axioms (L1)–(L5) is not independent, for instance:

(a) (L1) can be derived from the other axioms.

(b) (L5) (a) and (L1) (b) can be derived from the other axioms.

(c) (L5) (b) can be derived from the other axioms.

S2.22. Assume that $O \neq X \subset L$ possesses both an infimum and a supremum. Prove that

(a) $\quad \cap X \leqslant x$ for $x \in X$,

(b) $\quad x \leqslant \cup X$ for $x \in X$,

(c) $\quad \cap X \leqslant \cup X$.

S2.23. Prove that if a nonempty set $X \subset L$ possesses a supremum then the set $\{a \vee x : x \in X\}$ also possesses a supremum and the following holds:

$$a \vee \cup X = \cup \{a \vee x : x \in X\}.$$

S2.24. Under the assumptions of S2.23, prove that

$$a \wedge \cap X = \cap \{a \wedge x : x \in X\}.$$

S2.25. Let $\{a_t\colon t \in T\}$, $\{b_t : t \in T\}$ be two families of elements of the set L. Assume that there exist an infimum and a supremum of both of those sets. Finally, assume that $\bigwedge_{t \in T} a_t \leqslant b_t$. Prove that

a) $\bigcap_{t \in T} a_t \leqslant \bigcap_{t \in T} b_t$,

b) $\bigcup_{t \in T} a_t \leqslant \bigcup_{t \in T} b_t$,

where $\bigcap_{t \in T} a_t = \cap \{a_t : t \in T\}$, and $\bigcup_{t \in T} a_t = \cup \{a_t : t \in T\}$.

S2.26. Under the assumptions of S2.25, prove that if the families $\{a_t \vee b_t : t \in T\}$, $\{a_t \wedge b_t : t \in T\}$ possess both suprema and infima then:

(a) $\bigcup_{t \in T} (a_t \vee b_t) = \bigcup_{t \in T} a_t \vee \bigcup_{t \in T} b_t$,

(b) $\bigcap_{t \in T} (a_t \wedge b_t) = \bigcap_{t \in T} a_t \wedge \bigcap_{t \in T} b_t$,

(c) $\bigcap_{t \in T} a_t \vee \bigcap_{t \in T} b_t \leqslant \bigcap_{t \in T} (a_t \vee b_t)$.

S2.27. A lattice $\mathscr{L} = \langle L, \wedge, \vee \rangle$ is called *complete* if every nonempty subset of L possesses both a supremum and an infimum. Give examples of a complete lattice and of an incomplete lattice. Prove that every finite lattice is complete.

S2.28. An element $a \in L$ is called a *zero* (a *unity*) of the lattice $\langle L, \wedge, \vee \rangle$ if it satisfies the condition:

$$\bigwedge_x (a \wedge x = a) \, [\bigwedge_x (a \vee x = a)] .$$

(a) Prove that a lattice may possess at most one zero.

(b) Prove that a lattice may possess at most one unity.

(c) Find a lattice with a unity but without a zero.

(d) Find a lattice with zero but without a unity.

(e) Is it possible for an element of a lattice to be simultaneously both a zero and a unity in the lattice?

S2.29. Let $\mathscr{L}'$ be the lattice defined in S2.13.

(a) Prove that if $\mathscr{L}$ possesses a zero then $\mathscr{L}'$ possesses a unity.

(b) Prove the converse to (a).

(c) Prove an analogous theorem for the unity of $\mathscr{L}$.

(d) Prove that if $\mathscr{L}$ possesses both a zero and a unity then $\mathscr{L}'$ also possesses them.

(e) Is it true that if $\mathscr{L}$ is a distributive lattice then $\mathscr{L}'$ is distributive as well?

(f) Is it true that if $\mathscr{L}$ is a modular lattice then $\mathscr{L}'$ is modular as well?

S2.30. An *ideal* in the lattice $\mathscr{L}=\langle L, \wedge, \vee\rangle$ is a nonvoid subset $I \subset L$ satisfying the following conditions:

(I1) $a \in I \wedge b \in I \Rightarrow a \vee b \in I$,

(I2) $a \in I \wedge b \leqslant a \Rightarrow b \in I$.

Prove that L is an ideal in $\mathscr{L}$.

S2.31. Show that condition (I1) and (I2) are equivalent to the following $\mathrm{I}\frac{1}{2}$. $a \in I \wedge b \in I \Leftrightarrow a \vee b \in I$.

S2.32. Prove that if 0 is the zero of a lattice $\mathscr{L}$ then:

(a) $\{0\}$ is an ideal.

(b) 0 belongs to all the ideals of $\mathscr{L}$.

(c) $\{0\}$ is the common part of all the ideals in $\mathscr{L}$.

S2.33. Prove that if $a \in L$ then the set $(a)=\{b : b \leqslant a\}$ is an ideal in $\mathscr{L}$. It is called the *principal ideal* generated by a in $\mathscr{L}$.

S2.34. Prove that the common part of all ideals in $\mathscr{L}$ is nonempty iff $\mathscr{L}$ possesses a zero element.

S2.35. A *filter* in the lattice $\mathscr{L}=\langle L, \wedge, \vee\rangle$ is a nonempty subset $F \subset L$ satisfying the following conditions:

(F1) $a \in F \wedge b \in F \Rightarrow a \wedge b \in F$,

(F2) $a \in F \wedge a \leqslant b \Rightarrow b \in F$.

Prove that L is a filter in $\mathscr{L}$.

What is the meaning of the sign $\wedge$ in the predecessor and in the successor of the implication of (F1)?

S2.36. Prove that conditions (F1) and (F2) are equivalent to the following (F$1\frac{1}{2}$):

$$a \in F \wedge b \in F \Leftrightarrow a \wedge b \in F.$$

S2.37. Prove that if 1 is the unity of a lattice $\mathscr{L}$ then

(a) $\{1\}$ is a filter in $\mathscr{L}$.

(b) 1 belongs to all the filters of $\mathscr{L}$.

(c) $\{1\}$ is the intersection of all the filters in $\mathscr{L}$.

S2.38. Prove that if $a \in L$ then the set $[a]=\{b : a \leqslant b\}$ is a filter in $\mathscr{L}$. It is called the *principal filter* generated by a in $\mathscr{L}$.

S2.39. Prove that in a finite lattice every ideal and every filter are principal.

S2.40. Prove that if the derived ordering of the complete lattice $\mathscr{L} = \langle L, \wedge, \vee \rangle$ is a linear ordering then every ideal in $\mathscr{L}$ is principal. Give an example showing that the completeness condition cannot be omitted.

S2.41. Prove that every complete lattice possesses both a zero and a unity. Use this to show that every finite lattice possesses both a zero and a unity.

S2.42. Let $\mathscr{L} = \langle L, \wedge, \vee \rangle$ be a lattice and $\leqslant$ its derived ordering. Prove that if for every $X \subset L$, $\langle X, \leqslant \rangle$ is a lattice (i.e. X is closed under suprema and infima) then $\leqslant$ is a linear ordering. Prove that it is enough for finite sets to have the above property (indeed, two-element sets).

S2.43. Prove that if a is an element of the distributive lattice with a zero and a unity $\mathscr{L} = \langle L, \wedge, \vee, 0, 1 \rangle$ and

$$\bigvee_b [(a \wedge b = 0) \wedge (a \vee b = 1)],$$

then the element b of the above is uniquely determined by a.

A *Boolean algebra* is a relational system $\mathbf{a} = \langle A, \wedge, \vee, \sim, 0, 1 \rangle$ such that $\langle A, \wedge, \vee, 0, 1 \rangle$ is a distributive lattice with a zero and a unity and $\sim$ is a unary operation satisfying the following conditions:

$$a \wedge \sim a = 0, \qquad a \vee \sim a = 1.$$

Let $\mathfrak{a}$ be a Boolean algebra (Problems S2.44–S2.46):

S2.44. Prove that if $a \vee b = 1$ and $a \wedge b = 0$ then $b = \sim a$.

S2.45. Prove that $\sim \sim a = a$, $p \leqslant q \Leftrightarrow \sim q \leqslant \sim p$.

S2.46. Prove De Morgan's Laws:

$$\sim(p \wedge q) = \sim p \vee \sim q, \qquad \sim(p \vee q) = \sim p \wedge \sim q.$$

S2.47. Let $\mathfrak{a} = \langle A, \wedge, \vee, \sim, 0, 1 \rangle$ be a Boolean algebra. For the lattice $\mathfrak{a}_0 = \langle A, \wedge, \vee \rangle$ conduct the construction of S2.13. How should the operation $\sim'$ and the constants $0'$ and $1'$ be defined for the system $\mathfrak{a}' = \langle A, \wedge', \vee', \sim', 0', 1' \rangle$ to be a Boolean algebra? If I is an ideal in $\mathfrak{a}$, what is it in $\mathfrak{a}'$? What about a filter F of $\mathfrak{a}$?

S2.48. In the set $\mathbf{2} = \{0, 1\}$ define operations $\wedge$, $\vee$, and $\sim$ as follows:

$$x \wedge y = \min(x, y) \qquad x \vee y = \max(x, y) \qquad \sim x = 1 - x.$$

Prove that the system $\mathbf{2} = \langle 2, \wedge, \vee, \sim, 0, 1 \rangle$ is a Boolean algebra.

S2.49. Let $\langle X, \mathcal{O}\rangle$ be a topological space ($\mathcal{O}$ is the family of open subsets of the space), Cl the corresponding closure operation and Int the corresponding interior operation.

A *regular open* set is every set $A \subset X$ such that $A = \text{Int}\,[\text{Cl}(A)]$. In the family of all regular open sets introduce the operations

$$A \wedge B = A \cap B,$$

$$A \vee B = \text{Int}\,[\text{Cl}(A \cup B)],$$

$$\sim A = X - \text{Cl}\,A,$$

$$0 = O,$$

$$1 = X.$$

Prove that we get a Boolean algebra.

S2.50. A Boolean algebra $\langle A, \wedge, \vee, \sim, 0, 1\rangle$ is called *complete* iff the structure $\langle A, \wedge, \vee\rangle$ is a complete lattice. Prove that the infimum of the algebra of S2.49 is a complete Boolean algebra; in particular,

$$\bigcap_{t \in T} X_t = \text{Int}\,[\text{Cl}(\bigcap_{t \in T} X_t)], \qquad \bigcup_{t \in T} X_t = \text{Int}\,[\text{Cl}(\bigcup_{t \in T} X_t)],$$

where the symbols $\cap$ and $\cup$ denote on the left-hand sides the infimum and the supremum of the family of open sets and on the right-hand side the set-theoretical operations.

S2.51. Prove that if the supremum $\bigcup_{t \in T} a_t$ exists then the infimum $\bigcap_{t \in T} \sim a_t$ also exists and $\sim \bigcup_{t \in T} a_t = \bigcap_{t \in T} \sim a_t$. Prove an analogous result for infima.

S2.52. A *filter* F in the lattice $\mathscr{L} = \langle L, \wedge, \vee\rangle$ is said to be *prime* if the following is true $(a \vee b) \in F \Rightarrow a \in F \vee b \in F$. Prove that in the Boolean algebra the above condition is equivalent to the following: for every a, $a \in F$ or $\sim a \in F$. Introduce the notion of a *prime ideal* and prove an analogous equivalence.

S2.53. A filter F (an ideal I) of the lattice $\mathscr{L} = \langle L, \wedge, \vee\rangle$ is called a *proper filter* (a *proper ideal*) iff $F \neq L$ $(I \neq L)$. A *maximal filter* (*maximal ideal*) is a maximal element in the family of all proper filters (proper ideals). (A maximal filter is often called an *ultrafilter*.)

Prove that in the Boolean algebra $\mathfrak{a} = \langle A, \wedge, \vee, \sim, 0, 1\rangle$ a proper filter F is prime iff F is a maximal filter.

S2.54. Prove that every proper filter can be extended to a proper ultrafilter.

S2.55. Prove that for every element $a \in A$ and a filter F such that $a \notin F$ there exists an ultrafilter G such that $F \subset G$ and $a \notin G$.

S2.56. A *Boolean homormorphism* is a mapping f: $A \to A_1$ (where $\mathfrak{a} = \langle A, \wedge, \vee, \sim, 0, 1\rangle$ and $\mathfrak{a}_1 = \langle A_1, \wedge_1, \vee_1, \sim_1, 0_1, 1_1\rangle$ are Boolean algebras) which satisfies the following conditions:

$$f(a \wedge b) = f(a) \wedge_1 f(b), \quad f(a \vee b) = f(a) \vee_1 f(b),$$

$$f(\sim a) = \sim_1 f(a), \quad f(0) = 0_1, \quad f(1) = 1_1.$$

Prove that if f is a Boolean homomorphism then $f^{-1} * \{1_1\}$ is a filter and $f^{-1} * \{0_1\}$ is an ideal in $\mathfrak{a}$.

Let $\mathfrak{a} = \langle A, \wedge, \vee, \sim, 0, 1\rangle$ and $\mathfrak{a}_1 = \langle A_1, \wedge_1, \vee_1, \sim_1, 0_1, 1_1\rangle$ be two Boolean algebras. We say that $\mathfrak{a}$ is *isomorphic* with $\mathfrak{a}_1$ ($\mathfrak{a} \approx \mathfrak{a}_1$) if there exists a bijection f of A and A_1 which is a Boolean homomorphism.

Such a mapping is called a *Boolean isomorphism.*

S2.57. Prove that if F is an ultrafilter in a Boolean algebra $\mathfrak{a} = \langle A, \wedge, \vee, \sim, 0, 1\rangle$ and the mapping h of the set A into $\mathbf{2}$ is defined by

$$f(x) = \begin{cases} 1, & \text{if} \quad x \in F, \\ 0, & \text{if} \quad x \notin F, \end{cases}$$

then h is a Boolean homomorphism.

Prove the converse theorem.

S2.58 Prove that every Boolean algebra contains a subalgebra isomorphic to the algebra $\mathbf{2}$.

S2.59. Let φ be an expression of the propositional calculus and let T_φ be the set of the propositional variables occurring in formula φ. Given a mapping v of the set T_φ into a Boolean algebra $\mathfrak{a}$, extend v to φ inductively as follows:

$$v(p \wedge q) = v(p) \wedge v(q),$$

$$v(p \vee q) = v(p) \vee v(q),$$

$$v(\sim p) = \sim v(p),$$

$$v(p \Rightarrow q) = \sim v(p) \vee v(q),$$

$$v(p \Leftrightarrow q) = \big(v(p) \wedge v(q)\big) \vee \big(\sim v(p) \wedge \sim v(q)\big).$$

(such a mapping v is called a *valuation*).

Prove that formula φ is a tautology iff, for every valuation in any Boolean algebra, $v\varphi$ is equal to 1.

S2.60. Prove that if I is an ideal in $\mathfrak{a}$ then the set $\{x: \sim x \in I\}$ is a filter in $\mathfrak{a}$, and conversely, i.e. that if F is a filter in $\mathfrak{a}$ then $\{x: \sim x \in F\}$ is an ideal in $\mathfrak{a}$.

Prove an analogous result for prime ideals.

S2.61. Let F be a filter in $\mathfrak{a}$. Define a relation $\sim_F$ in A as follows

$$x \sim_F y \Leftrightarrow [(x \vee \sim y) \wedge (y \vee \sim x)] \in F.$$

Prove that $\sim_F$ is an equivalence relation.

Define the Boolean operations in $A/\sim_F$.

S2.62. Prove that if h is a Boolean epimorphism (i.e. a homomorphism and a surjection) then $h(\mathfrak{a}) \approx \mathfrak{a}/\sim_F$ where F is the counterimage of the unity of $h(\mathfrak{a})$.

S2.63. Let $\mathfrak{a} = \langle A, \wedge, \vee, \sim, 0, 1\rangle$, $\mathfrak{a}_1 = \langle A_1, \wedge_1, \vee_1, \sim_1, 0_1, 1_1\rangle$ be two Boolean algebras. In the Cartesian product $A \times A_1$ define the operations $\wedge'$, $\vee'$, $\sim'$, $0'$, $1'$ as follows:

$$\langle a, a_1\rangle \wedge' \langle b, b_1\rangle = \langle a \wedge b, a_1 \vee_1 b_1\rangle,$$
$$\langle a, a_1\rangle \vee' \langle b, b_1\rangle = \langle a \vee b, a_1 \wedge_1 b_1\rangle,$$
$$\sim' \langle a, a_1\rangle = \langle \sim a, \sim_1 a_1\rangle,$$
$$0' = \langle 0, 1_1\rangle, \quad 1' = \langle 1, 0_1\rangle.$$

Prove that $\langle A \times A_1, \wedge', \vee', \sim', 0', 1'\rangle$ is a Boolean algebra.

S2.64. Let $\{\mathfrak{a}_t\}_{t\in T}$ be an indexed family of Boolean algebras, $\mathfrak{a}_t = \langle A_t, \wedge_t, \vee_t, \sim_t, 0_t, 1_t\rangle$.

Define $\prod_{t\in T} \mathfrak{a}_t = \langle \underset{t\in T}{P} A_t, \wedge, \vee, \sim, 0, 1\rangle$, where the operations $\wedge$, $\vee$, $\sim$, 0 and 1 are defined as follows:

$$f \wedge g = h \Leftrightarrow \bigwedge_{t\in T} h(t) = f(t) \wedge_t g(t),$$
$$f \vee g = h \Leftrightarrow \bigwedge_{t\in T} h(t) = f(t) \vee_t g(t),$$
$$\sim f = g \Leftrightarrow \bigwedge_{t\in T} g(t) = \sim_t f(t),$$
$$0 = \{\langle t, 0_t\rangle : t \in T\},$$
$$1 = \{\langle t, 1_t\rangle : t \in T\}.$$

Prove that $\prod_{t\in T} \mathfrak{a}_t$ is a Boolean algebra.

S2.65. An element $a \in A$ is called an *atom* iff

$$\bigwedge_x [a \wedge x \neq a \Rightarrow a \wedge x = 0].$$

A Boolean algebra $\mathfrak{a}$ is said to be *atomic* iff for every $x \in A$ there exists an atom a such that $a \leqslant x$.

Prove that for every set X the algebra $\langle \mathscr{P}(X), \cap, \cup, -, O, X \rangle$ is an atomic Boolean algebra.

Give an example of an algebra which is not atomic.

S2.66. Prove that every finite Boolean algebra is atomic.

S2.67. Let $\mathfrak{a}$ be a complete and atomic Boolean algebra and At be its set of atoms. Prove that the mapping $h: A \to \mathscr{P}(\mathrm{At})$ defined as follows: $h(a) = \{x \in \mathrm{At} : x \leqslant a\}$ is an isomorphism of $\mathfrak{a}$ and the algebra $\langle \mathscr{P}(At), \cap, \cup, -, O, At \rangle$.

S2.68. Prove that if $\mathfrak{a}$ is a finite Boolean algebra then there exists a natural number n such that $\overline{\overline{A}} = 2^n$.

S2.69. Prove that a is an atom iff the filter $[a]$ is an ultrafilter.

S2.70. Prove that a is an atom iff, for every two filters F_1 and F_2, if $a \in F_1 \cap F_2$ then $F_1 = F_2$.

S2.71. Let $\mathfrak{a} = \langle A, \wedge, \vee, \sim, 0, 1 \rangle$ be a Boolean algebra and let X be the family of all ultrafilters in $\mathfrak{a}$. Given $a \in A$, let $X_a \subset X$ be the family of those filters F for which $a \in F$. Prove that:

(a) $X_a \cap X_b = X_{a \wedge b}$,

(b) $X_a \cup X_b = X_{a \vee b}$,

(c) $X - X_a = X_{\sim a}$.

S2.72. Define a topology in X of S2.71 as follows: The basis of the topology is the family consisting of X_a, $a \in A$. Prove that the space thus obtained is a compact, zero-dimensional Hausdorff space.

S2.73. Prove the following theorem of Stone: Every Boolean algebra is isomorphic to the algebra of all clopen subsets of a topological space.

S2.74. An algebra $\mathfrak{a}$ is said to be *dense* iff the derived ordering $\leqslant$ is a dense ordering. Prove that dense algebras are atomless.

S2.75. A *Boolean ring* is a commutative ring with unity in which every element is idempotent, i.e. $\bigwedge_a a^2 = a$. Let $\mathfrak{a} = \langle A, \wedge, \vee, \sim, 0, 1 \rangle$ be a Boolean algebra. Define an operation $\dot{-}$ in A as follows: $a \dot{-} b = (\sim a \wedge b) \vee (a \wedge \sim b)$. Prove that the system $\langle A, \wedge, \dot{-}, 0, 1 \rangle$ is a Boolean ring.

What is the subtraction operation in this ring? Check that Boolean rings satisfy the sentence $\bigwedge_a a+a=0$.

S2.76. Let $\mathscr{P}=\langle P, +, \cdot, 0, 1\rangle$ be a commutative ring with unity. Prove that the set W of the idempotent elements of $\mathscr{P}$ with the following operations $+'$ and $\cdot'$: $a \cdot' b = a \cdot b$, $a +' b = a+b-2ab$ is a Boolean ring.

S2.77. Let $\mathscr{P}=\langle P, +, \cdot, 0, 1\rangle$ be a Boolean ring.

Define the operations $\wedge, \vee, \sim$ as follows:

$$a \wedge b = a \cdot b, \quad a \vee b = a+b+a \cdot b, \quad \sim a = 1+a.$$

Prove that $\mathfrak{a}=\langle P, \wedge, \vee, \sim, 0, 1\rangle$ is a Boolean algebra.

S2.78. *Boolean polynomials* are defined inductively:

(a) Every variable is a polynomial; 0, 1 are polynomials.

(b) If f and g are polynomials then $f \wedge g$, $f \vee g$, $\sim f$ are also polynomials.

(c) Every polynomial arises from the variables, 0 and 1 by a finite number of applications of the operations from b).

A polynomial f is said to be *normal* if it is of the form $f=u_1 \vee \ldots \vee u_n$ where $u_i = v_{i_1} \wedge \ldots \wedge v_{i_k}$ and every v_j is of the form x_j or $\sim x_j$ where x_j is a variable or if f is 0 or 1.

Prove that for every polynomial f there exists a normal polynomial f_1 such that for every Boolean algebra $\mathfrak{b}=\langle B, \wedge, \vee, \sim, 0, 1\rangle$ and arbitrary $x_1, \ldots, x_n \in B$

$$f(x_1, \ldots, x_n) = f_1(x_1, \ldots, x_n).$$

S2.79. What is the connection between Problems S2.78 and 1.80?

S2.80. Let $\mathscr{L}=\langle L, \wedge, \vee, 0, 1\rangle$ be a distributive lattice with zero and unity. An element $a \in L$ is called a *Boolean element* of L iff there exists a $b \in L$ such that $a \wedge b = 0$, $a \vee b = 1$.

Prove that the sublattice of Boolean elements of $\mathscr{L}$ is a Boolean algebra.

S2.81. Let $\mathfrak{b}=\langle B, \wedge, \vee, \sim, 0, 1\rangle$ be a Boolean algebra, $\mathfrak{b}^n = \underbrace{\mathfrak{b} \times \ldots \times \mathfrak{b}}_{n}$, and finally $P=\{\langle x_1, \ldots, x_n\rangle : x_1 \geqslant x_2 \geqslant \ldots \geqslant x_n\}$.

Prove that P together with the operations inherited from $\mathfrak{b}^n$ is a distributive lattice with zero and unity.

Prove that the Boolean algebra of Boolean elements of P is isomorphic to $\mathfrak{b}$.

S2.82. Is a sublattice of a Boolean algebra necessarily a Boolean algebra?

SOLUTIONS

CHAPTER 1

PROPOSITIONAL CALCULUS

1.1. Consider the formula $\sim(\sim p \vee \sim q)$. We show that it may serve as a definition of the functor B_1; to see this we check the values taken by this formula, depending on the values of p and q:

p	q	$\sim p$	$\sim q$	$\sim p \vee \sim q$	$\sim(\sim p \vee \sim q)$	$p \wedge q$
0	0	1	1	1	0	0
0	1	1	0	1	0	0
1	0	0	1	1	0	0
1	1	0	0	0	1	1

Thus the formula $\sim(\sim p \vee \sim q)$ takes – under arbitrary valuation – the same values as $p \wedge q$ and so it may serve as a definition of $\wedge$ in terms of $\sim$ and $\vee$.

1.2. $p \vee q \Leftrightarrow \sim(\sim p \wedge \sim q)$.

1.3. $p \vee q \Leftrightarrow (\sim p \Rightarrow q)$.

1.4. Consider all functors definable by means of $\wedge$ and $\vee$.

1.5. Check the properties of the expressions $p \Leftrightarrow q$ and $p \Leftrightarrow \sim q$.

1.6. Prove that $\sim p \Leftrightarrow (p \Rightarrow A_0\, p)$, then use 1.2, 1.3, and the following: $p \Leftrightarrow q$ is equivalent to $(p \Rightarrow q) \wedge (q \Rightarrow p)$.

1.7. Notice first that $\sim p \Leftrightarrow B_8(p, p)$ and that $B_8(p, q) \Leftrightarrow (\sim p \wedge \sim q)$; thus $p \wedge q \Leftrightarrow B_8[B_8(p, p), B_8(q, q)]$. Since (cf. 1.2) the alternative is definable by means of conjunction and negation and similarly for implication $(p \Rightarrow q \Leftrightarrow \sim(p \wedge \sim q))$ and equivalence, we have $\sim$, $\wedge$, $\vee$, $\Rightarrow$, $\Leftrightarrow$ and thus all the other connectives are definable by means of B_8 only. A similar idea is used in the case of B_{14} $\sim p \Leftrightarrow B_{14}(p, p)$, $p \vee q \Leftrightarrow B_{14}[B_{14}(p, p), B_{14}(q, q)]$ etc.

1.8. Hint. The proof of the definability of all functors is obtained inductively. Assume that, for certain n, all n-ary functors are definable by means of B_8 (the proof for B_{14} is analogous). Let $f(x_1, \ldots, x_n, x_{n+1})$ be an $n+1$-ary functor. Then $f(x_1, \ldots, x_n, 0)$ and $f(x_1, \ldots, x_n, 1)$ are n-ary functors. Let $\Phi_1(x_1, \ldots, x_n)$, $\Phi_2(x_1, \ldots, x_n)$ be the respective definitions of $f(x_1, \ldots, x_n, 0)$, $f(x_1, \ldots, x_n, 1)$ in terms of B_8 only. Then $(x_{n+1} \wedge \Phi_1(x_1, \ldots, x_n)) \vee (\sim x_{n+1} \wedge \Phi_2(x_1, \ldots, x_n))$ is a definition of $f(x_1, \ldots, x_n, x_{n+1})$. Since we are able to express $\sim$, $\wedge$ and $\vee$ by means of B_8, $f(x_1, \ldots, x_n, x_{n+1})$ is also expressible by means of B_8 only.

1.9. 2^n.

1.19. If the expression under consideration is not a tautology, then there must exist a valuation under which the sides of the equivalence take different values. Therefore it is sufficient to prove that whenever $w(p \Rightarrow q)=0$ then $w(\sim q \Rightarrow \sim p)=0$ and, conversely, if $w(\sim q \Rightarrow \sim p)=0$ then $w(p \Rightarrow q)=0$. But if $w(p \Rightarrow q)=0$ then $w(p)=1$ and $w(q)=0$. Thus

$$w\,(\sim q)=1,\ w\,(\sim p)=0 \text{ and } w\,(\sim q \Rightarrow \sim p)=0.$$

Conversely, if $w(\sim q \Rightarrow \sim p)=0$ then $w(\sim q)=1$ and $w(\sim p)=0$. Thus $w(p)=1$ and $w(q)=0$ and so $w(p \Rightarrow q)=0$.

1.22. If under a valuation w we have $w[(\sim p \Rightarrow p) \Rightarrow p)]=0$ then $w(p)=0$ and $w(\sim p \Rightarrow p)=1$. But in that case $w(\sim p)=1$ and so $w(\sim p \Rightarrow p)=0$, which is absurd.

1.33. Yes. If $w(q)=0$ then $w[(p \vee q) \wedge \sim p]=w[(p \vee 0) \wedge \sim p]=w[p \wedge \sim p]=0$. Thus under no valuation does our formula take the value 0.

1.34. Yes.

1.35. Yes.

1.36. No. Consider a valuation w such that $w(p)=1$, $w(q)=0$.

1.37. No. Consider a valuation w such that $w(p)=0$, $w(q)=1$.

1.38. Yes.

1.39. Yes. Use the tautology: $\sim(p \wedge q) \Leftrightarrow \sim p \vee \sim q$.

1.40. Yes. Use the following: for every valuation w, $w(q \wedge \sim q)=0$.

1.41. No. Use the following: $w(p \vee q)=0 \Leftrightarrow w(p)=w(q)=0$.

1.42. No. Consider a valuation w such that $w(p)=1$, $w(q)=0$.

1.43. Yes. If $w(p \Rightarrow q)=0$ then $w(p)=1$ and $w(q)=0$, but in that case $w(p \wedge q)=0$.

1.44. Yes.

1.45. Yes. If $w(p \vee r \Rightarrow q \vee s)=0$ then $w(q)=w(s)=0$ and either $w(p)=1$ or $w(r)=1$. But in that case $w(p \Rightarrow q)=0$ or $w(r \Rightarrow s)=0$.

1.46. Yes.

1.47. Use the reasoning of 1.45.

1.48. No. Consider a valuation w such that $w(p)=w(q)=w(r)=0$.

1.49. Yes. Use 1.26 and the tautology $p \wedge q \Leftrightarrow q \wedge p$.

1.50. Yes. Compare with the tautology $(p \vee q) \Leftrightarrow (\sim p \Rightarrow q)$.

1.51. Yes. Use the following: $\sim(p \Rightarrow q) \Leftrightarrow (p \wedge \sim q)$, $p \wedge q \Rightarrow p$.

1.52. Yes. Notice that $w(q \vee r)=0$ implies $w(q)=w(r)=0$. Further, $w(p \wedge s)=1$ implies $w(p)=w(s)=1$, and finally $w(p \Rightarrow q)=0$.

1.53. Yes.

1.54. Yes. If $w(p\Rightarrow q)=0$ then $w(p)=1$ and $w(q)=0$ and so $w(p\wedge q)=0$. Finally, $w(p\wedge q\Leftrightarrow p)=0$. Conversely, if $w(p\wedge q\Leftrightarrow p)=0$ then $w(p)=1$ and $w(q)=0$, etc.

1.55. Yes.

1.56. Yes.

1.57. No. Consider a valuation w such that $w(p)=w(q)=0$, $w(r)=1$. Notice that looking for the counterexample we have to find a valuation w with the property $w(p\wedge q)=0$, $w(r)=1$.

1.58. Yes.

1.59. No. Consider valuation w such that $w(p)=w(s)=1$, $w(q)=w(r)=0$.

1.60. Yes.

1.61. Yes.

1.62. Yes. Consider the tautology $p\Rightarrow(\sim p\Rightarrow q)$.

1.63. Yes.

1.64. Yes.

1.65. No. The expression $(p\wedge q)\Rightarrow(p\Rightarrow r)$ is not a tautology.

1.66. Yes. Use the tautology $(p\wedge q)\Rightarrow(\sim q\Rightarrow\sim p)$.

1.67. Yes.

1.68. Yes.

1.69. Yes. Use the tautology $\sim p\Rightarrow(p\Rightarrow q)$.

1.70. Yes. Use the tautology $p\Rightarrow\sim\sim p$.

1.71. No.

1.75. The interpretation of the functors is as follows:

$$\sim p=1-p, \quad p\vee q=p\cdot q, \quad p\wedge q=p+q-pq, \quad p\Rightarrow q=(1-p)q.$$

1.76. Inductive proof works, however, the following simple observation is helpful: using the interpretation of 1.75, we get as the value of our expression $(1-w\Phi_1)\dots(1-w\Phi_n)\cdot(w\Psi)$. Since Ψ always takes value 0, our expression always takes value 0, and so is a tautology.

1.77. Use the interpretation of 1.75.

1.78. For even n. Prove it by induction or use the interpretation of 1.75.

1.79. Use the interpretation of 1.75.

1.80. Use the tautologies

$$(p \Leftrightarrow q) \Leftrightarrow [(p \Rightarrow q) \wedge (q \Rightarrow p)],$$
$$(p \Rightarrow q) \Leftrightarrow (\sim p \vee q),$$
$$p \wedge (q \vee r) \Leftrightarrow [(p \wedge q) \vee (p \wedge r)],$$
$$\sim (p \wedge q) \Leftrightarrow (\sim p \vee \sim q),$$
$$\sim (p \vee q) \Leftrightarrow (\sim p \wedge \sim q).$$

1.82. Use the tautologies

$$(p \Leftrightarrow q) \Leftrightarrow (q \Leftrightarrow p),$$
$$[(p \Leftrightarrow q) \Leftrightarrow r] \Leftrightarrow [p \Leftrightarrow (q \Leftrightarrow r)]$$

and the following: If Φ is a tautology then $\Phi \Leftrightarrow \Psi$ is a tautology if and only if Ψ is a tautology.

1.83. (a) $(p \wedge q) \vee (p \wedge \sim p \wedge r)$ (this can be simplified to $p \wedge q$),
(b) $(q \wedge \sim p) \vee p$,
(c) $\sim p \vee (q \wedge \sim p)$, this can be simplified to $\sim p$
(d) $p \vee \sim q \vee \sim p \vee (q \wedge \sim p) \vee (\sim q \wedge p)$ (this is a tautology)

1.84. (a) $p \wedge q$,
(b) $(q \vee p) \wedge (\sim p \vee p)$ (this can be simplified to $q \vee p$),
(c) $(\sim p \vee q) \wedge \sim p$ (this can be simplified to $\sim p$),
(d) there are a lot of equivalent formulas, e.g. $p \vee \sim p$.

1.85. Compare with 1.82.

1.86. Use induction on the degree of complexity of Φ using de Morgan's laws, i.e. 1.17 and 1.18.

1.87. Apply induction.

1.88. $[x-3>0] \vee [(x+3)<0] \vee [(x+2) \geqslant 0 \wedge (x-2) \leqslant 0]$.

Using the property $a/b \geqslant 0 \Leftrightarrow a \cdot b \geqslant 0 \wedge b \neq 0$, we notice that our inequality is equivalent to

$$(x^2-4)(x^2-9) \geqslant 0 \wedge x^2-9 \neq 0,$$

thus

$$(x-2)(x+2)(x-3)(x+3) \geqslant 0 \wedge (x-3) \neq 0 \wedge (x+3) \neq 0.$$

Using the property:

$$ab \geqslant 0 \Leftrightarrow [(a \geqslant 0 \wedge b \geqslant 0) \vee (a \leqslant 0) \wedge (b \leqslant 0)]$$

we get the following propositional function:

$$\{[(x-2) \geqslant 0 \wedge (x+2)(x-3)(x+3) \geqslant 0] \vee [(x-2) \leqslant 0 \wedge$$
$$\wedge (x+2)(x-3)(x+3) \leqslant 0]\} \wedge (x-3) \neq 0 \wedge (x+3) \neq 0.$$

By consecutive transformations we get

$$\{[(x-2)\geqslant 0\wedge(x+2)\geqslant 0\wedge(x-3)(x+3)\geqslant 0]\vee[(x-2)\geqslant 0\wedge$$
$$\wedge(x+2)\leqslant 0\wedge(x-3)(x+3)\leqslant 0]\vee[(x-2)\leqslant 0\wedge(x+2)\leqslant 0\wedge$$
$$\wedge(x-3)(x+3)\geqslant 0]\vee[(x-2)\leqslant 0\wedge(x+2)\geqslant 0\wedge$$
$$\wedge(x-3)(x+3)\leqslant 0]\}\wedge(x-3)\neq 0\wedge(x+3)\neq 0$$

and finally:

$$\{[(x-2)\geqslant 0\wedge(x+2)\geqslant 0\wedge(x-3)\geqslant 0\wedge(x+3)\geqslant 0]\vee$$
$$\vee[(x-2)\geqslant 0\wedge(x+2)\geqslant 0\wedge(x-3)\leqslant 0\wedge(x+3)\leqslant 0]\vee$$
$$\vee[(x-2)\geqslant 0\wedge(x+2)\leqslant 0\wedge(x-3)\geqslant 0\wedge(x+3)\leqslant 0]\vee$$
$$\vee[(x-2)\geqslant 0\wedge(x+2)\leqslant 0\wedge(x-3)\leqslant 0\wedge(x+3)\geqslant 0]\vee$$
$$\vee[(x-2)\leqslant 0\wedge(x+2)\geqslant 0\wedge(x-3)\geqslant 0\wedge(x+3)\leqslant 0]\vee$$
$$\vee[(x-2)\leqslant 0\wedge(x+2)\geqslant 0\wedge(x-3)\leqslant 0\wedge(x+3)\geqslant 0]\vee$$
$$\vee[(x-2)\leqslant 0\wedge(x+2)\leqslant 0\wedge(x-3)\geqslant 0\wedge(x+3)\geqslant 0]\vee$$
$$\vee[(x-2)\leqslant 0\wedge(x+2)\leqslant 0\wedge(x-3)\leqslant 0\wedge(x+3)\leqslant 0]\}\wedge$$
$$\wedge(x-3)\neq 0\wedge(x+3)\neq 0.$$

The second, third, fourth, fifth and seventh terms of the alternative are inconsistent. For instance we show that the third is inconsistent: indeed, the condition $x-2\geqslant 0$ implies $(x+3)\geqslant 0$. Using 1.72, (a) and (b), we find that our propositional function is equivalent to the following:

$$\{[(x-2)\geqslant 0\wedge(x+2)\geqslant 0\wedge(x-3)\geqslant 0\wedge(x+3)\geqslant 0]\vee$$
$$\vee[(x-2)\leqslant 0\wedge(x+2)\geqslant 0\wedge(x-3)\leqslant 0\wedge(x+3)\geqslant 0]\vee$$
$$\vee[(x-2)\leqslant 0\wedge(x+2)\leqslant 0\wedge(x-3)\leqslant 0\wedge(x+3)\leqslant 0]\}\wedge$$
$$\wedge(x-3)\neq 0\wedge(x+3)\neq 0.$$

Now, using the tautology of 1.52 several times, we get:

$$\{(x-3)\geqslant 0\vee(x+3)\leqslant 0\vee[(x+2)\geqslant 0\wedge(x-2)\leqslant 0]\}\wedge$$
$$\wedge(x+3)\neq 0\wedge(x-3)\neq 0.$$

Using the distributivity laws (1.48, 1.49) and 1.72 again, we get as the solution $3<x \vee -2 \leqslant x \leqslant 2 \vee x<-3$.

1.89. $(x-1) \geqslant 0$.

1.90. The inequality is valid for all x.

1.91. $(x+2)<0 \vee [(x-1) \geqslant 0 \wedge (x-2)<0] \vee (x-5) \geqslant 0$.

1.92. $[(x+1)>0 \wedge (x-1)<0] \vee (x-\sqrt[3]{2}) \geqslant 0$.

1.93. $(x+\sqrt{7}) \leqslant 0 \vee [(x+2)>0 \wedge (x-\sqrt{7}) \leqslant 0]$.

1.94. $[(x+3)>0 \wedge (x+1)<0] \vee [(x-3)>0 \wedge (x-4)<0]$.

1.95. $(x+4) \geqslant 0 \wedge (x-4) \leqslant 0$.

1.96. The inequality is valid for all x.

1.97. $[(x-1)>0 \wedge (x-2)<0] \vee (x-3)>0 \wedge (x-1)<0$.

1.98. Since $p_1 \vee p_2 \vee \ldots \vee p_n$ is true, at least one p_i must be true. Let i_0 be the least i for which p_i is true. Since $p_{i_0} \Rightarrow q_{i_0}$ is true, q_{i_0} is true (cf. 1.60). We conclude that, for $i \neq i_0$, q_i is false. But, for all i, $p_i \Rightarrow q_i$ is true, and so, for $i \neq i_0$ p_i must be false. Thus $q_{i_0} \Rightarrow p_{i_0}$ is true since both p_{i_0} and q_{i_0} are true and, for $i \neq i_0$, $q_i \Rightarrow p_i$ is true since both terms are false.

1.99. Use the following form of 1.98: Under the same assumptions $p_i \Leftrightarrow q_i (i=1, \ldots, n)$.

1.101. (a) p,

(b) $p \vee q$,

(c) $p \Leftrightarrow q$,

(d) $r \wedge s$.

1.102. (a) CpDqINpr. Indeed $p \wedge (q \vee (\sim p \Rightarrow r) = p \wedge (q \vee (\mathrm{N}p \Rightarrow r)$ $= p \wedge (q \vee \mathrm{IN}pr) = p \wedge \mathrm{D}q\mathrm{IN}pr = \mathrm{C}p\mathrm{D}q\mathrm{IN}pr$,

(b) IpIqIrp,

(c) EpCpNENqq,

(d) IIpqIIqrINrNp,

(e) ICDpqNpq,

(f) CDpqINpq.

1.103. (a) $[(p \wedge q) \vee (p \wedge \sim q)] \vee [(\sim p \wedge q) \vee (\sim p \wedge \sim q)]$. Indeed,

$$\begin{aligned}
&\mathrm{DDC}pq\mathrm{C}p\mathrm{N}q\mathrm{DCN}pq\mathrm{CN}p\mathrm{N}q \\
&\quad = \mathrm{DDC}pq\mathrm{C}p(\sim q)\,\mathrm{DC}(\sim p)q\mathrm{C}(\sim p)(\sim q) \\
&\quad = \mathrm{DD}(p \wedge q)(p \wedge \sim q)\,\mathrm{D}(\sim p \wedge q)(\sim p \wedge \sim q) \\
&\quad = \mathrm{D}[(p \wedge q) \vee (p \wedge \sim q)][(\sim p \wedge q) \vee (\sim p \wedge \sim q)] \\
&\quad = [(p \wedge q) \vee (p \wedge \sim q)] \vee [(\sim p \wedge q) \vee (\sim p \wedge \sim q)].
\end{aligned}$$

(b) $p \wedge (q \vee r) \Leftrightarrow [(p \wedge q) \vee (p \wedge r)]$,
(c) $(p \Rightarrow q) \Rightarrow (\sim q \Rightarrow \sim p)$,
(d) $(p \wedge q) \vee \sim\sim (q \vee \sim r)$,
(e) $\sim (p \wedge q) \Rightarrow (\sim p \wedge q)$.

1.104. Check all cases. N is the first connective only if the formula under consideration has the bracket form $\sim \Phi$ for a certain Φ.

1.105. A tautology takes value 1 under all valuations of its propositional variables. Thus, using the substitution rule, we get by 1.13 (p is substituted for r):

$$(*) \qquad [p \Rightarrow (q \Rightarrow p)] \Rightarrow [(p \Rightarrow q) \Rightarrow (p \Rightarrow q)].$$

Now we use the detachment rule (modus ponens). We know that $p \Rightarrow (q \Rightarrow p)$ holds (since it is one of the initial tautologies); by substitution we get (*) and thus by modus ponens we have:

$$(**) \qquad (p \Rightarrow q) \Rightarrow (p \Rightarrow p).$$

We use substitution again substituting in (**) $q \Rightarrow p$ for q, and get: $[p \Rightarrow (q \Rightarrow p)] \Rightarrow (p \Rightarrow p)$. Using detachment again we get $p \Rightarrow p$.

1.106. *Hint.* In 1.11 substitute 1.13 for p and $q \Rightarrow r$ for q. Then, after using detachment, make the substitutions: $q/(p \Rightarrow (q \Rightarrow r))$, $p/(q \Rightarrow r)$, $r/(p \Rightarrow q) \Rightarrow (p \Rightarrow r)$.

CHAPTER 2

ALGEBRA OF SETS

2.1. a, b, c.
2.2. a, b.
2.3. a.
2.4. This set has no elements.
2.5. $\{a\}$.
2.6. O.
2.7. $\{a, b\}, \{a\}$.
2.8. $\{\{a\}\}, \{a\}, a$.
2.9. $\{a, b, c\}, c$.
2.10. $\{a, b\}, \{\{a, b\}\}, O$.
2.11. 0, 1, 2.
2.12. 0, 1, 2.
2.13. 0, 1, 2.
2.14. 2.
2.15. 2, 3.
2.16. This set has no elements.
2.17. This set is empty.
2.18. 0, 1, 2, 3, ...
2.19. 2.
2.20. $-2, 2$.
2.21. This set is empty.
2.22. 1, 2, ..., 7. *Hint*: Solve the inequality $x^2-8x+1<0$ and then pick the natural numbers from the set of solutions.
2.23. 1, 2, 3, 4, 5.
2.24. -2.
2.25. $-\sqrt{2}, \sqrt{2}$.
2.26. -1.
2.27. All reals in $(-\sqrt{3}, \sqrt{3})$.
2.28. All real numbers.
2.29. We prove that $\sim(A\subset B)$. Indeed, if $A\subset B$ then, by definition, every element of A must also be an element of B. So consider the element b of A. Since $A\subset B$, we have $b\in B$ and so $(b=a)\vee(b=c)\vee(b=d)$. But

none of the components of this alternative is true since by assumption different letters denote different objects. This is a contradiction and so $A \subset B$ is false.

Note that $B \subset A$ holds.

2.30. $\sim(A \subset B)$, $\sim(B \subset A)$.

2.31. We show that $A \subset B$. We need to show the implication

$$x \in A \Rightarrow x \in B.$$

It is enough – in view of the tautology $(\alpha \Rightarrow \beta) \Leftrightarrow (\sim\beta \Rightarrow \sim\alpha)$ and the inference rules – to show that

$$x \notin B \Rightarrow x \notin A.$$

The last implication is true since the successor is true $\langle A = O \rangle$. On the other hand, $\sim(B \subset A)$ since $b \in B$ but $b \notin A = O$.

2.32. $A \subset B$, $\sim(B \subset A)$. *Remark*: From the proof of the inclusion $A \subset B$ as given in 2.31 and 2.32 it follows that, for every set B, $O \subset B$. Moreover, if $B \neq O$ then $\sim(B \subset O)$.

2.33. We show that $\sim(A \subset B)$. Indeed, if $A \subset B$ then every element of A is an element of B. Consider $\{a\}$, which is an element of A. Then $\{a\} \in B$ and, further, $\{a\} = a$. However, this implies that $a \in a$, which is false. On the other hand, it is easy to show that $B \subset A$.

2.34. $\sim(A \subset B)$, $B \subset A$.

2.35. $\sim(A \subset B)$, $\sim(B \subset A)$.

2.36. $A \subset B$, $B \subset A$. Remark: This implies $A = B$.

2.37. $\sim(A \subset B)$, $B \subset A$.

2.38. $A \subset B$, $B \subset A$.

2.39. $\sim(A \subset B)$, $B \subset A$.

2.40. $B \subset A$, $\sim(A \subset B)$.

Hint: To show $\sim(A \subset B)$ use the criteria of the equality of polynomials.

2.41. $A \subset B$, $B \subset A$.

2.42. $\sim(A \subset B)$, $B \subset A$.

2.43. $\sim(A \subset B)$, $B \subset A$.

Hint: Solve the equation $2x^3 - 5x^2 + 4x - 1 = 0$.

2.44. $A \subset B$. *Hint*: Use the theorem stating that the rational roots of a polynomial $x^n + a_{n-1}x^{n-1} + \ldots + a_0$ with integer coefficients are integers and they divide a_0.

2.45. $B \subset A$.

2.46. $\sim(A \subset B)$, $\sim(B \subset A)$.

Hint: Use the fact that A, being the set of solutions of an equation of an odd degree, is nonempty. Then use the result quoted in the hint to 2.44.

2.47. $\sim(A \subset B)$, $B \subset A$.

2.48. $A \subset B$, $\sim(B \subset A)$.

2.49. $A \subset B$, $\sim(B \subset A)$.

2.50. $\sim(A \subset B)$, $B \subset A$.

2.51. $d=b$ or $d=c$.

2.52. The equality holds for all a and b.

2.53. We show that our equality holds only when $a=c$ and $b=d$. Indeed, consider the set $\{\{a\}, \{a, b\}\}$ and its element $\{a\}$. From the assumed equality it follows that $\{a\} \in \{\{c\}, \{c, d\}\}$. Thus either $\{a\} = \{c\}$ or $\{a\} = \{c, d\}$.

If $\{a\} = \{c\}$ then $a=c$. On the other hand, if $\{a\} = \{c, d\}$ then $c=d$ since otherwise the set on the left hand side would consist of one element whereas that on the right hand side would consist of two. However, then $a=c=d$ and so in any case $a=c$.

Our equality implies also that $\{a, b\} \in \{\{c\}, \{c, d\}\}$; thus $\{a, b\} = \{c\}$ or $\{a, b\} = \{c, d\}$. Since we know that $a=c$, we have $\{c, b\} = \{c\}$ or $\{c, b\} = \{c, d\}$.

If $b \neq c$ then $\{c, b\} \neq \{c\}$ and so $\{c, b\} = \{c, d\}$; thus $b=c$ or $b=d$ and by our assumption $b \neq c$, simply $b=d$.

Thus only the case $b=c$ is left. Consider the set $\{c, d\}$. It is an element of the set $\{\{a\}, \{a, b\}\}$ and so $\{c,d\} = \{a\}$ or $\{c, d\} = \{a, b\}$. Since we know that $c=a$, we have $a=c=b$ and the equality $\{c, d\} = \{a\}$ implies $\{b, d\} = \{b\}$, and so $b=d$. Similarly, if $\{c, d\} = \{a, b\}$ then, by $a=c=b$, we get $\{b, d\} = \{b, b\}$, which implies $d=b$. Thus finally:

$$a=c \quad \text{and} \quad b=d.$$

2.54. $b=a$.

2.55. $a=b=d$.

2.56. The sets under consideration cannot be equal.

2.57. $A \cup B = \{a, b, c, d\}$, $A \cap B = \{c\}$, $A - B = \{a, b\}$, $B - A = \{d\}$.

2.58. $A \cup B = \{\{a, b\}, c, d\}$, $A \cap B = \{c\}$, $A - B = \{\{a, b\}\}$, $B - A = \{d\}$.

2.59. $A \cup B = \{x, y, \{z\}, a\}$, $A \cap B = \{x, y\}$, $A - B = \{\{z\}\}$, $B - A = \{a\}$.

2.60. $A \cup B = \{\{a, \{a\}\}, a, \{a\}\}, A \cap B = \{a\}, A - B = \{\{a, \{a\}\}\},$
$B - A = \{\{a\}\}.$

2.61. $A \cup B = \{a, \{a\}, \{b\}\}, A \cap B = B, A - B = \{a\}, B - A = O.$

2.62. It is easy to see that $A \cup B = \{\{a, \{b\}\}, c, \{c\}, \{a, b\}, \{b\}\}$. We show that

$$A \cap B = \{\{a, b\}, c\}.$$

Consider the consecutive elements of A: If $\{a, \{b\}\}$ belongs to $A \cap B$ then it must be one of the elements $\{a, b\}$, c, $\{b\}$. Our assumptions imply that $\{a, \{b\}\} \neq c$ (since c is not a set). Further, $\{a, \{b\}\} \neq \{b\}$ since otherwise $b = a$ or $b = \{b\}$. The first possibility is excluded by the assumption and the other implies that $b \in b$, which is absurd. Similarly, $\{a, \{b\}\} \neq \{a, b\}$. Now we show that $\{c\} \notin A \cap B$. Indeed, if $\{c\} = \{a, b\}$ then $c = a$ or $c = b$, which is absurd. Similarly, $\{c\} \neq \{b\}$ (since $c \neq b$) and $\{c\} \neq c$ because $\{c\} = c$ implies $c \in c$.

On the other hand,

$$A - B = \{\{a, \{b\}\}, c\},$$

since none of the listed elements belongs to B. Finally

$$B - A = \{\{b\}\},$$

which follows from the fact that $\{b\} \neq \{a, \{b\}\}$, $(a \neq b)$ $\{b\} \neq c$ (c is not a set), $\{b\} \neq \{c\}$ $(b \neq c)$, and finally $\{b\} \neq \{a, b\}$ $(b \neq a)$.

2.63. $A \cup B = \mathcal{N}$, $A \cap B = O$, $A - B = A$, $B - A = B$.

2.64. $A \cup B = \{2\}$, $A \cap B = O$, $A - B = O$, $B - A = B$.

2.65. $A \cup B = A$, $A \cap B = B$, $A - B = O$, $A - B$ is the set of all reals smaller than the number 1 (with the exception of 0), $B - A = O$.

2.66. $A \cup B = B$, $A \cap B = A$, $A - B = O$, $B - A = \langle 1, 2)$.

2.67. (a) $(A \cap B) \cap C = O$,

(b) $(A \cap -B) \cap C$ is the set of triangles with the angles equal to $\frac{1}{4}\pi, \frac{1}{4}\pi, \frac{1}{2}\pi$.

(c) $(-A) \cap (B \cap C) = O$,

(d) $(-A) \cap (C \cap -B)$ is the set of triangles with one of the angles equal to $\frac{1}{2}\pi$ and the other angles different from $\frac{1}{4}\pi$.

(e) $(A \cap B) \cap -C$ is the set of equilateral triangles.

2.81. From the definition of equality of sets it follows that it is enough to prove the equivalence

$$x \in (A \cup B) - C \Leftrightarrow x \in (A - C) \cup (B - C).$$

Using the definitions of set operations, we get:

$$x \in (A \cup B) - C \Leftrightarrow x \in A \cup B \wedge x \notin C \Leftrightarrow (x \in A \vee x \in B) \wedge \sim x \in C,$$

$$x \in (A - C) \cup (B - C) \Leftrightarrow x \in A - C \vee x \in B - C$$

$$\Leftrightarrow (x \in A \wedge x \notin C) \vee (x \in B \wedge x \notin C)$$

$$\Leftrightarrow (x \in A \wedge \sim x \in C) \vee (x \in B \wedge \sim x \in C).$$

Consider the following formula of the propositional calculus:

$$(\alpha \vee \beta) \wedge \gamma \Leftrightarrow (\alpha \wedge \gamma) \vee (\beta \wedge \gamma).$$

It is easy to see that it is a tautology. But if, for α, β, γ we substitute $x \in A$, $x \in B$, $x \notin C$, respectively, then we get the equivalence

$$(x \in A \vee x \in B) \wedge \sim x \in C \Leftrightarrow (x \in A \wedge \sim x \in C) \vee$$

$$\vee (x \in B \wedge \sim x \in C),$$

which finishes the proof.

2.90. Yes.

2.91. Yes.

2.92. No. Consider $A = \{1, 2\}$, $B = \{1\}$. Then

$$A \cap (A \cup B) = \{1, 2\} \neq \{1\} = B.$$

2.93. No. Consider $A = C = \{1, 2\}$, $B = O$. Then

$$(A \cup B \cup C) - (A \cup B) = O \neq C.$$

2.94. Let $A \subset B$ and $C \subset D$ (otherwise the antecedent of the implication is false and the whole implication is true). Then, for any x,

$$x \in A \Rightarrow x \in B \quad \text{and} \quad x \in C \Rightarrow x \in D.$$

Now it is enough to show that for an arbitrary x

$$x \in A \cap C \Rightarrow x \in B \cap D.$$

Use the following tautology of the propositional calculus:

$$[(\alpha \Rightarrow \beta) \wedge (\gamma \Rightarrow \delta)] \Rightarrow [\alpha \wedge \gamma \Rightarrow \beta \wedge \delta].$$

Substituting for α, β, γ, δ, respectively, $x \in A$, $x \in B$, $x \in C$ and $x \in D$ we get

$$[(x \in A \Rightarrow x \in B) \wedge (x \in C \Rightarrow x \in D)] \Rightarrow [(x \in A \wedge x \in C)$$

$$\Rightarrow (x \in B \wedge x \in D)].$$

By our assumptions the antecedent of the implication is true, and so the consequent must be also true (the modus ponens rule). Thus $x \in A \wedge \wedge x \in C \Rightarrow x \in B \wedge x \in D$. Using the definition of the set intersection, we can write this implication in the desired form:

$$x \in A \cap C \Rightarrow x \in B \cap D.$$

2.101. Our equality holds if and only if

$$x \in (A \cap B) \cup (C \cap B) \Leftrightarrow x \in B.$$

But $x \in (A \cap B) \cup (C \cap B)$, implies $(x \in A \wedge x \in B) \vee (x \in C \wedge x \in B)$, and in any case $x \in B$.

So we just need to check when the converse implication holds. Let x be an arbitrary element of B. If x does not belong to either A or C, then $x \notin A \cap B$ and $x \notin C \cap B$, and so

$$\sim(x \in A \cap B \vee x \in B \cap C).$$

Thus, if $x \in B$, then x must belong to either A or C. Finally, we find that our equality holds if and only if $B \subset A \cup C$.

2.102. $C \cap A \subset B$.

2.103. $C \cap A \subset B$.

2.104. $B \cap C = O$.

2.105. $A \cup B \subset C$, $A \cap B = O$.

2.106. $A \cap B \subset C \subset A$.

2.110. A cannot consist of more than $2^n - 1$ sets. If none of the sets A_i is empty and they are pairwise disjoint, then A has exactly $2^n - 1$ elements.

Hint: Use the combinatorial interpretation of $\binom{n}{k}$ and

$$\binom{n}{0} + \binom{n}{1} + \ldots + \binom{n}{n} = 2^n.$$

2.111. A cannot have more than $2^{2^n} - 1$ elements. If none of the sets A_i' s is empty and they are pairwise disjoint, then A has exactly $2^{2^n} - 1$ elements.

Hint: Note that $X - Y = X \cap (V - Y)$, where $V = A_1 \cup \ldots \cup A_n$, and so our problem is reduced to that of 2.108.

2.119. Yes.

2.120. Yes.

2.121. No. Consider $A=C\neq O$, $B=O$, then

$$A\cup(B\dot{-}C)=A, \quad (A\cup B)\dot{-}(A\cup C)=O.$$

2.122. *Hint*: To each constituent $A_1^{i_1}\cap\ldots\cap A_n^{i_n}$ adjoin the zero-one sequence $i_1,\ldots,i_n$ and note that there are exactly 2^n such sequences.

2.124. The sum of the constituents is the whole X.

2.127. *Hint*: Use induction.

2.128. $A\times B=\{\langle 0,1\rangle,\langle 0,2\rangle,\langle 1,1\rangle,\langle 1,2\rangle\}$,
$B\times A=\{\langle 1,0\rangle,\langle 1,1\rangle,\langle 2,0\rangle,\langle 2,1\rangle\}$.

2.129. $A\times B=\{\langle 0,0\rangle,\langle 0,2\rangle,\langle 0,3\rangle,\langle 1,0\rangle,\langle 1,2\rangle,\langle 1,3\rangle,$
$\langle 2,0\rangle,\langle 2,2\rangle,\langle 2,3\rangle\}$.

2.130. $A\times B=\{\langle 1,1\rangle,\langle 1,2\rangle,\langle 1,3\rangle,\langle 1,4\rangle,\langle 1,5\rangle\}$.
$B\times A=\{\langle 1,1\rangle,\langle 2,1\rangle,\langle 3,1\rangle,\langle 4,1\rangle,\langle 5,1\rangle\}$.

2.131. $A\times B=O$, $B\times A=O$.

2.134. $A\times(B\times C)=\{\langle 0,\langle 1,2\rangle\rangle,\langle 0,\langle 1,3\rangle\rangle,\langle 1,\langle 1,2\rangle\rangle,$
$\langle 1,\langle 1,3\rangle\rangle\}$.
$(A\times B)\times C=A\times B\times C=\{\langle 0,1,2\rangle,\langle 0,1,3\rangle,\langle 1,1,2\rangle,$
$\langle 1,1,3\rangle\}$.

Note: $A\times(B\times C)\neq A\times B\times C$.

2.135. Fig. 1 and Fig. 2.

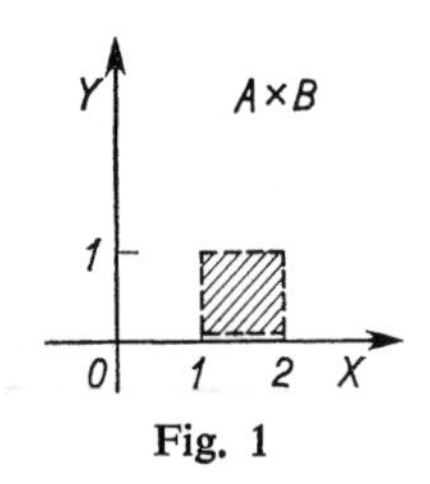

Fig. 1

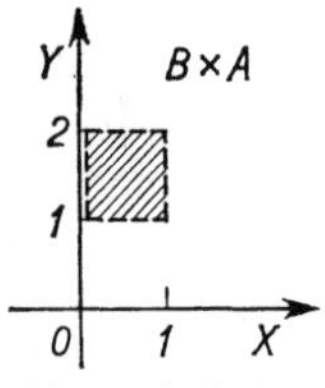

Fig. 2

2.136. Fig. 3.

2.137. Fig. 4 and Fig. 5.

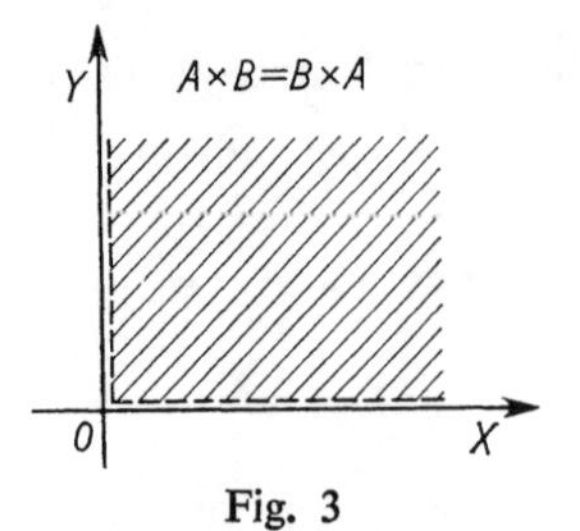

Fig. 3

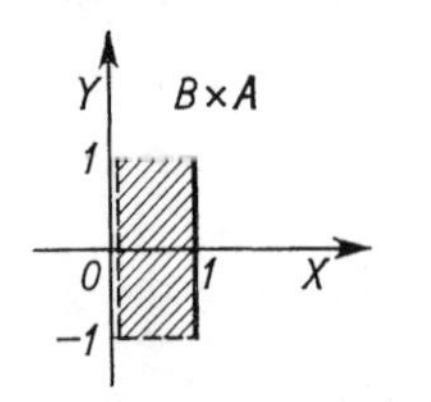

Fig. 4

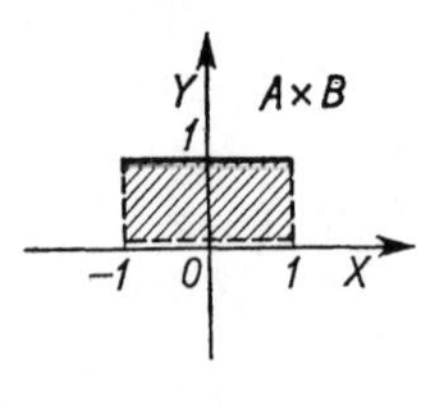

Fig. 5

2.138. Fig. 6 and Fig. 7.

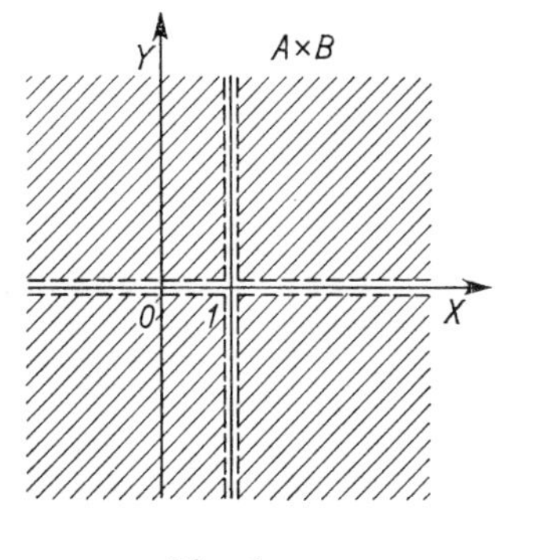

Fig. 6

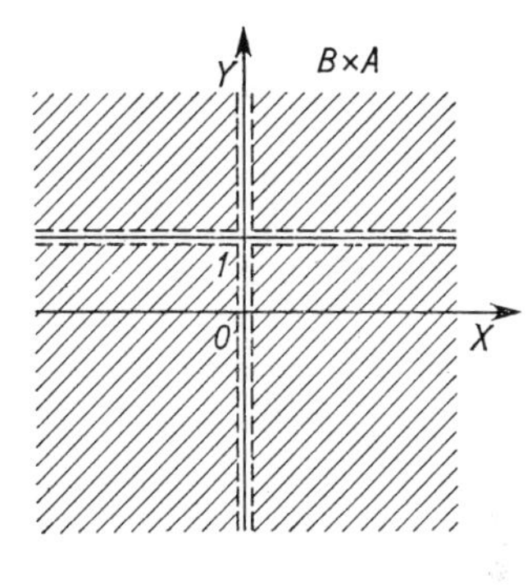

Fig. 7

2.139. Fig. 8 and Fig. 9.

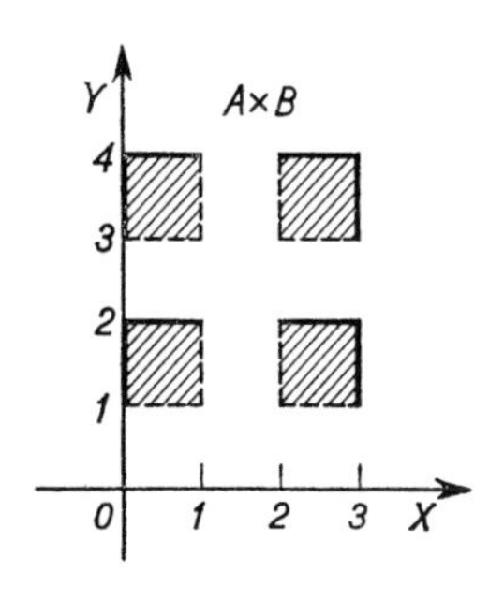

Fig. 8

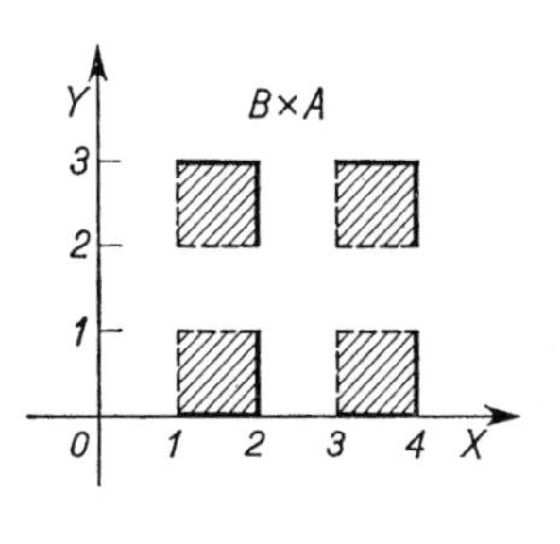

Fig. 9

2.140. Fig. 10.

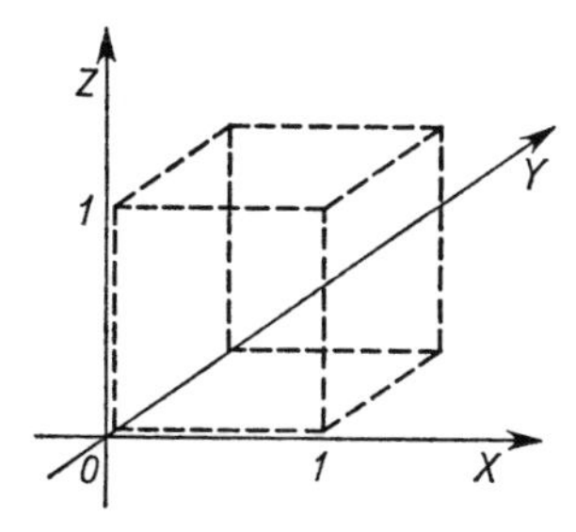

Fig. 10

2.141. Fig. 11.

2.142. Fig. 12.

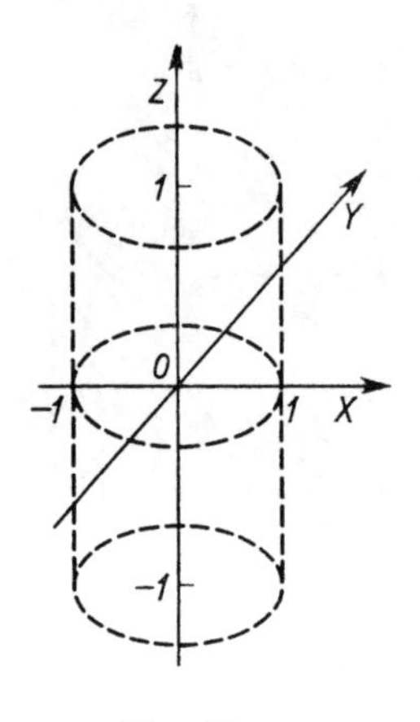

Fig. 11

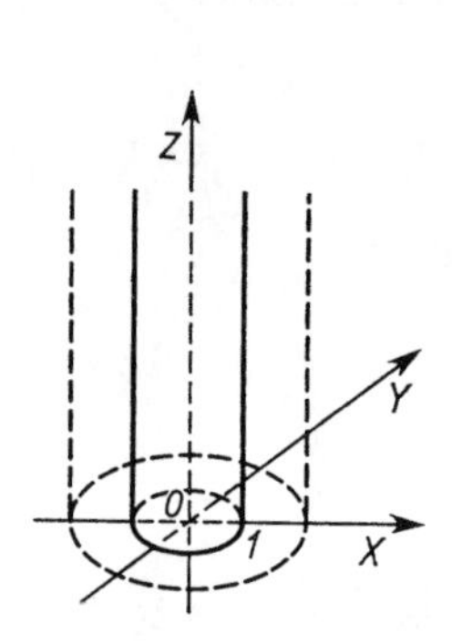

Fig. 12

2.143. Let $\langle x, y\rangle \in A\times(B\cup C)$. Then $x\in A\wedge y\in B\cup C$, and so $x\in A\wedge(y\in B\vee y\in C)$. This implies that $(x\in A\wedge y\in B)\vee(x\in A\wedge y\in C)$, and so

$$\langle x, y\rangle \in A\times B \quad \text{or} \quad \langle x, y\rangle \in A\times C,$$

which means $\langle x, y\rangle \in (A\times B)\cup(A\times C)$.

Conversely,

$$\begin{aligned}\langle x, y\rangle \in (A\times B)\cup(A\times C) &\Rightarrow \langle x, y\rangle \in A\times B\vee\langle x, y\rangle\in A\times C\\ &\Rightarrow (x\in A\wedge y\in B)\vee(x\in A\wedge y\in C)\\ &\Rightarrow (x\in A\wedge y\in B\cup C)\Rightarrow\\ &\Rightarrow \langle x, y\rangle\in A\times(B\cup C).\end{aligned}$$

So the equality is true.

2.144. Yes.

2.145. No.

2.146. No.

2.157. $\mathscr{P}(A)=\{O, \{a\}, \{b\}, \{c\}, \{a, b\}, \{a, c\}, \{b, c\}, \{a, b, c\}\}$.

2.158. $\mathscr{P}(A)=\{O\}$.

2.159. $\mathscr{P}(A)=\{O, \{O\}\}$.

2.160. $\mathscr{P}(A)=\{O, \{a\}, \{\{a\}\}, \{\{\{a\}\}\}, \{a, \{a\}\}, \{a, \{\{a\}\}\}, \{\{a\}, \{\{a\}\}\}, \{a, \{a\}, \{\{a\}\}\}\}$.

2.163. By the definition of $\mathscr{P}(A)$, $A \in \mathscr{P}(A)$. If $A=\mathscr{P}(A)$, then $A \in A$, which is absurd.

2.164. No. Consider $X=\{O, \{O\}, \{\{O\}\}, \ldots\}$.

CHAPTER 3

PROPOSITIONAL FUNCTIONS. QUANTIFIERS

3.1. $Z=(-\infty, -1\rangle \cup \langle 1, \infty)$.

3.2. $Z=\mathcal{N}$.

3.3. $Z=O$.

3.4. $Z=\{1\}$.

3.5. $Z=O$.

3.6. If the complex numbers are interpreted as points of the plane, then Z is the part of the plane outside the unit disc.

3.7. $Z=\{-\frac{1}{2}\}$.

3.8. $Z=\mathcal{N}$.

3.9. $Z=\{-1, 1\}$.

3.10. $Z=\langle -2, -1\rangle$.

3.11. $Z_1 \cap Z_2$.

3.12. $Z_1 \cup Z_2$.

3.13. $X-Z_1$.

3.14. $(X-Z_1)\cup Z_2$.

3.15. $(Z_1 \cap Z_2)\cup[(X-Z_1)\cap(X-Z_2)]$.

3.16. $(X-Z_1)\cap(X-Z_2)$.

3.17. $\Phi_1(x)\wedge\Phi_2(x)$.

3.18. $\Phi_1(x)\vee\Phi_2(x)$.

3.19. $\Phi_1(x)\wedge\sim\Phi_2(x)$.

3.21. $\Phi(x)$ is true in X.

3.22. $\sim\Phi(x)$ is true in X.

3.23. (a) The set X. (b) If $Z=Z_1=Z_2$ is the graph of $\Phi_1(x)$ (thus also of $\Phi_2(x)$), then Z is the graph of $\Phi_1(x)\wedge\Phi_2(x)$ as well.
(c) The graph of the propositional function $\Phi_1(x)\vee\sim\Phi_2(x)$ is X.

3.24. Let Z be the graph of Φ_2. (a) Z, (b) X, (c) O, (d) Z.

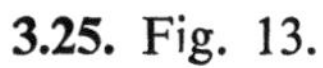

3.25. Fig. 13.
3.26. Fig. 14.

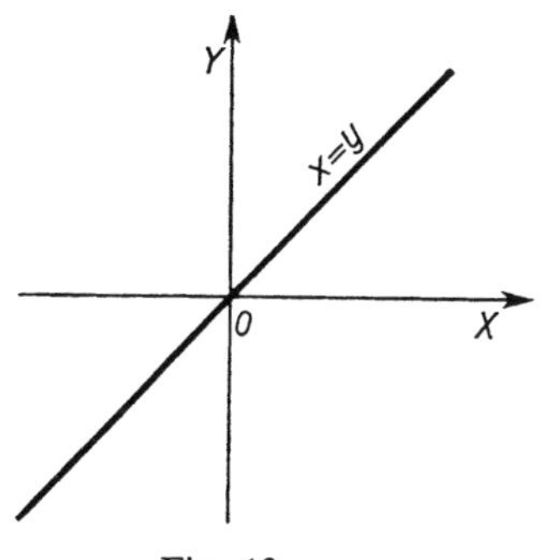

Fig. 13

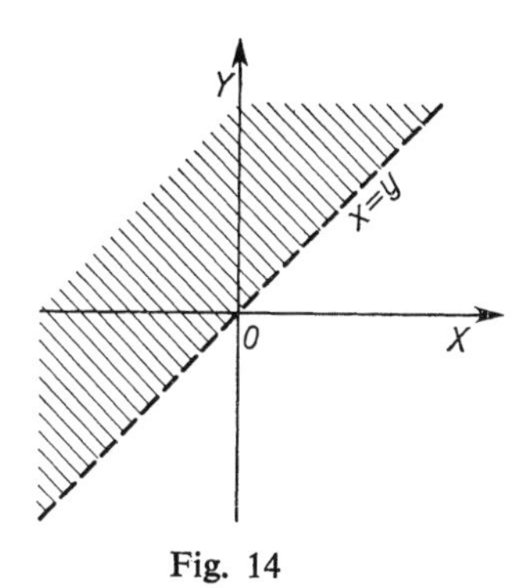

Fig. 14

3.27. Fig. 15.
3.28. Fig. 16.

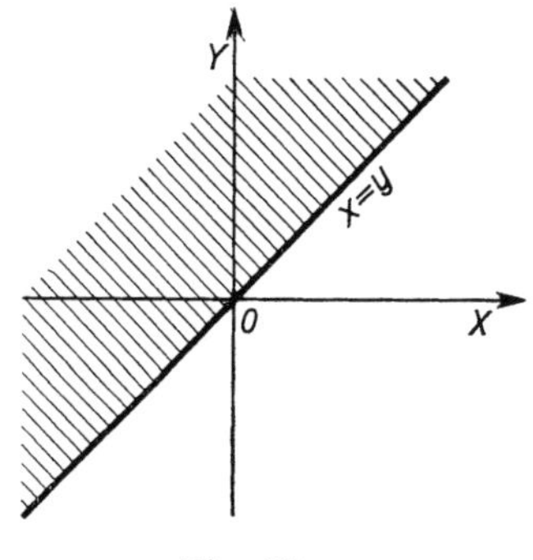

Fig. 15

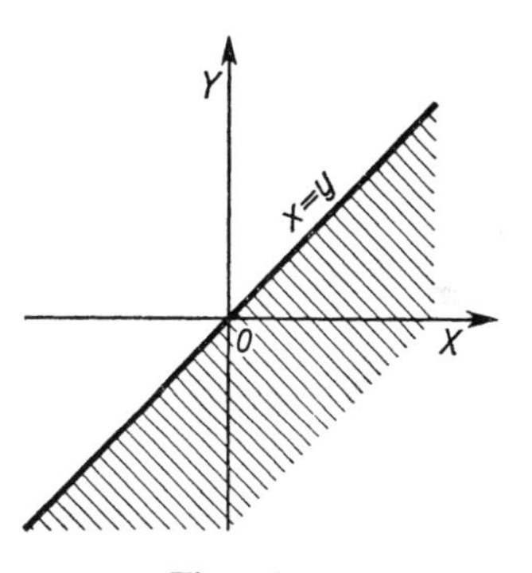

Fig. 16

3.29. Fig. 17.
3.30. Fig. 18.

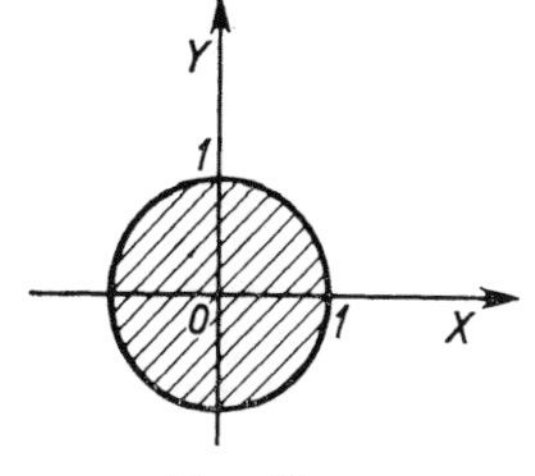

Fig. 17

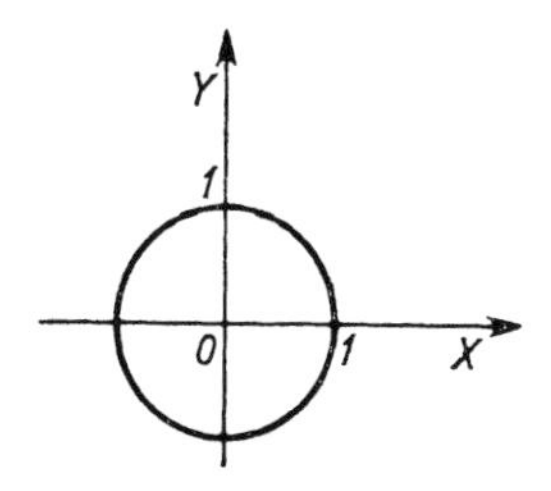

Fig. 18

3.31. Fig. 19. The angle α is determined by $\operatorname{tg}\alpha=\frac{a}{b}$. *Note*: If $b=0$ then $\alpha=\frac{1}{2}\pi$ and the line meets axis X at the point $x=-c$.

3.32. The plane without the diagonal.

3.33. The graph consists of the point (0, 0).

3.34. Fig. 20.

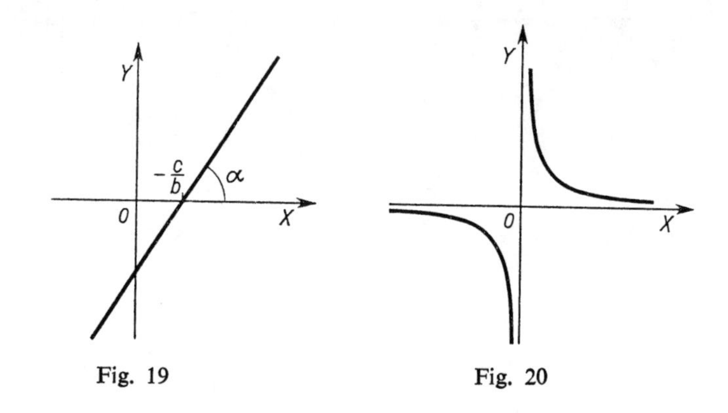

Fig. 19 Fig. 20

3.35. Fig. 21. *Note*: The graph consists of the axes OX and OY.

3.36. Fig. 22.

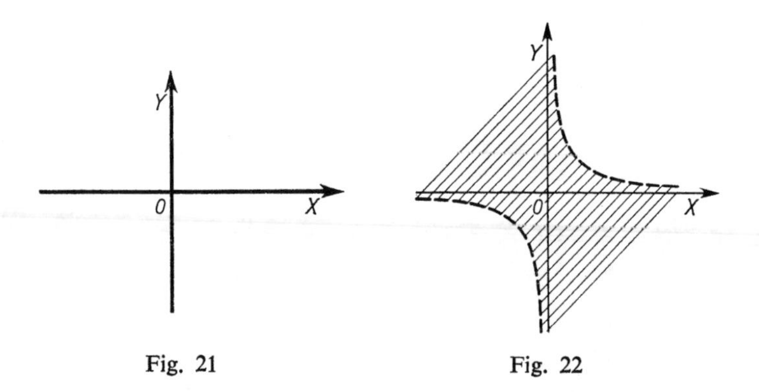

Fig. 21 Fig. 22

3.37. Fig. 23. *Note*: We give the solution for the case $a<0$. If $a=0$ the graph is a line, and in the case of $a>0$ the arms of the parabola are directed downwards.

3.38. Fig. 24.

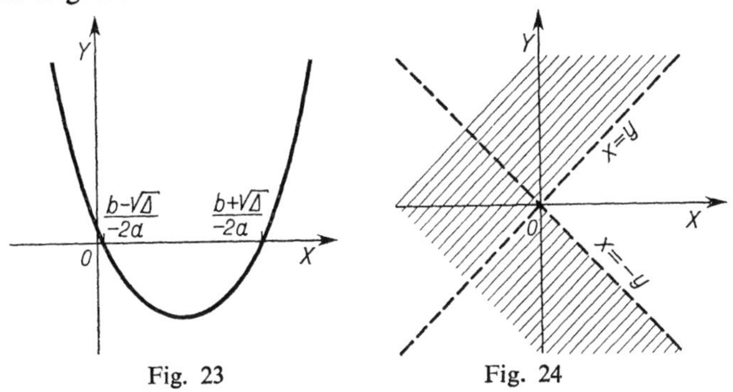

Fig. 23 Fig. 24

3.39. Fig. 25.

3.40. Fig. 26. *Note*: This is the case where $a \neq 0 \neq b$.

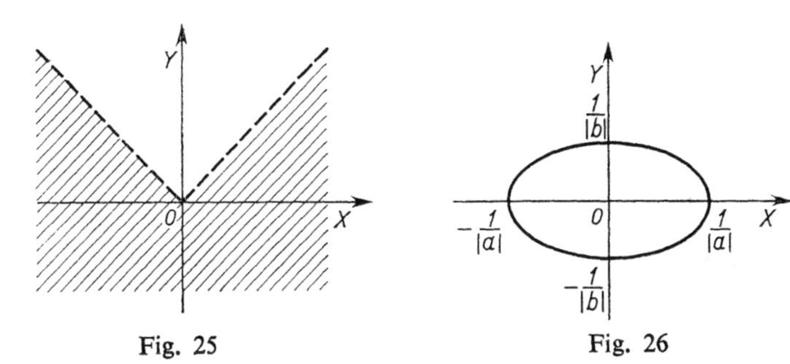

Fig. 25 Fig. 26

3.41. Whole plane.

3.42. The graph is empty.

3.43. Fig. 27.

3.44. Fig. 28.

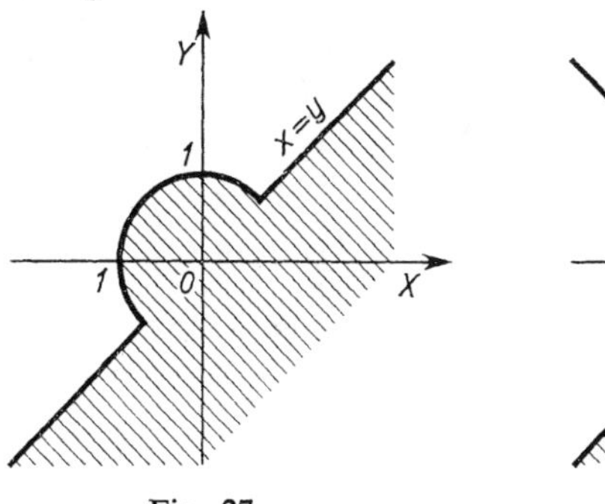

Fig. 27

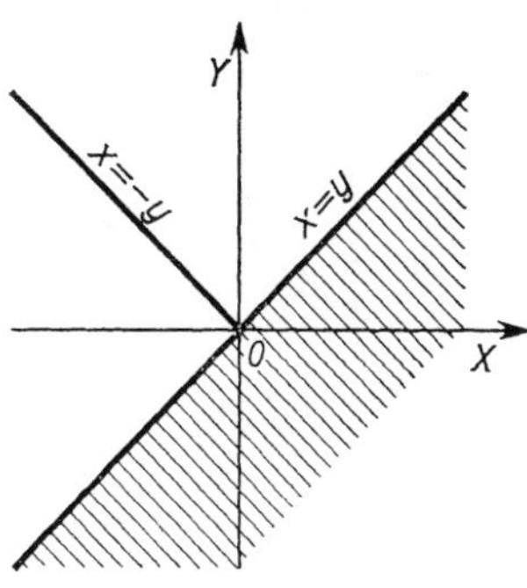

Fig. 28

3.45. Whole plane without the line $x=y$ but with the point $(\frac{1}{2}, \frac{1}{2})$ included.

3.46. The graph consists of the point (0 ,0).

3.47. Fig. 29.

3.48. Compare Fig. 29.

3.49. Whole plane.

3.50. Fig. 30.

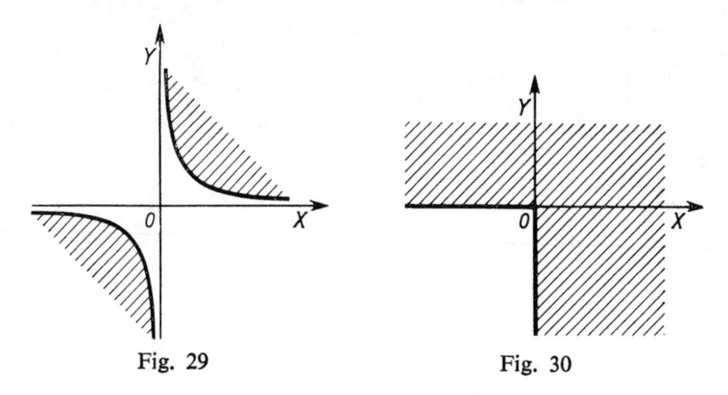

Fig. 29 Fig. 30

3.51–3.53. Whole plane.

3.54. Fig. 31.

3.55. Fig. 32.

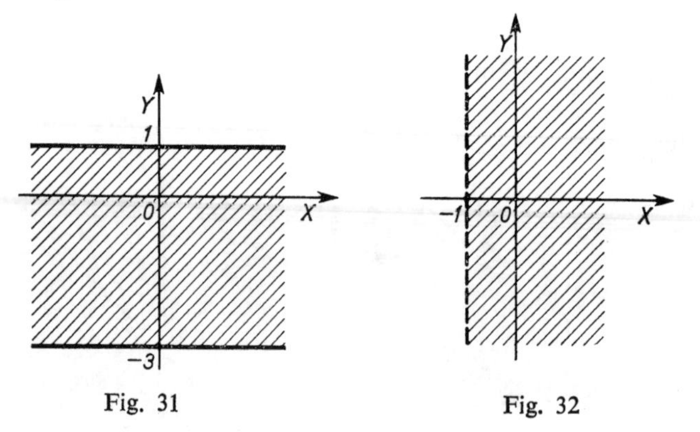

Fig. 31 Fig. 32

3.56. Fig. 33. Note: The graph is the whole unit ball (i.e. the interior and the boundary).

3.57. Fig. 34. Note: The graph is the plane which forms with the axis Oz an angle of $\frac{1}{4}\pi$ and contains the line $x=-y$.

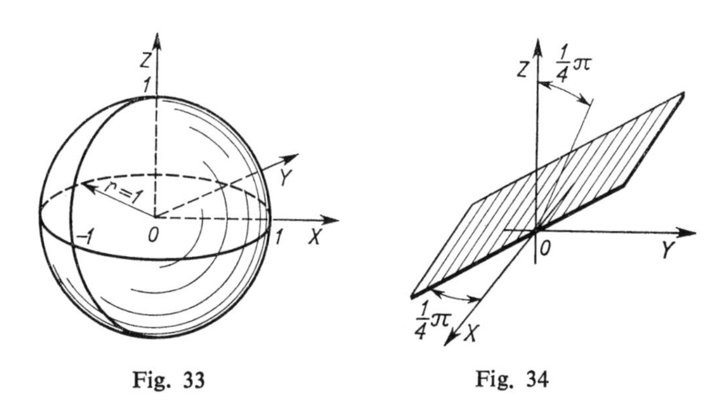

Fig. 33 Fig. 34

3.58. Fig. 35. *Note*: The graph is the plane perpendicular to the plane XY which meets the axis Y at the point $Y=1$ and the axis X at the point $X=1$.

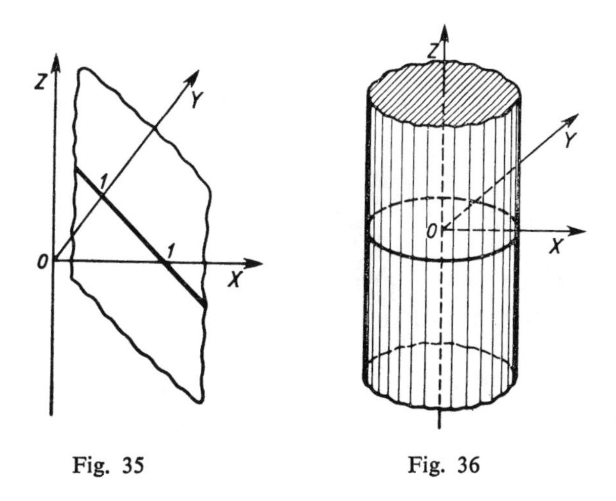

Fig. 35 Fig. 36

3.59. Fig. 36. *Note*: The graph is an infinite cylinder, the interior included.

3.60. Fig. 37. *Note*: The graph is the interior of the cylinder.

3.61. Fig. 38. *Note*: The graph is the interior of the cube.

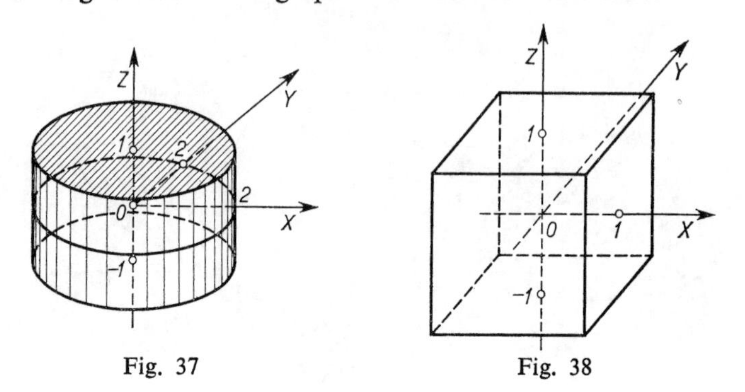

Fig. 37

Fig. 38

3.62. Fig. 39. The graph is the paraboloid (without the interior).

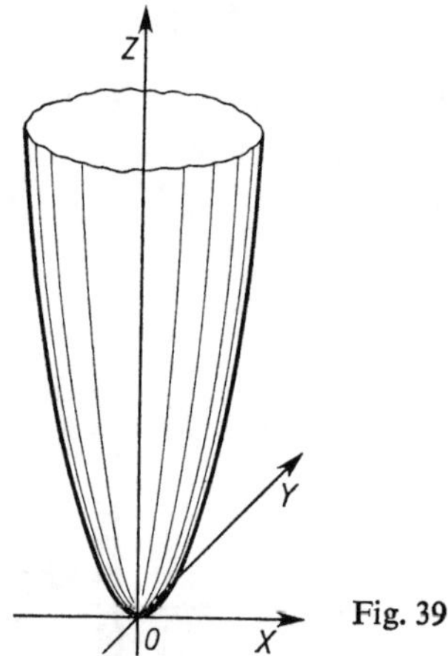

Fig. 39

3.65. Our assumption implies that, for $\langle x_1, \ldots, x_n\rangle \in X_1 \times \ldots \times X_n$, $\Phi(x_1, \ldots, x_n)$ holds. Since $X_i \neq 0$, we have $X_1 \times \ldots \times X_n \neq 0$. Pick $\langle a_1, \ldots$ $\ldots, a_n\rangle \in X_1 \times \ldots \times X_n$. Then $\Phi(a_1, \ldots, a_n)$ holds. Thus Φ is satisfiable.

The assumption $X_i \neq 0$ cannot be omitted. Indeed, if, for a certain i, $X_i = 0$, then $X_1 \times \ldots \times X_n = 0$ and Φ is not satisfiable since there is no element in $X_1 \times \ldots \times X_n$ which satisfies Φ (there being no elements in $X_1 \times \ldots \times X_n$).

3.67. If $\Phi(x)$ and $\Psi(y)$ are satisfiable, then there are elements a and b such that $\Phi(a)$ and $\Psi(b)$ hold. Thus there exists a pair $\langle a, b\rangle$ satisfying the function $\Theta(x, y) = \Phi(x) \wedge \Psi(y)$.

On the other hand, if the function $\Theta(x, y) = \Phi(x) \wedge \Psi(y)$ is satisfiable, then there exists a pair $\langle a, b \rangle$ usch that $\Theta(a, b)$ holds, i.e. $\Phi(a) \wedge \Psi(b)$ holds. Using the tautologies $\alpha \wedge \beta \Rightarrow \alpha$, $\alpha \wedge \beta \Rightarrow \beta$ and modus ponens, we get $\Phi(a)$ and $\Psi(b)$, i.e. Φ and Ψ are satisfiable.

The above property is not valid for the propositional function of the form $\Phi(x)$ and $\Psi(x)$. One implication holds: if $\Phi(x) \wedge \Psi(x)$ is satisfiable then there exists an a such that $\Phi(a) \wedge \Psi(a)$ and so $\Phi(a)$ and $\Psi(a)$ hold.

The converse implication may fail: Take $x > 0$ as Φ and $x < 0$ as Ψ. Both propositional functions are satisfiable (in reals) but their conjunction is false.

3.68. Assume that $\Phi(x) \Rightarrow \Psi(y)$ is not satisfiable. Thus, for an arbitrary pair $\langle a, b \rangle$, $\Phi(a) \Rightarrow \Psi(b)$ is false. Therefore, for arbitrary a and b, $\Phi(a)$ is true and $\Psi(b)$ is false.

On the other hand if $\Phi(x)$ is true whereas $\Psi(y)$ is false, then, for an arbitrary pair $\langle a, b \rangle$, $\Phi(a)$ is true whereas $\Psi(b)$ is false; thus $\Phi(a) \Rightarrow \Psi(b)$ is false and so $\Phi(x) \Rightarrow \Psi(y)$ is not satisfiable.

For functions of the form $\Phi(x) \Rightarrow \Psi(x)$ an analogous property holds.

3.69. $\sim \Phi(x)$ is satisfiable if and only if $\Phi(x)$ is not true.

In the answers to problems 3.73–3.85 the variables underlined are bounded and those with a line on top are free.

3.73. $\bigwedge_{x} \Phi(\underline{x}, \overline{y}, \overline{z})$.

3.74. $\bigwedge_{x} \bigwedge_{y} \Phi(\underline{x}, \underline{y}, \overline{z})$.

3.75 $\bigvee_{z} \Phi(\overline{x}, \overline{y}, \underline{z})$.

3.76. $\bigvee_{x} \Phi(\underline{x}, \overline{y}, \overline{z})$.

3.77. $[\bigvee_{x} \bigwedge_{y} \Phi(\underline{x}, \underline{y}, \overline{z})] \Rightarrow \Psi(\overline{x}, \overline{y}, \overline{z})$.

3.78. $\bigwedge_{x} \bigwedge_{y} \Phi(\underline{x}, \underline{y}, \overline{z}) \wedge \bigvee_{z} \Psi(\overline{x}, \overline{y}, \underline{z})$.

3.79. $\bigwedge_{x} \Phi(\underline{x}, \overline{y}, \overline{z}) \Rightarrow \{\bigvee_{z} [\bigvee_{y} \Psi(\overline{x}, \underline{y}, \underline{z}) \wedge \bigwedge_{z} \Theta(\overline{x}, \overline{y}, \underline{z})]\}$.

3.80. $\bigvee_{x} [\Phi(\underline{x}, \overline{y}, \overline{z}) \Rightarrow \Psi(\underline{x}, \underline{x}, \overline{y})]$

$$\Rightarrow \{\bigvee_{x} \bigvee_{z} [\Phi(\underline{x}, \underline{x}, \overline{y}) \wedge \Theta(\underline{x}, \overline{y}, \overline{y})]\}.$$

3.81. $\bigvee_{x} (\underline{x} < \overline{y} \vee \underline{x} < \overline{z})$.

3.82. $\bigvee_{x} \bigwedge_{y} [(\underline{x} < \underline{y}) \Rightarrow (\underline{x} < \overline{z} \wedge \overline{z} < \underline{y})]$.

3.83. $\bigwedge_{x} (\underline{x} | \overline{y} \wedge \underline{x} | \overline{z} \Rightarrow \underline{x} | \overline{z})$.

3.84. $(\bigwedge_{x} \bigvee_{y} \underline{x} < \underline{y}) \vee (\overline{x} < \overline{z})$.

3.85. $\bigvee_{x} (\underline{x} < \underline{x} \vee \underline{x} < \overline{z})$.

3.88. $(-1, +1)$.

3.89. $\mathscr{R} - \{0\}$.

3.90. O.

3.91. O.

3.92. $\mathscr{R}$.

3.93. O.

3.94. $\mathscr{R}$.

3.95. O.

3.112. Let the graph A of the function $\Phi(x, y)$ be a subset of the plane V. Then:

(a) The graph of $\bigwedge_{x} \Phi(x, y)$ is obtained as follows: Take $V - A$, project it to the axis $0y$ and take the complement.

(b) The graph of $\bigvee_{x} \Phi(x, y)$ is the projection of A to $0y$.

(c) The graph of $\bigwedge_{y} \Phi(x, y)$ is obtained by the procedure described in (a) with the axis $0x$ instead of $0y$.

(d) The graph of $\bigvee_{y} \Phi(x, y)$ is the projection of A to Ox.

3.114. $\bigvee_{y} x = 2y$.

3.115. $\bigvee_{y_1} \bigvee_{y_2} x = y_1 y_1 + y_2 y_2$.

3.116. $\bigwedge_{y} \bigwedge_{z} (x = yz \Rightarrow y = 1 \vee y = x)$.

3.117 $\bigvee_{y} \bigvee_{z} [(y \neq x) \wedge (y \neq 1) \wedge x = yz]$.

3.118. $\bigvee_{y_1}\bigvee_{z_1}(x=yy_1)\wedge(x=zz_1)\wedge$
$$\wedge\bigwedge_{u}[\bigvee_{y_1}\bigvee_{z_1}(u=yy_1\wedge u=zz_1)\Rightarrow\bigvee_{x_1}x_1x=u].$$

3.119. $\bigvee_{x_1}\bigvee_{x_2}(y=xx_1\wedge z=xx_2)\wedge$
$$\wedge\bigwedge_{u}[\bigvee_{u_1}\bigvee_{u_2}(y=uu_1\wedge z=uu_2)\Rightarrow\bigvee_{u_0}uu_0=x].$$

Note: Using in addition the propositional function $x<y$, we can get different forms of the functions from Problems 3.118 and 3.119. For instance, in the case of 3.119 we get the following:

$$\bigvee_{x_1}\bigvee_{x_2}xx_1=y\wedge xx_2=z\wedge\bigwedge_{u}[\bigvee_{u_1}\bigvee_{u_2}(uu_1=y\wedge uu_2=z)\Rightarrow u\leqslant x].$$

3.120. $\bigwedge_{y}\bigwedge_{z}[(x=4y+z\wedge z<4)\Rightarrow(z=1\vee z=2)]$.

3.121. $\bigwedge_{x}\bigwedge_{y}\bigwedge_{z}[(x=2y+z\wedge z<2)\Rightarrow(z=0\vee z=1)]$.

3.122. $\bigwedge_{n}\bigvee_{p}[\bigwedge_{y}\bigwedge_{z}(p=yz\Rightarrow y=1\vee y=p)\wedge(n\leqslant p\wedge p\leqslant 2n)]$.

Note: Compare with 3.116.

3.123. $\bigwedge_{u}(\bigvee_{u_1}uu_1=x\Rightarrow\bigvee_{u_2}uu_2=y)$.

3.124. $\bigwedge_{u}\{2u+1>3\Rightarrow[\bigvee_{p_1}\bigvee_{p_2}\bigwedge_{y}\bigwedge_{z}((p_1=yz\Rightarrow p_1=y\vee 1=y)\wedge$
$$\wedge(p_2=yz\Rightarrow p_2=y\vee 1=y))\Rightarrow 2u+1=p_1+p_2]\}.$$

3.125. $\bigwedge_{z}\bigwedge_{y}\bigwedge_{x}\{\bigvee_{x_1}\bigvee_{x_2}(xx_1=z\wedge xx_2=y)\wedge$
$$\wedge\bigwedge_{u}[\bigvee_{u_1}\bigvee_{u_2}(uu_1=z\wedge uu_2=y)\Rightarrow u\leqslant x]\}.$$

Note: Compare with 3.119.

3.126. $\bigwedge_{z}\bigwedge_{y}\bigwedge_{u}\bigvee_{x}\{\bigvee_{z_1}\bigvee_{y_1}\bigvee_{u_1}(z\cdot z_1=x\wedge y\cdot y_1=x\wedge u\cdot u_1=x)\wedge$
$$\wedge\bigwedge_{t}[\bigvee_{z_1}\bigvee_{y_1}\bigvee_{u_1}(z\cdot z_1=t\wedge y\cdot y_1=t\wedge u\cdot u_1=t)\Rightarrow x\leqslant t]\}.$$

3.127. $\bigwedge_{x}\bigvee_{y}x<y$.

3.128. $\bigwedge_{x}\bigvee_{t}\{[\bigwedge_{y}\bigwedge_{z}x=y\cdot z\Rightarrow y=1\vee y=x]$
$$\Rightarrow[(\bigwedge_{y}\bigwedge_{z}t=y\cdot z\Rightarrow y=1\vee y=t)\wedge x<t]\}.$$

3.129. $\sim \bigvee_{x} x^2 < 0.$

3.130. $\bigvee_{x} \{f(x)=0 \wedge \bigwedge_{y} [f(y)=0 \Rightarrow x=y]\}.$

3.131. $\bigwedge_{x} \bigwedge_{y} \bigvee_{z} (x<y \Rightarrow x<z \wedge z<y).$

3.132. $\sim \bigvee_{x} \bigwedge_{y} y \leqslant x.$

3.133. $\sim \bigvee_{y} y^2 = x.$

3.134. $\bigvee_{x} (z=y \vee z^2=y \vee z^3=y).$

3.135. $\bigwedge_{x} \bigwedge_{y} [x<y \Rightarrow f(x)>f(y)].$

3.136. $\bigwedge_{n \in \mathcal{N}} \bigwedge_{m \in \mathcal{N}} (n<m \Rightarrow a_n < a_m).$

3.137. $\bigwedge_{n \in \mathcal{N}} 0 < a_n.$

3.138. $\bigwedge_{\varepsilon>0} \bigvee_{n \in \mathcal{N}} \bigwedge_{m_1 \in \mathcal{N}} \bigwedge_{m_2 \in \mathcal{N}} (m_1 > n \wedge m_2 > n \Rightarrow |a_{m_1} - a_{m_2}| < \varepsilon).$

3.139. $\bigvee_{x} \bigwedge_{n \in \mathcal{N}} |a_n| < x.$

3.140. $\bigvee_{n \in \mathcal{N}} \bigwedge_{m_1 \in \mathcal{N}} \bigwedge_{m_2 \in \mathcal{N}} (n<m_1 \wedge n<m_2 \Rightarrow a_{m_1} = a_{m_2}).$

3.141. If Φ_1 is the formula from 3.140 and Φ_2 the formula from 3.138, then our result is $\Phi_1 \Rightarrow \Phi_2$.

3.142. $\bigvee_{x} \bigwedge_{n \in \mathcal{N}} |a_n| < x \Rightarrow \bigwedge_{\varepsilon>0} \bigvee_{n \in \mathcal{N}} \bigwedge_{m \in \mathcal{N}} \bigvee_{k_1 \in \mathcal{N}} \bigvee_{k_2 \in \mathcal{N}} [n<m$

$\Rightarrow (m<k_1 \wedge m<k_2 \wedge |a_{k_1} - a_{k_2}| < \varepsilon)].$

3.143. $\bigwedge_{\varepsilon>0} \bigvee_{\delta>0} \bigwedge_{x} (|x-x_0| < \delta \Rightarrow |f(x) - f(x_0)| < \varepsilon).$

3.144. $\bigwedge_{x_0 \in \langle a, b \rangle} \bigwedge_{\varepsilon>0} \bigvee_{\delta>0} \bigwedge_{x \in \langle a, b \rangle} (|x-x_0| < \delta \Rightarrow |f(x) - f(x_0)| < \varepsilon)$

$\Rightarrow \bigvee_{y} \bigwedge_{x \in \langle a, b \rangle} |f(x)| < y.$

3.145. $\bigwedge_{\varepsilon>0} \bigvee_{\delta>0} \bigwedge_{x_1 \in \langle a, b \rangle} \bigwedge_{x_2 \in \langle a, b \rangle} (|x_1 - x_2| < \delta \Rightarrow |f(x_1) - f(x_2) < \varepsilon).$

3.146. $\bigwedge_{x \in \mathcal{R}} [x \leqslant a \wedge \bigwedge_{y} (y<a \Rightarrow \bigvee_{x \in \mathcal{R}} y \leqslant x)].$

3.147. $\bigwedge_{x \in \mathcal{R}} [a \leqslant x \wedge \bigwedge_{y} (a<y \Rightarrow \bigvee_{x \in \mathcal{R}} x \leqslant y)].$

3.148. If $\Phi_1(x_0)$ is the formula from 3.143 and $\Psi_1(a)$ and $\Psi_2(a)$ are formulas from 3.146 and 3.147, respectively, then the required formula is:

$$\bigwedge_{x \in \langle a, b\rangle} \Phi(x) \Rightarrow \bigvee_{x \in \langle a, b\rangle} \Psi_1[f(z)] \wedge \bigvee_{z \in \langle a, b\rangle} \Psi_2[f(z)].$$

3.149. $$[\bigwedge_{x} \bigwedge_{\varepsilon>0} \bigvee_{\delta>0} \bigwedge_{x'} |x-x'|<\delta$$
$$\Rightarrow(|f(x)-f(x')|<\varepsilon \wedge |g(x)-g(x')|<\varepsilon)]$$
$$\Rightarrow[\bigwedge_{x} \bigwedge_{\varepsilon>0} \bigvee_{\delta>0} \bigwedge_{x'} |x-x'|<\delta \Rightarrow |f(x)g(x)-f(x')g(x')|<\varepsilon].$$

3.150. $$[\bigwedge_{\varepsilon>0} \bigvee_{\delta>0} \bigwedge_{x_1} \bigwedge_{x_2} |x_1-x_2|<\delta$$
$$\Rightarrow(|f(x_1)-f(x_2)|<\varepsilon \wedge |g(x_1)-g(x_2)|<\varepsilon)]$$
$$\Rightarrow[\bigwedge_{\varepsilon>0} \bigvee_{\delta>0} \bigwedge_{x_1} \bigwedge_{x_2} |x_1-x_2|<\delta$$
$$\Rightarrow|f(x_1)+g(x_1)-f(x_2)-g(x_2)|<\varepsilon].$$

3.151. Consider the proposition $\sim\bigvee_x \Phi(x)$. It says that there is no x such that Φ. Take an arbitrary x_0. It does not possess the property Φ, i.e. it possesses the property $\sim\Phi$; thus $\sim\Phi(x_0)$ is true.

Our reasoning was valid for an arbitrary x_0, and thus, for an arbitrary x_0, $\sim\Phi(x_0)$ is true. Thus from $\sim\bigvee_x \Phi(x)$ we proved $\bigwedge_x \sim\Phi(x)$.

On the other hand, if $\bigwedge_x \sim\Phi(x)$ then, for an arbitrary x_0, $\sim\Phi(x_0)$ holds, i.e. x_0 having the property $\sim\Phi$ does not possess the property Φ. If there were an x_1 with the property Φ, then by the above reasoning it would not possess the property $\sim\Phi$. The contradiction obtained shows $\sim\bigvee_x \Phi(x)$.

3.161. Let Φ and Ψ be propositional functions in $x, y, \ldots, z$ and $u, \ldots, t$. Consider an arbitrary tuple $y_0, \ldots, z_0, u_0, \ldots, t_0$. If for such a tuple the statement.

$$\bigwedge_x \Phi(x, y_0, \ldots, z_0) \vee \Psi(u_0, \ldots, t_0)$$

is false, then the whole implication is true irrespective of the value of the successor of the implication. So let $y_0, \ldots, z_0, t_0, \ldots, u_0$ be a tuple for which the statement

$$\bigwedge_x [\Phi(x, y_0, \ldots, z_0) \vee \Psi(u_0, \ldots, t_0)]$$

is true.

Note that $\Psi(u_0, \ldots, t_0)$ is a proposition since it has no free variables and thus has a logical value. If it is true, then the formula $\Phi(x, y_0, \ldots \ldots, z_0) \vee \Psi$ is true, and so it remains true when we insert the universal quantifier in front of it. In that case, however, the alternative

$$\bigwedge_x \Phi(x, y_0, \ldots, z_0) \vee \Psi(u_0, \ldots, t_0)$$

is clearly true. If $\Psi(u_0, \ldots, t_0)$ is false, then the propositional function $\Phi(x, y_0, \ldots, z_0)$ must be true. Thus the sentence $\bigwedge_x \Phi(x, y_0, \ldots, \ldots, z_0)$ is true, and so the alternative

$$\bigwedge_x \Phi(x, y_0, \ldots, z_0) \vee \Psi(u_0, \ldots, t_0)$$

is also true.

Thus for any tuple $y_0, \ldots, t_0$ our implication holds, and so the implication

$$\bigwedge_x [\Phi(x) \vee \Psi] \Rightarrow \bigwedge_x \Phi(x) \vee \Psi$$

holds.

3.167. No. As $\Phi(x)$ take $\bigvee_z x < z$.

3.168. No. As $\Phi(x)$ and $\Psi(x)$ take $x < 0$ and $x \geqslant 0$, respectively.

3.169. No. As $\Phi(x, y)$, take $x < y$ (in reals).

3.170. No. As $\Phi(x)$ and $\Psi(x)$ take $x < 0$ and $x > 0$, respectively.

3.171. No. As $\Phi(x)$ and $\Psi(x)$ take $x^2 > 1$ and $x > 1$, respectively.

3.172. No. As $\Phi(x)$ and $\Psi(x)$ take $2|x$ and $3|x$, respectively.

3.173. No. As $\Phi(x)$ and $\Psi(x)$ take $x < 0$ and $x^2 < 0$, respectively.

3.174. No. Take as $\Phi(x, y)$ the formula $x = x \wedge \bigvee_x y < x$.

3.175. No. Take as $\Phi(x, y)$ the formula $y < x$.

3.176. No. Take as $\Phi(x)$ the formula $x < 0$.

3.177. No. Take as $\Phi(x)$ the formula $x < 0$.

3.179. We give a formal proof of the formula

$$\sim \bigvee_x \Phi(x) \Leftrightarrow \bigwedge_x \sim \Phi(x).$$

According to our convention, in which formulas with free variables are treated as universal sentences, axiom (g) can be written down as

$$\bigwedge_x [\Phi(x) \Rightarrow \bigvee_x \Phi(x)].$$

By the use of the tautology $(\alpha \Rightarrow \beta) \Leftrightarrow (\sim \beta \Rightarrow \sim \alpha)$, the formula bounded by the quantifier $\bigwedge_x$ may be changed to $\sim \bigvee_x \Phi(x) \Rightarrow \sim \Phi(x)$, thus we get:

$$\bigwedge_x [\sim \bigvee_x \Phi(x) \Rightarrow \sim \Phi(x)].$$

Using axiom (a), we get:

$$\bigwedge_x \sim \bigvee_x \Phi(x) \Rightarrow \bigwedge_x \sim \Phi(x).$$

But the variable x is not free in the formula $\bigvee_x \Phi(x)$, and thus we can use axiom (e); we have

$$\sim \bigvee_x \Phi(x) \Rightarrow \bigwedge_x \sim \bigvee_x \Phi(x) \quad \text{and} \quad \bigwedge_x \sim \bigvee_x \Phi(x) \Rightarrow \bigwedge_x \sim \Phi(x).$$

Using the transitivity of the implication, we finally get

$$\sim \bigvee_x \Phi(x) \Rightarrow \bigwedge_x \sim \Phi(x).$$

Consider the formula $\bigwedge_x [\bigwedge_x \sim \Phi(x) \Rightarrow \sim \Phi(x)]$, following from axiom (b). Using the transposition rule (as above), we can replace it by the following formula:

$$\bigwedge_x [\Phi(x) \Rightarrow \sim \bigwedge_x \sim \Phi(x)].$$

By the distribution (axiom (f)) we get

$$\bigvee_x \Phi(x) \Rightarrow \bigvee_x \sim \bigwedge_x \sim \Phi(x).$$

But x is not free in $\sim \bigwedge_x \sim \Phi(x)$ and so we can use the axiom (j), which gives

$$\bigvee_x \Phi(x) \Rightarrow \sim \bigwedge_x \sim \Phi(x).$$

Using transposition again, we finally get

$$\bigwedge_x \sim\Phi(x) \Rightarrow \sim\bigvee_x \Phi(x),$$

which finishes the proof of equivalence.

3.180. *Hint*: Note that the axioms of the predicate calculus are true in all domains whereas the statement $\bigwedge_x \Phi(x) \Rightarrow \bigvee_x \Phi(x)$ is false in the empty domain.

3.184. In the following proof, besides of axioms and theorems of logic, we will use the following theorems of arithmetic:

(1) $$\bigwedge_x \bigwedge_y \bigwedge_z x<y \Rightarrow x+z<y+z,$$

(2) $$\bigwedge_x \bigwedge_y \bigwedge_z [y>0 \Rightarrow (x<z \Rightarrow xy<zy)]$$

and the properties of the relation $=$ and $<$.

From theorem (1), using (d) and (e) we immediately get

$$\bigwedge_x \bigwedge_y x<y \Rightarrow 2x<y+x, \quad \bigwedge_x \bigwedge_y x<y \Rightarrow y+x<2y,$$

$$\bigwedge_x \bigwedge_y y<x \Rightarrow 2y<y+x, \quad \bigwedge_x \bigwedge_y y<x \Rightarrow y+x<2x,$$

and so using (2) and substituting for y the number $\frac{1}{2}$, we get

$$\bigwedge_x \bigwedge_z [\tfrac{1}{2}>0 \Rightarrow (x<z \Rightarrow \tfrac{1}{2}x<\tfrac{1}{2}z)].$$

Using axiom (a) and the tautology from 3.160, we get

$$\tfrac{1}{2}>0 \rightarrow \bigwedge_x \bigwedge_z (x<z \Rightarrow \tfrac{1}{2}x<\tfrac{1}{2}z).$$

After using modus ponens and changing the variables, we get

(3) $$\bigwedge_x \bigwedge_y (x<y \Rightarrow \tfrac{1}{2}x<\tfrac{1}{2}y),$$

and similarly:

(4) $$\bigwedge_x \bigwedge_y (y<x \Rightarrow \tfrac{1}{2}y<\tfrac{1}{2}x),$$

By substitution we get

$$(5) \qquad \bigwedge_x \bigwedge_y [2y<x+y \Rightarrow y<\tfrac{1}{2}(x+y)],$$

$$(6) \qquad \bigwedge_x \bigwedge_y [2y<y+x \Rightarrow y<\tfrac{1}{2}(x+y)].$$

We now form the conjunctions of (5) and (3) and of (6) and (4); using the distributivity of the universal quantifier with respect to conjunction and the transitivity of implication, we get

$$\bigwedge_x \bigwedge_y \left(x<y \Rightarrow x<\frac{x+y}{2}\right), \quad \bigwedge_x \bigwedge_y \left(y<x \Rightarrow y<\frac{x+y}{2}\right)$$

and similarly

$$\bigwedge_x \bigwedge_y \left(x<y \Rightarrow \frac{x+y}{2}<y\right), \quad \bigwedge_x \bigwedge_y \left(y<x \Rightarrow \frac{x+y}{2}<x\right).$$

Thus

$$\bigwedge_x \bigwedge_y \left(x<y \Rightarrow x<\frac{x+y}{2}<y\right), \quad \bigwedge_x \bigwedge_y \left(y<x \Rightarrow y<\frac{x+y}{2}<x\right).$$

But $x<\frac{x+y}{2} \wedge \frac{x+y}{2}<y \Rightarrow \bigvee_z x<z \wedge z<y$, and so by axiom (*g*) we get

$$\bigwedge_x \bigwedge_y (x<y \Rightarrow \bigvee_z x<z<y),$$

and similarly

$$\bigwedge_x \bigwedge_y (y<x \Rightarrow \bigvee_z y<z<x).$$

Using axiom (b), we get:

$$x<y \Rightarrow \bigvee_z x<z<y, \quad y<x \Rightarrow \bigvee_z y<z<x.$$

But then the alternative of predecessors implies the alternative of successor, which is:

$$x<y \vee y<x \Rightarrow \bigvee_z (x<z<y \vee y<z<x).$$

But $=$ and $<$ have the following property:

$$x \neq y \Rightarrow x<y \vee y<x,$$

so we finally get

$$\bigwedge_x \bigwedge_y x \neq y \Rightarrow \bigvee_z (x < z < y \vee y < z < x).$$

3.190. The simultaneously conjunctive and disjunctive form is

$$\bigwedge_x \bigwedge_y [\Phi(x) \vee \Psi(y)].$$

3.191. The simultaneously conjunctive and disjunctive form is

$$\bigvee_x \bigvee_y [\Phi(x) \wedge \Psi(y)].$$

3.192. The simultaneously conjunctive and disjunctive form is

$$\bigwedge_x \bigwedge_y [\sim \Phi(x) \vee \Psi(y)].$$

3.193. The conjunctive form is

$$\bigwedge_x \bigwedge_t \bigvee_u \bigvee_s \{[\sim \Phi(x) \vee \Psi(t, y)] \wedge [\sim \Psi(s, y) \vee \Phi(u)]\}.$$

The disjunctive form is

$$\bigwedge_x \bigwedge_t \bigvee_u \bigvee_s \{[\sim \Phi(x) \wedge \sim \Psi(s, y)] \vee [\sim \Phi(x) \wedge \Phi(u)] \vee$$
$$\vee [\Psi(t, y) \wedge \sim \Psi(s, y)] \vee [\Psi(t, y) \wedge \Phi(u)]\}.$$

3.194. The simultaneously conjunctive and disjunctive form is

$$\bigwedge_z \bigwedge_y [\sim \Phi(z) \vee \sim \Psi(x, y)].$$

3.195. $\sim \bigwedge_x \{\bigvee_y \Phi(x, y) \Rightarrow \sim \bigwedge_y [\Psi(x, y, z) \vee \bigwedge_z \Theta(x, y, z)]\}$

$$\Leftrightarrow \bigvee_x \{\bigvee_y \Phi(x, y) \wedge \bigwedge_y [\Psi(x, y, z) \vee \bigwedge_z \Theta(x, y, z)]\}$$

$$\Leftrightarrow \bigvee_x \bigvee_y \{\Phi(x, y) \wedge \bigwedge_u [\Psi(x, u, z) \vee \bigwedge_z \Theta(x, u, z)]\}$$

$$\Leftrightarrow \bigvee_x \bigvee_y \bigwedge_u \{\Phi(x, y) \wedge [\Psi(x, u, z) \vee \bigwedge_t \Theta(x, u, t)]\}$$

$$\Leftrightarrow \bigvee_x \bigvee_y \bigwedge_u \bigwedge_t \{\Phi(x, y) \wedge [\Psi(x, u, z) \vee \Theta(x, u, t)]\}.$$

This is the conjunctive normal form. The disjunctive form is

$$\bigvee_{x}\bigvee_{y}\bigwedge_{u}\bigwedge_{t}\{[\Phi(x,y)\wedge\Psi(x,u,z)]\vee[\Phi(x,y)\wedge\Theta(x,u,t)]\}.$$

Note: We use here the fact that Φ, Ψ, Θ did not contain – in the original formula – the variables u and t.

3.196. The conjunctive form is

$$\bigvee_{x}\bigwedge_{u}\bigwedge_{t}\bigvee_{z}\{[\Phi(u,y)\vee\sim\Psi(x,t)]\wedge[\Phi(u,y)\vee\sim\Theta(z,x)]\}.$$

The disjunctive form is

$$\bigvee_{x}\bigwedge_{u}\bigwedge_{t}\bigvee_{z}\{\Phi(u,y)\vee[\sim\Psi(x,t)\wedge\sim\Theta(z,x)]\}.$$

3.197. *Hint*: Use 3.190 and 3.191.

3.198. *Hint*: Reason as in 3.197.

3.199. No. Different form result from the freedom of choice of the variable while we change bounded variables and also from the possibility of taking quantifiers into prenex in a different order.

CHAPTER 4

RELATIONS. EQUIVALENCES

4.1. $D(R)=\{a,b\}$, $D^*(R)=\{b,c\}$.
4.2. $D(R)=\{a\}$, $D^*(R)=\{a,b,c\}$.
4.3. $D_1(R)=\{a\}$, $D_2(R)=\{b,c,d\}$, $D_3(R)=\{b,c\}$.
4.4. $D(R)=\mathcal{N}$, $D^*(R)=\mathcal{N}-\{0\}$.
4.5. $D_1(R)=\{-3,-2,-1,0,1,2,3\}=D_2(R)$, $D_3(R)=\{0,1,2,3\}$.
4.6. $D_1(R)=\mathcal{R}-\{0\}=D_2(R)$, $D_2(R)=\mathcal{N}$.

4.8. R has none of the considered properties in the set X. However, R restricted to $\{a, b\}$ is reflexive, transitive, connected and weakly antisymmetric.

4.9. R is reflexive, symmetric, and transitive in X. In the set $\{a, b\}$ R is in addition connected, in the set $\{c, d\}$ is asymmetric.

4.10. R is antireflexive. In the set $\{a, b, c\}$ it is antireflexive, symmetric and connected, in the set $\{a, c, d\}$ it is antireflexive and connected.

4.11. R has none of the properties in question in X. In the set $\{a, b, c\}$ it is reflexive, symmetric, transitive, and connected.

4.12. If $\langle x, y\rangle \in R$ then $x \in D(R)$ and $y \in D^*(R)$, i.e.

$$\langle x,y\rangle \in D(R)\times D^*(R).$$

4.13. Let $x \in X$. Then $\langle x, x\rangle \in R \wedge \langle x, x\rangle \in S$, and thus $\langle x, x\rangle \in R \cap \cap S$, which – x being arbitrary – proves the theorem.

4.14. Let $I_{D(R)\cup D^*(R)} \subset R$; then $\bigwedge_{x\in D(R)\cup D^*(R)} \langle x,x\rangle \in R$, which just means that R is reflexive.

Conversely, if R is reflexive, then, by definition, it is reflexive in $D(R) \cup D^*(R)$ i.e.

$$\bigwedge_{x\in D(R)\cup D^*(R)} \langle x,x\rangle \in R, \quad \text{or} \quad I_{D(R)\cup D^*(R)} \subset R.$$

4.15. If R is antireflexive, then

$$\bigwedge_{x\in D(R)\cup D^*(R)} \langle x,x\rangle \notin R, \quad \text{or} \quad \bigwedge_{u} (u \in I_{D(R)\cup D^*(R)} \Rightarrow u \notin R);$$

thus

$$I_{D(R)\cup D^*(R)} \cap R = O.$$

Now note that all the implications are in fact equivalences.

4.16. Note that $I_{D(R)\cup D^*(R)} \cap R = O \Leftrightarrow I_{D(R)\cup D^*(R)} \subset X^2 - R$, and the set-theoretical union or intersection of an arbitrary family of sets satisfying the above condition also satisfies it.

4.17. Yes.

4.18. Yes. Assume $\langle x, y\rangle \in R - I_X$, we show that $\langle y, x\rangle \notin R - I_X$. Indeed, suppose $\langle y, x\rangle \in R - I_X$; thus $\langle y, x\rangle \in R$ (and similarly $\langle x, y\rangle \in R$). By the weak antisymmetry of R, $y = x$, which contradicts our assumption.

The converse is also true. If $\langle x, y\rangle \in R$, $\langle y, x\rangle \in R$, we show that $x = y$. Indeed, if $x \neq y$ then $\langle x, y\rangle \in R - I_X$ and $\langle y, x\rangle \in R - I_X$, which contradicts the fact that $R - I_X$ is asymmetric.

4.19. No.

4.23. Yes. Just add I_X.

4.24. No. For instance $\{\langle a, a\rangle, \langle a, b\rangle\} \subset \{a, b\}^2$ cannot be extended.

4.25. Yes.

4.26. No.

4.27. No.

4.28. Yes.

4.29. Yes.

4.30. Recall the symmetry condition:

$$\bigwedge_{x,y \in X} (\langle x, y\rangle \in R \Rightarrow \langle y, x\rangle \in R)$$

or

$$\bigwedge_{x,y \in X} (\langle x, y\rangle \in R \Rightarrow \langle x, y\rangle \in R^{-1}),$$

which was to be proved.

4.31. We show first that $R \subset S \Rightarrow R^{-1} \subset S^{-1}$. Assume $R \subset S$ and $\langle x, y\rangle \in R^{-1}$; then $\langle y, x\rangle \in R$, $\langle y, x\rangle \in S$ and finally $\langle x, y\rangle \in S^{-1}$. On the other hand, $(R^{-1})^{-1} = R$. If $\langle x, y\rangle \in (R^{-1})^{-1}$ then $\langle y, x\rangle \in R^{-1}$ and $\langle x, y\rangle \in R$. The converse inclusion is equally clear. Now, if $R \subset R^{-1}$ then $R^{-1} \subset (R^{-1})^{-1}$, i.e. $R^{-1} \subset R$, which gives $R = R^{-1}$. The converse implication is obvious since $X = Y \Rightarrow X \subset Y$.

4.32. (a) Yes. Let $\langle x, y\rangle \in (R \cup S)^{-1}$. Then $\langle y, x\rangle \in R \cup S$, which is equivalent to $\langle y, x\rangle \in R \vee \langle y, x\rangle \in S$, which in turn implies $\langle x, y\rangle \in R^{-1} \vee \langle x, y\rangle \in S^{-1}$ and finally $\langle x, y\rangle \in R^{-1} \cup S^{-1}$.

(b) Yes.

(c) Yes.

(d) Yes.

4.33. Yes. This is, in fact, an other formulation of the connectedness property.

4.34. Yes.

4.35. Yes. Assume that R and S are transitive, and let $\langle x, y\rangle \in R \cap S$ and $\langle y, z\rangle \in R \cap S$. Then $\langle x, y\rangle \in R$, $\langle x, y\rangle \in S$, $\langle y, z\rangle \in R$, $\langle y, z\rangle \in S$, and so $\langle x, y\rangle \in R \wedge \langle y, z\rangle \in R$ and $\langle x, y\rangle \in S \wedge \langle y, z\rangle \in S$. Thus $\langle x, z\rangle \in R$ and $\langle x, z\rangle \in S$; thus finally $\langle x, z\rangle \in R \cap S$.

4.36. No. It is enough to give a counterexample. Let R be the relation $\leqslant$ in $\mathcal{N}$ and let S be $\geqslant$ in $\mathcal{N}$. Then both R and S are connected in $\mathcal{N}$, but $R \cap S = I_{\mathcal{N}}$ and $I_{\mathcal{N}}$ is not connected.

4.37. Yes. Let $\langle x, y\rangle \in X^2$. Then $\langle x, y\rangle \in R \cup R^{-1} \cup I_X$ and $\langle x, y\rangle \in S \cup S^{-1} \cup I_X$; thus

$$\begin{aligned}\langle x, y\rangle \in R \cup R^{-1} \cup I_X \cup S \cup S^{-1} \cup I_X \\ &= R \cup S \cup R^{-1} \cup S^{-1} \cup I_X \\ &= (R \cup S) \cup (R \cup S)^{-1} \cup I_X .\end{aligned}$$

(The last passage uses 4.32 (a) and the condition used in 4.33).

4.38. No. It is enough to provide a counterexample. Let $X = \{1, 2, 3\}$, $R = I_X$ and $S = R \cup \{\langle 1, 2\rangle \cup \langle 2, 3\rangle\}$. Then R is a transitive relation, but S is not. Indeed, $\langle 1, 2\rangle \in S$ and $\langle 2, 3\rangle \in S$ but $\langle 1, 3\rangle \notin S$.

4.39. Assume that R is transitive, i.e.

$$\bigwedge_{x,y,z} (xRy \wedge yRz \Rightarrow xRz),$$

and let $\langle x, z\rangle \in R \circ R$. Then there exists a y such that $\langle x, y\rangle \in R$ and $\langle y, z\rangle \in R$, and so, by transitivity, $\langle x, z\rangle \in R$. Conversely, if $R \circ R \subset R$ then since $\langle x, y\rangle \in R$ and $\langle y, z\rangle \in R$ implies $\langle x, z\rangle \in R \cap R$ and $R \circ R \subset R$, we have $\langle x, z\rangle \in R$, which fnishes the proof.

4.40. Let $\langle u, v\rangle \in I_X \circ I_Y$; then there is an element t such that $\langle u, t\rangle \in I_X$ and $\langle t, v\rangle \in I_Y$, but in that case $u = t$ and $u \in X$; moreover $t = v$ and $v \in Y$, and thus $u = t = v$ and $t \in X \cap Y$, and finally $\langle u, v\rangle \in I_{X \cap Y}$.

Conversely, if $\langle u, v\rangle \in I_{X\cap Y}$ then $u=v$ and $u \in X \cap Y$; thus $\langle u, v\rangle \in I_X$ and $\langle v, v\rangle \in I_Y$ and so $\langle u, v\rangle \in I_{X\circ Y}$.

4.41. Let $\langle x, t\rangle \in R\circ(S\circ T)$. Then there exists a y such that $xRy\wedge$ $\wedge\, yS\circ Tt$. Thus there is a z such that $xRy\wedge(ySz\wedge zTt)$. By the tautology $[p\wedge(q\wedge r)]\Leftrightarrow[(p\wedge q)\wedge r]$ we get $(xRy\wedge ySz)\wedge zTt$, which means $x[(R\circ S)\circ T]t$. The converse inclusion is proved analogously.

4.42. Assume $\langle x, z\rangle \in (R\circ S)^{-1}$; then $\langle z, x\rangle \in R\circ S$ and so there exists a y such that $\langle z, y\rangle \in R$ and $\langle y, x\rangle \in S$. But then $\langle y, z\rangle \in R^{-1}$ and $\langle x, y\rangle \in S^{-1}$; thus $\langle x, y\rangle \in S^{-1}$, which gives $\langle x, z\rangle \in S^{-1}\circ R^{-1}$. As all the passages were in fact equivalences we get the desired equality.

4.43. Let $X\in D(R)$. Then there exists a y such that $\langle x, y\rangle \in R$ and thus $\langle x, y\rangle \in R^{-1}$. But then $\langle x, x\rangle \in R\circ R^{-1}$, which shows (since x is arbitrary) that $I_{D(R)}\subset R\circ R^{-1}$. The other part of the problem is proved analogously.

4.44. (a) The relations are interpreted as the subsets of the Cartesian plane.

(b) $D(R)$ is the projection of R onto the axis Ox and $D^*(R)$ the projection of R onto the axis Oy.

(c) The diagonal (i.e. the straight line described by the equation $y=x$) is included in R.

(d) The diagonal is a symmetry axis of R.

(e) The set-theoretical union of R, the diagonal, and the symmetric image of R with respect to the diagonal is the whole plane.

(f) The diagonal is disjoint from R.

(g) R does not contain a pair of points symmetric with respect to the diagonal.

(h) The only points of R which have in R images which are symmetric (with respect to the diagonal) belong to the diagonal.

4.45. Reflexivity, symmetry, transitivity.

4.46. Reflexivity, symmetry, transitivity. Indeed, for every x, $2|2x$. If $2|x+y$ then $2|y+x$ (since $x\neq y=y\neq x$). Finally, if $2|x+y$ and $2|y+z$, then $2|x+2y+z$ and $2|2y$, and so $2|x+2$.

4.47. Asymmetry and transitivity. Note that R is not reflexive, in particular $\langle 0, 0\rangle \notin R$. R is not connected since $\langle 3, 5\rangle \notin R$, $\langle 5, 3\rangle \notin R$, $3\neq 5$.

4.48. Antireflexivity, asymmetry, transitivity.

4.49. Antireflexivity, asymmetry, transitivity. Indeed, assume xRy

and yRz; then $x=2$ and $y=3$ ($R=\{\langle 2, 3\rangle\}$) and $y=2$ and $2=3$, which is absurd. Thus the whole implication $xRy \wedge yRz \Rightarrow xRz$ is valid.

4.50. Transitivity, weak antisymmetry. Indeed, $xRy \wedge yRx \Rightarrow x=1 \wedge y=1$ (as $R=\{\langle 1, 1\rangle\}$), thus $x=y$.

4.51. Reflexivity, symmetry, transitivity.

4.52. Antireflexivity, symmetry.

4.53. Reflexivity, symmetry, transitivity.

4.54. Weak antisymmetry. Indeed, if xRy and yRx then $x^3=y^2$ and $y^3=x^2$; thus $x=0$ and $y=0$ or $x=1$ and $y=1$, and so always $x=y$.

4.55. Antireflexivity, asymmetry, transitivity. Note that R is not connected; indeed, $2\neq -2$ but $\sim(2R-2) \wedge \sim(-2R2)$.

4.56. Antireflexivity, symmetry.

4.57. Reflexivity, symmetry. Note that R is not transitive. Indeed, $3R1$ and $1R0$ but $\sim 3R0$. We infer from this that there are reflexive and symmetric relations which are not transitive. In other words, the property of being reflexive and symmetric is not sufficient to imply transitivity.

4.58. Symmetry.

4.59. Reflexivity, symmetry, transitivity. Indeed, if $x \in \mathcal{N}$ then $x \leqslant 5$ or $x>5$. If $x \leqslant 5$ then xRx since $x=x$, and if $x>5$ then $2|x+x$. Similarly, if xRy then $(x \leqslant 5 \wedge y \leqslant 5 \wedge x=y)$ or $(x>5 \wedge y>5 \wedge 2|x+y)$. In case $y=x$ and so yRx; in the other case $2|y+x$. In any case yRx.

4.60. Antireflexivity, symmetry, connectedness.

4.61. Reflexivity, symmetry, transitivity.

4.62. Antireflexivity, symmetry.

4.63. Antireflexivity.

4.64. Antireflexivity.

4.65. Asymmetry.

4.66. Weak antisymmetry.

4.67. Antireflexivity, asymmetry.

4.68. None.

4.69. Reflexivity, weak antisymmetry, transitivity. Note that $x-x=0+0i$. If $x-y=a+bi$, $y-x=c+di$ where $a, b, c, d \in \mathcal{N}$, then $x-y+{}+y-x=0=(a+c)+(b+d)i$, and so $a+c=b+d=0$. As $a \in \mathcal{N}$ and $c \in \mathcal{N}$, therefore $a=c=0$. Similarly $b=d=0$. So $x-y=0$, i.e. $x=y$.

4.70. Reflexivity, symmetry. Note that our relation is not transitive. Indeed $\langle 1, 1\rangle R \langle 0, 0\rangle$ and $\langle 0, 0\rangle R \langle 1, 2\rangle$ but $\sim\langle 1, 1\rangle R \langle 1, 2\rangle$.

4.71. Symmetry.

4.72. Symmetry.

4.73. Reflexivity, symmetry, transitivity. Indeed, $X \dot{-} X = 0$, and O is finite. Then $X \dot{-} Y = Y \dot{-} X$. Finally, $X \dot{-} Z \subset (X \dot{-} Y) \cup (Y \dot{-} Z)$, and so if $X \dot{-} Y$ and $Y \dot{-} Z$ are finite then $X \dot{-} Z$ is finite.

4.74. Reflexivity, symmetry, transitivity.

4.75. Symmetry. Relation R is not transitive. Indeed, let $X = \text{Par}$, $Y = \{1\}$, $Z = \text{Par}$. Then $X \cap Y = O = Y \cap Z$ and obviously $O \subset \mathcal{N} - \text{Par}$; thus $XRY \wedge YRZ$ but $X \cap Z = \text{Par}$ and $\sim(X \cap Z \subset \mathcal{N} - \text{Par})$.

4.76. Reflexivity, symmetry, transitivity. Since this problem is important in applications, particularly those connected with model theory, we suggest that the reader, if not able to solve it himself, should read the answer carefully.

Reflexivity. Let $Z \in \mathcal{P}(X)$. Then $Z \dot{-} Z = O$. According to condition 1°, $Z \dot{-} Z \in I$ and so ZRZ.

Symmetry. Since $Z \dot{-} T = T \dot{-} Z$ therefore if $Z \dot{-} T \in I$ then also $T \dot{-} Z \in I$.

Transitivity. If TRY and YRZ then $T \dot{-} Y \in I$ and $Y \dot{-} Z \in I$, and so $(T \dot{-} Y) \cup (Y \dot{-} Z) \in I$ (according to 2°). Since $T \dot{-} Z \subset (T \dot{-} Y) \cup (Y \dot{-} Z)$, we have $T \dot{-} Z \cup [(T \dot{-} Y) \cup (Y \dot{-} Z)] = (T \dot{-} Y) \cup (Y \dot{-} Z)$ and so $(T \dot{-} Z) \cup \cup [(T \dot{-} Y) \cup (Y \dot{-} Z)] \in I$; hence (by 2° again) $X \dot{-} Z \in I$.

Note that Problem 4.73 is a particular case of our problem, namely, the finite subsets of the set $\mathcal{N}$ form an ideal.

4.77. Antireflexivity, asymmetry, transitivity. Note that our relation s not connected: $\sim\{1/n\}_{n \in N} R \{1/n^2\}_{n \in N}$, $\sim\{1/n^2\}_{n \in N} R \{1/n\}_{n \in N}$ and $\sim\{1/n\}_{n \in N} = \{1/n^2\}_{n \in N}$.

4.78. Symmetry.

4.79. Reflexivity, symmetry, transitivity.

4.80. Reflexivity, transitivity. Note that our relation is not antisymmetric, since the equality of domains does not imply the equality of relations.

4.81. Reflexivity, symmetry, transitivity. This is again a very important problem: check the solution carefully.

$$\{x : f(x) \neq f(x)\} = O \in I \quad \text{(by our assumption)},$$

$$\{x : f(x) \neq g(x)\} = \{x : g(x) \neq f(x)\},$$

$$\{x : f(x) \neq h(x)\} \subset \{x : f(x) \neq g(x)\} \cup \{x : g(x) \neq h(x)\}.$$

Since each of the sets on the right hand side of the inclusion belongs to the ideal I, so does their union, and finally a subset of that union, $\{x : f(x) \neq h(x)\}$, belongs to I.

4.82. *Proof*: (a) $\Rightarrow$. Note that $x \in [x]_R$ and so $x \in [y]_R$

$\Leftarrow$. xRy implies $\bigwedge_z zRx \Leftrightarrow zRy$ (by transitivity and symmetry), which means $[x]_R = [y]_R$.

(b) $\Leftarrow$. We have

$$xRy \Rightarrow [x]_R = [y]_R$$

(by (a)) and so $[x]_R \cap [y]_R = [x]_R$. But $x \in [x]_R$ and so $[x]_R \cap [y]_R \neq O$.

$\Rightarrow$. Assume that $[x]_R \cap [y]_R \neq O$ and so there is a z such that zRx and zRy. By symmetry, xRz and zRy and thus xRy.

4.83. Reflexivity. Since $\bigcup_{i \in I} X_i = X$, we have, for given x, $\bigvee_{i \in I} x \in X_i$ and so $\bigvee_{i \in I} (x \in X_i \wedge x \in X_i)$.

Symmetry. If $\bigvee_{i \in I} (x \in X_i \wedge y \in X_i)$, then $\bigvee_{i \in I} (y \in X_i \wedge x \in X_i)$.

Transitivity. If $x \in X_{i_0} \wedge y \in X_{i_0}$ and $y \in X_{i_1} \wedge z \in X_{i_1}$ then $i_0 = i_1$. Since $y \in X_{i_0} \cap X_{i_1}$ and $X_{i_0} \cap X_{i_1} \neq O$. Thus $x \in X_{i_0}$ and $z \in X_{i_0}$, and so $\bigvee_{i \in I} (x \in X_i \wedge z \in X_i)$.

4.84. R_X is not reflexive.

4 85. The necessary and sufficient condition is as follows:

$$X_i \cap X_j \neq O \Leftrightarrow X_i = X_j.$$

4.86. Yes. There are three equivalence classes:

$$A_0 = \{6n : n \in \mathcal{N}\},$$

$$A_1 = \{6n+2 : n \in \mathcal{N}\},$$

$$A_3 = \{6n+4 : n \in \mathcal{N}\}.$$

4.87. Yes. If we interpret complex numbers as points in the plane (i.e. the number $x+yi$ as $\langle x, y\rangle$, then the equivalence classes are the lines parallel to the axis $0y$.

4.88. No, it is not reflexive.

4.89. Yes. There are two equivalence classes: $A_0 = \text{Par}$, $A_1 = \mathcal{N} - \text{Par}$.

4.90. No. The relation R is not symmetric.

4.91. No. The relation R is not transitive; namely $1R3 \wedge 3R2$ but $\sim 1R2$.

4.92. Yes. An equivalence class is determined by the coefficients of the powers greater than or equal to 3.

4.93. Yes. An equivalence class is determined by a polynomial in which all nonzero coefficients are equal to 1. Adding to it consecutively polynomials with even coefficients, we get all polynomials which are in the relation R with our representative.

4.94. Yes. Using the geometric interpretation of the complex numbers, we see that one of the equivalence classes, namely $[0]_R$, is a one-element set, $[0]_R = \{0\}$. All the other classes are of the form of a straight line without a point. Namely, if $x \neq 0$ then $[x]_R$ arises from the line passing through x and 0 by deleting 0.

4.95. Yes. Note that if $a \in R$ then $(x+yi)^2 = ai$ means that $x^2 - y^2 = 0$ i.e. $x^2 = y^2$, and this relation is clearly an equivalence. Every equivalence class (apart from the class $[0]_R$, which is a one-element set) is two element, consisting of x and $-x$.

4.96. Yes. Note that all the polynomials form one of the equivalence classes.

4.97. Yes. Compare with 4.92.

4.98. No. It is not reflexive.

4.99. Yes. There are two equivalence classes: $A_1 = \{1, 3, 5, \ldots, 15\}$, $A_2 = \{2, 4, \ldots, 16\}$.

4.100. Yes. An equivalence class is determined by a real number, which is the common limit of all elements of the class.

4.101. Yes. This problem is, in fact, another formulation of Problem 4 100. If we define a relation $\lim a_n - b_n = 0$ between the converging sequences of rationals $\{a_n\}_{n \in \mathcal{N}}$ and $\{b_n\}_{n \in \mathcal{N}}$ as

$$\bigwedge_n \bigvee_k \bigwedge_m \left[m > k \Rightarrow |(a_m - b_m)| < \frac{1}{n} \right],$$

then the equivalence classes of our relation may be treated as real numbers. The above method of the construction of real numbers is due to G. Cantor.

4.102. No. For instance $\sim 1R1$.

4.103. Yes. This is, in fact, a construction of the ring Z_k (remainders modulo k). There are k classes of equivalence. Marking them successively the numbers $0, \ldots, k-1$, we have $A_i = \{(k \cdot n) + i : n \in \mathcal{N}\}$.

4.104. Yes. $A_r = \{\mathbf{B} : \mathrm{Det}\, \mathbf{B} = r\}$, where $r \in \mathcal{R}$.

4.105. Yes. An equivalence class is determined by an element of the form:

$$\begin{bmatrix} a & b \\ c & 0 \end{bmatrix}.$$

4.106. Yes. $[0]_R = \mathcal{R}[t]$.

4.107. No. R is not reflexive.

4.108. No. R is not transitive.

4.109. No R is not reflexive.

4.110. Yes. The only problem is to show transitivity. Assume $xy = t^2$ and $yz = u^2$. Then $xzy^2 = t^2u^2$ and so $y^2 | (tu)^2$ and $y | tu$. Let $w = tu | y$. Then $w \in \mathcal{N}$ and $xz = w^2$. In order to find the equivalence classes recall the following fact: Every natural number greater than one has a unique representation of the form $p_1^{\alpha_1} \ldots p_n^{\alpha_n}$ where p_i are prime numbers and $i < j \Rightarrow p_i < p_j$, moreover, all α_i are natural numbers, $\alpha_i \neq 0$. Given number $a = p_1^{\alpha_1} \ldots p_n^{\alpha_n}$, consider $b = p_1^{\varepsilon_1} \ldots p_n^{\varepsilon_n}$, where ε_i is 0 or 1 depending on whether α_i is even or odd. Every number b of this form determines an equivalence class; b is the least number in its class.

4.111. Yes. An equivalence class is determined by a positive rational number, namely by one in which the numerator is the first element of the pair and the denominator is the second element of the pair.

4.112. Yes. The equivalence classes are determined by integers. This is one of the methods of constructing integers.

4.113. Yes.

4.114. No. Our relation is not symmetric.

4.115. Yes. There are two kinds of equivalence classes, namely $[x]_R = \{x\}$ for x' es even and three other classes $A_0 = \{6n+1 : n \in \mathcal{N}\}$, $A_1 = \{6n+3 : n \in \mathcal{N}\}$, $A_2 = \{6n+5 : n \in \mathcal{N}\}$.

4.116. Yes. There are eight equivalence classes: $A_0 = \{6n : n \in \mathcal{N}\}$, $A_1 = \{6n+2 : n \in \mathcal{N}\}$, $A_2 = \{6n+4 : n \in \mathcal{N}\}$, $A_3 = \{10n+1 : n \in \mathcal{N}\}$, $A_4 = \{10n+3 : n \in \mathcal{N}\}$, $A_5 = \{10n+5 : n \in \mathcal{N}\}$, $A_6 = \{10n+7 : n \in \mathcal{N}\}$, $A_7 = \{10n+9 : n \in \mathcal{N}\}$. It is easy to see that $A_0 \cup \ldots \cup A_7 = \mathcal{N}$.

4.117. No. The relation R is not transitive. Indeed, let a, b be different elements of the set X. Then $\{a\} RO$ and $OR\{b\}$ but $\sim \{a\} R \{b\}$.

4.118. No. It is not reflexive.

4.119. Yes. There are 11 equivalence classes: nine one-element classes determined by $z \in \mathscr{Z}$ such that $|z| < 5$, and, moreover, two infinite classes, one consisting of the remaining even numbers and one consisting of the remaining odd numbers.

4.120. Yes. Note that it is a particular case of problem 4.76. Indeed if X is a set and C a fixed subset of X, then $\mathscr{P}(C)$ is an ideal in $\mathscr{P}(X)$. Every equivalence class is determined by a subset of the set $X - C$.

4.121. Yes. There are two equivalence classes, one consisting of the polynomials of even degrees and the other consisting of the polynomials of odd degrees.

Hint: Use the following: $\deg(f \cdot g) = \deg f + \deg g$.

4.122. No. It is not reflexive.

4.123. Yes. There are two one-element equivalence classes $[0]_R$ and $[1]_R$. All the other equivalence classes are infinite and of the form $\{a^n : n \in \mathscr{N}\}$, where a is a natural number not of the form b^k where $b, k \in \mathscr{N}$.

4.124. Yes. The equivalence classes are of the form $A \times B$, where A is an equivalence class of R_1 and B is an equivalence class of R_2. Thus we prove in this problem the following theorem: The Cartesian product of equivalences is an equivalence. One can prove that the Cartesian product of partial orderings is again a partial ordering. An analogous statement for connected relations is false.

4.125. (a) Yes. The relation $R_1 \cap R_2$ is reflexive (compare Problem 4.13) and transitive (compare Problem 4.35). If $\langle x, y\rangle \in R_1 \cap R_2$ then $\langle x, y\rangle \in R_1$ and $\langle x, y\rangle \in R_2$, and so $\langle y, x\rangle \in R_1$ and $\langle y, x\rangle \in R_2$; thus $\langle y, x\rangle \in R_1 \cap R_2$, which shows that $R_1 \cap R_2$ is symmetric.

(b) No. It is not necessarily transitive. It is enough to give an example. Let $X = \mathscr{N} - \{0\}$ and let R_1, R_2 be defined, respectively, as follows: $xR_1 y \Leftrightarrow 3|x-y$, $xR_2 y \Leftrightarrow 2|x-y$. Then

$$\langle 1,4\rangle \in R_1 \cup R_2, \quad \langle 4,6\rangle \in R_1 \cup R_2, \quad \langle 1,6\rangle \notin R_1 \cup R_2.$$

(c) No. It is not reflexive.

The equivalence classes of the relation $R_1 \cap R_2$ are nonempty sets of the form $A \cap B$ where A is an equivalence class of R_1 and B is an equivalence class of R_2.

4.126. (a) Yes. The proof is analogous to that used in 4.125 (a).

(b) No. Example: $R_1=\{\langle a,a\rangle, \langle a,b\rangle, \langle b,a\rangle, \langle b,b\rangle\}$, $R_2=\{\langle b,b\rangle, \langle b,c\rangle, \langle c,b\rangle, \langle c,c\rangle\}$.

4.127. (b) The function f must be one-to-one.

4.128. Let $E[x]$ be the entier of x, i.e. the largest integer smaller than or equal to x. Define $aRb \Leftrightarrow E[a]=E[b]$. Note that R is just the relation $\sim_E$ (compare 4.127.).

4.129. $xRy \Leftrightarrow 2 \mid x+y$.

4.130. $\langle x,y\rangle R(z,t\rangle \Leftrightarrow (xy=zt=0) \vee (xy\neq 0 \wedge zt\neq 0 \wedge xz>0 \wedge yt>0)$.

4.131. $\langle x,y\rangle R\langle z,t\rangle \Leftrightarrow (x^2+y^2=z^2+t^2)$.

Note: A point is being considered as a degenerate circle of radius 0.

4.132. Let $R_1=R_{\mathscr{A}}$ and $R_2=R_{\mathscr{B}}$; then

$$\bigwedge_{x,y}(xR_1 y \Rightarrow xR_2 y).$$

Every set of the family $\mathscr{B}$ is the union of some subfamily of $\{A_t\}_{t\in T}$.

4.133. *Hint*: Compare 4.132.

4.134. Fig. 40.

4.135. Fig. 41.

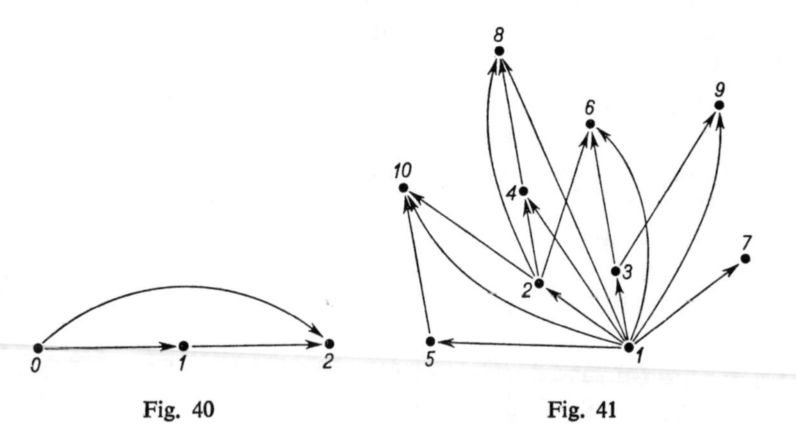

Fig. 40 Fig. 41

4.136. Fig. 42 and Fig. 43.

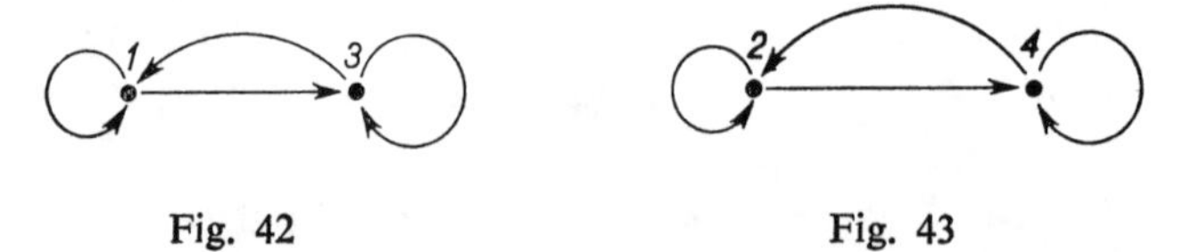

Fig. 42 Fig. 43

4.137. Every point of the diagram of a reflexive relation is the starting point of an arrow which ends at the same point.

4.138. The diagram of a transitive relation has the following property: If it is possible to pass from point a to point b along the arrows of the diagram, then there is an arrow with the origin at a and the end-point at b.

4.139. There are no cycles in the diagram, i.e. for no point is it possible to leave it and come back to it along the arrows.

4.140. If there is an arrow with the origin at a and the end-point at b, then there is also an arrow with the origin at b and the end-point at a.

4.141. For arbitrary points a and b, there is either an arrow starting at a and ending at b or conversely.

4.142. There is no arrow with the same origin and the end-point.

4.143. If there is an arrow staring at a and ending at b, then there is no arrow starting at b and ending at a.

4.144. If there is a path in the direction of the arrows from a to b, then there is also such path in the opposite direction.

4.145. The diagram has no cycle.

4.146. The diagram can be imbedded in the straight line.

4.147. No arrow ends at that point.

4.148. No arrow starts at that point.

4.149. For every point x of the diagram, there is a path leading from that point to x.

4.150. For every point x of the diagram, there is a path from x to that point.

CHAPTER 5

FUNCTIONS

5.1. Let $f\colon X\to Y$, $A\subset X$, $A\neq O$. By the last of our assumptions, there is an x in A. By the definition, $f(x)\in f*A$ and so $f*A\neq O$.

5.2. Let $f:\mathscr{R}\to\mathscr{R}$ be defined by the equation $f(x)=x^2$. Let $A=\langle -2, -1\rangle$. Then $A\neq O$ but $f^{-1}*A=O$ since, for all x, $f(x)\in\mathscr{R}^+$ and $\langle -2, -1\rangle\cap\mathscr{R}^+=O$.

5.3. According to the definition $I_X=\{\langle x, x\rangle : x\in X\}$. One has to prove that: I_X is a function, the domain of I_X is the whole of X, the range of I_X is again the whole of X, and, finally, I_X is one-to-one. If the pair $\langle x, y\rangle\in I_X$ then $y=x$, and so, if $\langle x, y\rangle\in I_X$ and $\langle x, z\rangle\in I_X$, then $y=x$ and $z=x$ and so $y=z$.

Thus we have proved that I_X is a function. Since for all $x\in X$, $\langle x, x\rangle\in I_X$, the domain of I_X is the whole of X. By the same reason $Rf=X$. Thus $f\colon X\to X$. Our function is one-to-one since $\langle y, x\rangle\in I_X\Rightarrow y=x$ and so $\langle y, x\rangle\in I_X$ and $\langle z, x\rangle\in I_X$ imply $y=x$ and $z=x$, and thus $y=z$.

5.4. No. We need to show two pairs $\langle x, y_1\rangle$ and $\langle x, y_2\rangle$ such that $y_1\neq y_2$ but $\langle x, y_1\rangle\in R$ and $\langle x, y_2\rangle\in R$. Take $x=2$, $y_1=2$, $y_2=-2$.

5.5. Yes. Indeed, $x^2=y^2$ implies $|x|=|y|$. But in $\mathscr{R}^+$, $|x|=x$, and so $x=y$. Actually $R=I_{\mathscr{R}^+}$.

5.6. Yes.

5.7. No. The pairs $\langle 1, -1\rangle$ and $\langle 1, 1\rangle$ both belong to R, but $1\neq -1$.

5.8. No. The pairs $\langle 1, i\rangle$ and $\langle 1, 2i\rangle$ both belong to R, but $i\neq 2i$.

5.9. (a) Every straight line parallel to the axis $0y$ meets R at at most one point.

(b) Every straight line parallel to the axis $0y$ meets R in (at most) one point.

(c) Conjunction of (a) and (b).

(d) For every set R in the plane there exists a subset $S\subset R$ such that the projection of S on to $0x$ is the same as the projection of R on to $0x$ and is such that every straight line parallel to the axis $0y$ meets S at (at most) one point.

5.10. The mapping f is an injection but it is not a surjection. $Rf=\mathscr{R}^+-\{0\}$.

5.11. The mapping f is a bijection.

5.12. The mapping f is neither an injection nor a surjection. $Rf=\mathscr{Z}$, $f(\frac{1}{2})=f(\frac{1}{3})=0$.

5.13. The mapping f is neither an injection nor a surjection. $Rf=\mathscr{R}-\{2\}$, $f(1)=f(-\frac{1}{2})=0$.

5.14. The mapping f is a bijection. Use the fact that the derivative of f exists and is always positive, and finally the Darboux property.

5.15. The mapping f is neither an injection nor a surjection.

$$Rf=\left\langle f\left(\frac{\ln\ln 2-\ln\ln 3}{\ln 3-\ln 2}\right), \infty\right\rangle.$$

5.16. The mapping f is neither an injection nor a surjection. $Rf=(-1,1), f(2)=f(\frac{1}{2})=\frac{4}{5}$.

5.17. The mapping f is not an injection but it is a surjection, $f(1)=f(0)=0$.

5.18. The mapping f is not an injection but it is a surjection, $f(1)=f(-1)=0$.

Note that f is continuous and has 3 roots,

$$\lim_{x\to\infty} f(x)=+\infty, \qquad \lim_{x\to-\infty} f(x)=-\infty,$$

$$\lim_{x\to 0^+} f(x)=0=\lim_{x\to 0} f(x)$$

and use the Darboux property.

5.19. The mapping f is not a bijection.

5.20. The mapping f is neither an injection nor a surjection, $f(1)=f(-1)=1$, $Rf=\mathscr{R}^+$.

5.21. The mapping f is neither an injection nor a surjection, $f(1)=f(-1)=0$, $Rf=\langle -\frac{9}{4}, \infty)$.

5.22. The mapping f is neither an injection nor a surjection,

$$Rf=\left\langle -\frac{1}{\ln 2}\cdot 2^{-\frac{1}{\ln 2}-1}-1, +\infty\right).$$

5.23. The mapping f is neither an injection nor a surjection, $f(\pi)=f(0)=0$, $Rf=\langle -1, 1\rangle$.

5.24. The mapping f is a bijection. Though $f(x)=\dfrac{x}{x+1}$ is not continuous at the point $x=-1$, we have defined for this point the value

$f(-1)=1$. On the other hand, $\lim_{x\to+\infty} f(x)=\lim_{x\to-\infty} f(x)=1$. The graph of f consists of two parts: for $x<-1$ and for $x>-1$. In the first part f is monotone, and that part of the graph is below the line $y=1$; in the second part f is again monotone and that part of the graph is above the line $y=1$.

5.25. First note that $f\circ g\,(x)=f[g(x)]$. The equality $f\circ g=f$ means that, for all x, $f\circ g(x)=f(x)$ i.e. $f[g(x)]=f(x)$. But f is an injection and so $g(x)=x$. Thus $g=I_X$. The assumption that f is an injection cannot be omitted; Let $g:\mathscr{R}\to\mathscr{R}$ and $f:\mathscr{R}\to\mathscr{R}$ be mappings defined as follows: $g(x)=|x|$, $f(x)=x^2$. Since $|x|^2=x^2$, we have $f\circ g=f$, but–obviously–g is not an identity.

5.26. As before, note that $g\circ f(x)=g[f(x)]$. Thus the equality $g\circ f=f$ reduces to $g[f(x)]=f(x)$. Since f is a surjection, each element $t\in X$ is of the form $f(x)$, and so, for each $x\in X$, $g(x)=x$. The example given in 5.25 may be used to show that the assumption that f is a surjection cannot be omitted.

5.27. The sentence which we want to prove is of the form

$$(p\wedge q)\Rightarrow(r\wedge s).$$

Thus it is enough to show that the statement $p\wedge q\wedge(\sim r\vee\sim s)$ leads to a contradiction. In other words, under our assumption it is impossible for f either not to be an injection or not to be a surjection.

Assume that f is not an injection. Then there are x_1 and x_2 such that $x_1\neq x_2$ but $f(x_1)=f(x)_2$. Let y be the common value of f in x_1 and x_2. By our assumption $g\circ f=IX$ i.e., for an arbitrary x, $g[f(x)]=x$. Thus $g[f(x_1)]=x_1$ and $g[f(x_2)]=x_2$, and so $g(y)=x_1$ and $g(y)=x_2$. Since g is a mapping, $x_1=x_2$, contradicting our assumption.

Assume now that g is not a surjection, i.e. $g*Y\to X$. As $g:Y\to X$; thus $g*Y\subset X$. We conclude that $X-g*Y\neq O$. Let $x\in X-g*Y$. By the assumption $g\circ f(x)=x$, i.e. $g[f(x)]=x$, thus x belongs to the range of g (because it is the value of g at the point $f(x)$), which is a contradiction.

5.28. (a) We need to show equivalence, i.e. two implications. We first prove that if the left converse exists, then f is an injection. Indeed, otherwise there would be x_1 and x_2 such that $x_1\neq x_2$ but $f(x_1)=f(x_2)$. However, then $g\circ f(x_1)=g[f(x_1)]=g[f(x_2)]=g\circ f(x_2)$. By the assumption $g \circ f = I_X$ we have $x_1 = x_2$, contradicting the choice of x_1 and x_2.

If f is an injection and X is nonempty, let x_0 be a fixed element of X. We define g as follows: If $y \notin f{*}X$ then $g(y) = x_0$. If $y \in f{*}X$, i.e. $y = f(x)$ for a certain x, then we get $g(y) = x$. Since f is one-to-one; the above definition is sound.

(b) Hint for the construction of the right converse in the case of f being a surjection:

Given $y \in Y$ let $U_y = \{x : f(x) = y\}$ (i.e. $U_y = f^{-1}{*}\{y\}$). Let g be a map with the domain Y, such that $g(y) \in U_y$. Check that g is the required right converse.

Note: The above construction requires so-called *axiom of choice* (c.f. Chapter 10, Cardinal and ordinal arithmetic).

5.29. First note that for an arbitrary mapping $f: X \to Y$ we have $f \circ I_X = I_Y \circ f = f$. Further, the composition of mappings is associative, i.e. for arbitrary f, g, h we have $f \circ (g \circ \mathrm{h}) = (f \circ g) \circ h$. Apply this to the composition $g_1 \circ (f \circ g_2)$. Since $f \circ g_2 = I_Y$ we have $g_1 \circ (f \circ g_2) = g_1 \circ I_Y = g_1$. On the other hand, $g_1 \circ (f \circ g_2) = (g_1 \circ f) \circ g_2 = I_X \circ g_2 = g_2$. Thus $g_1 = g_2$.

5.30. Assume $g \circ f(x_1) = g \circ f(x_2)$, i.e. $g[f(x_1)] = g[f(x_2)]$. Since g is an injection, we have $f(x_1) = f(x_2)$, and so $x_1 = x_2$ holds.

5.31. Assume that g is not an injection. Then there exist y_1 and y_2 such that $y_1 \neq y_2$ but $g(y_1) = g(y_2)$. Since f is a surjection, there are x_1 and x_2 such that $f(x_1) = y_1$ and $f(x_2) = y_2$. Clearly $x_1 \neq x_2$ (since f is a function). Consider now the values of $g \circ f$ in x_1 and in x_2. $g \circ f(x_1) = g[f(x_1)] = g(y_1) = g(y_2) = g[f(x_2)] = g \circ f(x_2)$. The above equalities imply $g \circ f(x_1) = g \circ f(x_2)$, which, since $g \circ f$ is an injection, gives $x_1 = x_2$, a contradiction.

Example: $f: \mathscr{R}^+ \to \mathscr{R}$, $g: \mathscr{R} \to \mathscr{R}$ are mappings defined as follows: $f(x) = x$, $g(x) = x^2$, $g \circ f$ is an injection, but g is not an injection.

5.32. Assume, on the contrary, that f is not a surjection. Then $f{*}X \neq Y$, i.e. $Y - f{*}X \neq 0$. Consider $y_0 \in Y - f{*}X$ and the value of g in y_0, say z_0. Since g is an injection, for every $y \neq y_0$ we have $g(y) \neq z_0$. This holds, in particular, for every $y \in f{*}X$, and so z_0 is not a value of $g \circ f$, contradicting the assumption that $g \circ f$ is a surjection.

5.33. *Example*: $X = \mathscr{N}$. Since every natural number n can be uniquely represented in the form $2^l(2m+1) - 1$, we can map n to the expotent l

such that $n=2^l(2m+1)-1$. This mapping is obviously a surjection but it is not an injection (for instance $f(5)=f(9)=1$). $Rf=\mathcal{N}$ since $f(2^l-1)=l$. An example of this sort cannot be found among finite sets. The proof follows by induction on the cardinality of the set X.

5.34. *Example*: $X=\mathcal{N}$, $f:\mathcal{N}\to\mathcal{N}$ is defined as $f(n)=2n$. As in Problem 5.33 an example cannot be found among finite sets.

5.35. *Hint*: Use the solutions of 5.17 and 5.18.

5.36. Assume that f is a surjection. Let $y\in Y$. Then there exists an x such that $f(x)=y$. By definition this element belongs to the set $f^{-1}*\{y\}$. Similarly, each element x of $f^{-1}*\{y\}$ has the property $f(x)=y$. Thus, if $f^{-1}*\{y\}$ is nonempty, then y is a value of f.

5.37. *Hint*: Use the definition.

5.38. $f*\{a, b\}=a$. The result shows that it is necessary to distinguish between $f(X)$ and $f*X$.

5.39. f is an injection but is not a surjection, for instance $3\notin f*\mathcal{N}$.

5.40. It is enough to snow that f is an injection and a surjection. The mapping f is continuous (in the sense of the calculus) and its derivative f is positive. Thus f, being continuous and increasing, is an injection. Since $\lim_{x\to-\infty} f(x)=0$, $\lim_{x\to+\infty} f(x)=+\infty$ and $f(x)>0$, f is also a surjection. The inverse mapping is $g:\mathcal{R}^+-\{0\}\to\mathcal{R}$ defined by $g(x)=\log_2 x$.

5.41. f does not possess an inverse mapping since it is not a surjection. Compare with Problem 5.36.

5.42. (a) Yes.

(b) Yes.

(c) $\mathcal{R}_0[t]$.

(d) $\mathcal{R}_2[t]$. Generally note that for every $n\in\mathcal{N}$ we have $\varphi*\mathcal{R}_n[t]=\varphi^{-1}*\mathcal{R}_n[t]=\mathcal{R}_n[t]$.

5.43. (a) Yes. Indeed, if $f\in C_\infty\langle 0, 1\rangle$, then in particular f is continuous in $\langle 0, 1\rangle$. Thus if we define an auxiliary function g by $g(x)=\int_0^x f(t)\,dt$, then $f=\varphi[g(x)+x]$.

(b) No. $\varphi(x)=\varphi(x+1)=0$.

(c) $\{-1\}$, the set consisting of one constant function.

(d) $\mathcal{R}[t]$.

(e) It is the set of polynomials of the form $g(t)+k$ where $g\in\mathcal{Q}[t]$ and $k\in\mathcal{R}$.

(f) It is the set of functions of the form $-\cos x+x+k$ or $\sin x+x+k$, where $k\in\mathscr{R}$.

(g) $\{e^x+k : k\in\mathscr{R}\}$.

5.44. (a) $\langle 0, 2\rangle$. Use the Darboux property and the continuity and the differentiability of f.

(b) $\langle 6, 12\rangle$.

(c) O. Note that f possesses at the point $x=\frac{3}{2}a$ unique minimum equal to $-\frac{1}{4}$, f is continuous and differentiable,

$$\lim_{x\to-\infty} f(x)=\lim_{x\to+\infty} f(x)=+\infty,$$

and so $f*\mathscr{R}=\langle -\frac{1}{4}, +\infty)$.

(d) O.

e) $\{0\}$.

5.45. (a) $\langle 0, 2\rangle$.

(b) $\{1\}$.

(c) $\left\{\frac{3}{2}, \frac{\sqrt{2}+2}{2}, 2\right\}$.

(d) This is the union of segments of the form

$$(-\tfrac{1}{6}\pi+2k\pi, \tfrac{7}{6}\pi+2k\pi),$$

i.e.

$$f^{-1}*(\tfrac{1}{2},\infty)=\bigcup_{k\in\mathscr{Z}}(-\tfrac{1}{6}\pi+2k\pi, \tfrac{7}{6}\pi+2k\pi).$$

(e) $\bigcup_{k\in\mathscr{Z}}[(2k+1)\pi, (2k+2)\pi]$.

(f) $\{\frac{3}{2}\pi+k\pi : k\in\mathscr{Z}\}$.

5.46. (a) $\mathscr{R}_0^+$.

(b) $\mathscr{R}_2[t]$.

(c) $\{t+1, -t-1\}$.

(d) O.

(e) $\{t-1, 1-t, t^2-1, 1-t^2\}$.

5.47. (a) No. $f(t^2+2)=f(1)=1$.

(b) Yes. $z=f(\mathrm{Im}(z)t+\mathrm{Re}(z))$.

(c) It is the set of all polynomials divisible by the binomial t^2+1, i.e. the principal ideal generated by this polynomial.

(d) It is the set of all polynomials of the form $\varphi(t)(t^2+1)+k$, where $k\in\mathscr{R}$, i.e. $\{\varphi(t)\ (t^2+1)+k : \varphi(t)\in\mathscr{R}[t]\wedge k\in\mathscr{R}\}$.

(e) $\mathscr{R}[t]$.

5.48. (a) Yes, indeed $f(\langle\varphi, 1\rangle)=\varphi$.

(b) $\mathscr{R}_3[t]$.

(c) $\{0\}\times(\mathcal{N}-\{0\})$.

(d) $\{\langle 1/n(t^2+t+1), n\rangle : n\in\mathcal{N}-\{0\}\}$.

5.49. (a) Yes. If ε is a root of unity of order n, then there is ε_1 of order $n\cdot k$ such that $\varepsilon_1^k=\varepsilon$.

(b) No. Let ε_1, ε_2 be two different roots of unity of order k, then $\varepsilon_1^k=\varepsilon_2^k=1$.

(c) It is the set of roots of unity of order k.

(d) $\{-1, 1\}$.

(e) It is the set of roots of unity of order $4k$.

5.50. (a) Yes.

(b) No.

(c) $\mathscr{R}$.

5.51. (a) No. For instance the vector [0, 0, 1] is not a value of our function since it is not self-orthogonal.

(b) No. $f([0, 0, 1])=f([0, 0, 2])=[0, 0, 0]$.

(c) $\{[0, 0, k]\colon k\in\mathscr{R}\}$.

(d) $\{[s, 0, 0] : s\in\mathscr{R}\}$.

(e) O.

5.52. (a) No. $\varphi * X=\mathscr{R}^+$.

(b) No. For every vector s we have: $\varphi(\mathbf{s})=\varphi(-\mathbf{s})$.

(c) O.

(d) The vector $[0, \ldots, 0]$.

(e) It is the unit sphere in $\mathscr{R}^n$.

5.53. (a) Yes, as $z=\varphi(z-1)$.

(b) $\langle -1, 0)$.

(c) $(-\infty, -1)$.

(d) $\mathcal{N}-\{0\}$.

(e) $\{2, 3\}$.

5.54. (a) No; $0\notin\varphi*\mathcal{N}^2$.

(b) No; $\varphi(\langle 0, 1\rangle)=\varphi(\langle 1, 0\rangle)=2$.

(c) $\mathcal{N}-\{0, 1\}$.

(d) O.

(e) $[\mathrm{Par}\times(\mathcal{N}-\mathrm{Par})]\cup[(\mathcal{N}-\mathrm{Par})\times\mathrm{Par}]$.

5.55. (a) Yes; $k=\varphi(\langle 1, k\rangle)$.

(b) Par.

(c) $(\mathcal{N}\times\{0\})\cup(\{0\}\times\mathcal{N})$.

(d) $(\mathrm{Par}\times\mathcal{N})\cup(\mathcal{N}\times\mathrm{Par})$.

(e) Let T be our set i.e. $T=\{2^n : n\in\mathcal{N}\}$. Then $\varphi^{-1}*T=T\times T$.

5.56. (a) Yes; $k=\varphi(\langle 1, k\rangle)$.

(b) No;$\varphi(\langle a, b\rangle)=\varphi(\langle -a, b\rangle)=a^2b$.

(c) $(\{0\}\times\mathcal{Z})\cup(\mathcal{Z}\times\{0\})$.

(d) $\{\langle 1,1\rangle,\langle -1,1\rangle\}$,

(e) $\mathcal{Z}\times\mathcal{N}$

(f) It is the set of those natural numbers which are squares, i.e. $\{n : \bigvee_k n=k^2\}$.

5.57. (a) Yes; $\langle k, l\rangle=\varphi((k+l)/2,\ (k-l)/2)$.

(b) Yes. If $\varphi(\langle x, y\rangle)=\varphi(\langle x_1, y_1\rangle)$, then $x+y=x_1+y_1$ and $x-y=x_1-y_1$, and so $x=x_1$ and $y=y_1$, i.e. $\langle x, y\rangle=\langle x_1, y_1\rangle$.

(c) It is the straight line $y=-1$.

(d) $\{\langle 0, 0\rangle\}$.

5.58. (a) No, $3\notin f*\mathcal{N}^2$.

(b) No, $f(\langle 7, 1\rangle)=f(\langle 1, 7\rangle)=50$.

(c) $\{\langle 0, 0\rangle\}$.

(d) O.

5.59. (a) No, $2\notin f*\mathcal{N}^2$.

(b) No, $f(\langle 7, 5\rangle)=f(\langle 5, 1\rangle)=24$.

(c) $\{\langle x,x\rangle : x\in\mathcal{N}\}$, i.e. $I_{\mathcal{N}}$.

(d) $\mathcal{N}$ – Par.

(e) $\{k : \bigvee_l k=l^2\}$.

5.60. (a) Yes, $x=f(\langle x, x\rangle)$.

(b) No, $f(\langle 1, 2\rangle)=f(\langle 2, 2\rangle)=2$.

(c) $\{\langle 0, 0\rangle\}$.

(d) $(\{k\}\times\{1,2,\dots,k\})\cup(\{1,2,\dots,k\}\times\{k\})$.

(e) $\mathcal{N}$.

5.61. (a) Yes.

(b) Yes.

(c) $\mathcal{R}$.

(d) $\{x-i : x\in\mathcal{R}^+\}$.

(e) X.

5.62. (a) $\langle -2\tfrac{1}{4}, +\infty)$.

(b) $\langle -2, +\infty)$.

(c) $(-2, 1)$.

(d) $\left\langle -2, \frac{-1-\sqrt{5}}{2}\right\rangle \cup \left\langle \frac{-1+\sqrt{5}}{2}, 1\right\rangle$.

(e) $\{-2, 1\}$.

5.63. (a) $\langle -2\frac{1}{4}, +\infty)$. Indeed, our function is continuous. Its graph is a parabola. Moreover, our function is differentiable on the whole rea line and its derivative has just one root, namely $x=2\frac{1}{2}$. This is clearly the global minimum of our function since it has no other minima and tends to infinity in both infinities. The value of our function at $2\frac{1}{2}$ is $-2\frac{1}{4}$. Thus $Rf=\langle -2\frac{1}{4}, +\infty)$.

(b) $(4, +\infty)$.

(c) $(-\infty, 1) \cup (4, +\infty)$.

(d) $\left\langle \frac{5-\sqrt{24}}{2}, \frac{5-\sqrt{17}}{2}\right\rangle \cup \left\langle \frac{5+\sqrt{17}}{2}, \frac{5+\sqrt{24}}{2}\right\rangle$.

(e) $\{1, 4\}$.

5.64. (a) $\langle 1, +\infty)$.

(b) $\{0\}$.

(c) $\langle 1, 2\rangle$. Indeed, $f(0)=1$, $f(1)=2$. In the entire open segment $(0, 1)$ our function is differentiable and its derivative $2^x \ln 2$ is positive; thus f is increasing and so $f*\langle 0,1\rangle=\langle 1, 2\rangle$.

(d) $\langle -\log_2 3, -1\rangle \cup \langle 1, \log_2 3\rangle$.

Hint: Note that f is even and so its graph is symmetric w.r.t. $0y$; thus the counteirmage of any set possesses a symmetry center at the origin. Finally apply the reasoning of (c).

(e) $\{-2, 2\}$.

5.65. (a) Yes. For every $x \in \langle 0, 1)$, $x=f(x)$.

(b) $\mathscr{Z}$. Indeed, $\mathscr{Z}$ consist of points x such that $E[x]=x$.

(c) $\mathscr{R}$.

(d) $\mathscr{R}-\mathscr{Z}$.

5.66. (a) Yes. Our function is continuous in $\mathscr{R}$, $\lim_{x\to-\infty} f(x)=-\infty$, $\lim_{x\to\infty} f(x)=+\infty$.

(b) Note that $f(x)=(x+1)(x-1)^2$, and $f'(x)=3x^2-2x-1$ $=(3x+1)(x-1)$. It follows that, for $x \in \langle 3, \infty)$, $f'(x)>0$ holds. Since $\lim_{x\to\infty} f(x)=+\infty$, we have $f*\langle 3, \infty)=\langle f(3), +\infty)=\langle 16, \infty)$.

(c) $\mathscr{R}$.

(d) $\{-1, 1\}$.

(e) $\{0, 3\}$.

5.67. (a) No; for instance $3 \notin f*\mathcal{Q}$, otherwise $\sqrt{2}$ is rational.

(b) No; for instance $\frac{3}{2} \notin f*\mathcal{Q}$.

(c) $\{-\frac{2}{3}, \frac{2}{3}\}$.

(d) $\{0\}$.

(e) $\langle\frac{1}{3}, 1\rangle \cap \mathcal{Q}$.

5.68. (a) Yes. Note that $f(3)=0$, $\lim_{x\to+\infty} f(x)=+\infty$ and that in the whole half-line $(3, +\infty)$ f is differentiable.

(b) $\{2, 3\}$.

(c) Since $\langle 3, \infty)\subset\langle 2, \infty)$, and so also $f*\langle 3, \infty)\subset f*\langle 2, \infty)$, but $f*\langle 3, \infty)=\mathcal{R}^+$, we have $f*\langle 2, \infty)=\mathcal{R}^+$.

(d) $\left(\dfrac{5-\sqrt{5}}{2}, \dfrac{5+\sqrt{5}}{2}\right)-\{2,3\}$.

(e) $\{0, 2, 6\}$.

5.69. (a) No. For instance $1 \notin Rf$.

(b) $\{3, 4\}$.

(c) $\left\langle\dfrac{7-\sqrt{5}}{2}, \dfrac{7+\sqrt{5}}{2}\right\rangle$.

(d) $\{-6, -2, 0\}$.

(e) $(-\infty, \frac{1}{4}\rangle$.

5.70. To show this it is necessary to prove that 1° $(f*A)\cup(f*B)\subset f*(A\cup B)$ and 2° $f*(A\cup B)\subset(f*A)\cup(f*B)$.

1° If $y\in(f*A)\cup(f*B)$, then $y\in f*A$ or $y\in f*B$. If the former is true, then there exists an $x\in A$ such that $y=f(x)$. Since $x\in A$, we have $x\in A\cup B$ and $y=f(x)\in f*(A\cup B)$. In the latter case we proceed similarly.

2° If $y\in f*(A\cup B)$, then there exists an $x\in A\cup B$ such that $y=f(x)$. By definition $x\in A$ or $x\in B$, i.e. $y\in f*A$ or $y\in f*B$.

5.71. If $y\in f*A\cap B$, then there exists an $x\in A\cap B$ such that $y=f(x)$. By definition $x\in A$ and $x\in B$. Thus $f(x)\in f*A$ and $f(x)\in f*B$, and so $f(x)\in(f*A)\cap(f*B)$; thus

$$y\in(f*A)\cap(f*B).$$

Example: Let $f: \mathcal{R}\to\mathcal{R}$ be a mapping defined by $f(x)=x^2$. Let $A=\mathcal{R}-\mathcal{R}^+$, $B=\mathcal{R}^+$. $A\cap B=O$, and so $f*(A\cap B)=O$. But

$f*A=\mathcal{R}^+-\{0\}, f*B=\mathcal{R}^+$, and thus

$$(f*A)\cap(f*B)=\mathcal{R}^+-\{0\}.$$

5.72. If $y\in(f*A)-(f*B)$, then $y\in f*A$ and $y\notin f*B$ and so there exists an $x\in A$ such that $f(x)=y$. Clearly $x\notin B$ (since then $f(x)\in f*B$); thus $x\in A-B$ and $f(x)\in f*(A-B)$.

Example: The same mapping as in the example of 5.71. Take $A=\mathcal{R}$, $B=\mathcal{R}^+$.

5.73. *Hint*: Use 5.70 and the following: $u\subset v\Leftrightarrow u\cup v=v$.

5.74. Let $x\in A$. Then $f(x)\in f*A$ and so $x\in f^{-1}*(f*A)$.

Example: As in 5.71. Take $A=\mathcal{R}^+$.

5.75. *Hint*: Use the reasoning of 5.71.

5.76. We need to show: 1° $(f^{-1}*A)\cap(f^{-1}*B)\subset f^{-1}*(A\cap B)$. 2° $f^{-1}*(A\cap B)\subset(f^{-1}*A)\cap(f^{-1}*B)$.

1° If $x\in(f^{-1}*A)\cap(f^{-1}*B)$, then $x\in f^{-1}*A$ and $x\in f^{-1}*B$, i.e. $f(x)\in A$ and $f(x)\in B$, and so $f(x)\in(A\cap B)$, and finally $x\in f^{-1}*$ $*(A\cap B)$.

2° If $x\in f^{-1}*(A\cap B)$, then $f(x)\in A\cap B$, i.e. $f(x)\in A$ and $f(x)$ $\in B$, and so $x\in f^{-1}*A$ and $x\in f^{-1}*B$; thus $x\in(f^{-1}*A)\cap(f^{-1}*B)$.

5.77. *Hint*: Reason as in 5.76.

5.78. Assume that $A\subset B$, $A\cup B=B$ and $f^{-1}*(A\cup B)=f^{-1}*B$. According to 5.75 we have $f^{-1}*(A\cup B)=(f^{-1}*A)\cup(f^{-1}*B)$, and so $(f^{-1}*A)\cup(f^{-1}*B)=f^{-1}*B$, i.e. $f^{-1}*A\subset f^{-1}*B$.

5.79. Two inclusions have to be shown: 1° $f*(f^{-1}*A)\subset A\cap(f*X)$ and 2° $A\cap(f*X)\subset f*(f^{-1}*A)$.

1° $y\in f*(f^{-1}*A)$ implies $y\in f*X$ (since $f^{-1}*A\subset X$) and $y\in A$, since $y\in f*(f^{-1}*A)$, means that there exists an $x\in f^{-1}*A$ such that $y=f(x)$ and so $y\in A$.

2° $y\in A\cap(f*X)$ means that $y\in A$ and that y is a value of f. Thus there exists an $x\in X$ such that $y=f(x)$. But then $x\in f^{-1}*A$ and so $y\in f^{-1}*A$.

5.80. *Hint*: Use the arguments of 5.70 and 5.71.

5.81. *Hint*: Use the arguments of 5.75 and 5.76.

5.82. *Hint*: You have to show that if f is not an injection, then – in each of the four possible cases – one can construct a suitable counterexample. On the other hand, if f is an injection, then each of the missing inclusions holds.

5.83. It is enough to show that 1°: If f is a surjection then the counterimage of a nonempty subset is nonempty, and 2°: If f is not a surjection then there exists a nonempty set whose counterimage is empty.

1° Let $f\colon X \to Y$ be a surjection. $A \subset Y$, $A \neq O$. Then there exists y such that $y \in A$. Since f is a surjection, there is an x such that $f(x)=y$. But then this x belongs to $f^{-1}*A$.

2° Assume f is not a surjection. Then $f*X \neq Y$, and so $Y-f*X \neq O$. The set $Y-f*X$ is nonempty but its counterimage is empty.

Note: Our reasoning corresponds to the following tautology: $(p \Leftrightarrow q) \Leftrightarrow [(p \Rightarrow q) \wedge (\sim p \Rightarrow \sim q)]$.

5.84. *Hint*: Use 5.79 and 5.83.

5.85. (a) We need to show: 1° $f*[A \cap (f^{-1}*B)] \subset (f*A) \cap B$ and 2° $(f*A) \cap B \subset f*(A \cap f^{-1}*B)$.

1° $f*[A \cap (f^{-1}*B)] \subset (f*A) \cap f*(f^{-1}*B) = (f*A) \cap (B \cap f*X) = (f*A) \cap (f*X) \cap B = (f*A) \cap B$. Indeed, $A \subset X \Rightarrow f*A \subset f*X$ and so $(f*A) \cap (f*X) = f*A$.

2° If $y \in (f*A) \cap B$, then $y \in f*A$ and $y \in B$. So there exists an $x \in A$ such that $f(x)=y$. Since $f(x)=B$, we have $x \in f^{-1}*B$, and so $x \in (A \cap f^{-1}*B)$, and $y \in f*(A \cap f^{-1}*B)$.

(b) We have $A \subset f^{-1}*(f*A)$, and so $A \cap f^{-1}*B \subset f^{-1}*(f*A) \cap f^{-1}*B$, i.e. $A \cap f^{-1}*B \subset f^{-1}*[(f*A) \cap B]$.

5.86. *Hint*: The very fact that f possesses an inverse implies that f is a bijection.

5.88: *Note*: We use this to show $\aleph_0 \cdot \aleph_0 = \aleph_0$. A function with the property as in our problem is called a *pairing function*.

5.89. $(g \circ f)(x) = E[x+\frac{2}{3}]$, $(g \circ f)*A = \{0,1\}$, $(g \circ f)*\mathcal{R} = \mathcal{N}$.

5.90. $(g \circ f)(\langle x,y \rangle) = \sqrt{x^2+y^2+1}$, $(g \circ f)*A = (1, 1+\sqrt{2})$, $(g \circ f)*B = \langle 1, 1+\sqrt{2} \rangle$.

5.91. $(g \circ f)(n) = n^2-n$, $(g \circ f)*A = \{2, 12, 30, 56, 90\}$, $(g \circ f)*B = \{0\}$.

5.92. $(g \circ f)(n) = e^{n^2}$, $(g \circ f)*A = \{e^{n^2} : n \in \mathcal{N}\}$, $(g \circ f)*B = \{e^{4k^2} : k \in \mathcal{N}\}$.

CHAPTER 6

GENERALIZED SET-THEORETICAL OPERATIONS

6.2. We show that $\bigcup_{t \in \mathcal{N}} A_t = \mathcal{R}^+$. For every nonnegative real number x, there exists a natural number n_x such that $n_x \leqslant x < n_x + 1$. As n_x take $E[x]$. Thus, for an arbitrary $x \in \mathcal{R}^+$ there exists an n such that $x \in A_n$, i.e. $x \in \mathcal{R}^+ \Rightarrow x \in \bigcup_{n \in \mathcal{N}} A_n$.

To see the converse implication note that if $x \in \bigcup_{t \in \mathcal{N}} A_t$, then for a certain t_0 we have $x \in A_{t_0}$. Since for an arbitrary t, $A_t \subset \mathcal{R}^+$, the proof is complete.

Similarly one can show that $\bigcup_{t \in \mathcal{R}^+} A_t = \mathcal{R}^+$.

We now show that $\bigcap_{t \in \mathcal{N}} A_t = O$. Indeed, assume $x \in \bigcap_{t \in \mathcal{N}} A_t$ and consider n_x as above. Since $x < (n_x + 1)$, we have $x \notin A_{n_x+1}$, and thus x is not in some of the sets A_t, which completes the proof.

By the above $\bigcap_{t \in \mathcal{R}^+} A_t = O$.

6.3. $\bigcup_{t \in \mathcal{N}} A_t = \mathcal{R}$, $\quad \bigcap_{t \in \mathcal{N}} A_t = \{0\}$,

$\bigcup_{t \in \mathcal{R}^+} A_t = \mathcal{R}$, $\quad \bigcap_{t \in \mathcal{R}^+} A_t = \{0\}$,

6.4. $\bigcup_{t \in \mathcal{N}} A_t = \langle 0, +\infty)$, $\quad \bigcap_{t \in \mathcal{N}} A_t = O$,

$\bigcup_{t \in \mathcal{R}^+} A_t = \langle 0, +\infty)$, $\quad \bigcap_{t \in \mathcal{R}^+} A_t = O$.

6.5. $\bigcup_{t \in \mathcal{N}} A_t = \langle 0, +1 \rangle$, $\quad \bigcap_{t \in \mathcal{N}} A_t = \{0\}$,

$\bigcup_{t \in \mathcal{R}^+} A_t = \langle 0, +1)$, $\quad \bigcap_{t \in \mathcal{R}^+} A_t = \{0\}$,

6.6. $\bigcup_{t \in \mathcal{N}} A_t = (-1, 1\rangle$, $\quad \bigcap_{t \in \mathcal{N}} A_t = \{0\}$,

$\bigcup_{t \in \mathcal{R}^+} A_t = (-1, 1\rangle$, $\quad \bigcap_{t \in \mathcal{R}^+} A_t = \{0\}$.

6.7. $\bigcup_{t \in \mathcal{N}} A_t = \{0\} \cup \langle 1, +\infty)$, $\quad \bigcap_{t \in \mathcal{N}} A_t = O$,

$\bigcup_{t \in \mathcal{R}^+} A_t = \mathcal{R}^+$, $\quad \bigcap_{t \in \mathcal{R}^+} A_t = O$.

6.8. We show that $\bigcup_{t\in\mathcal{N}} A_t = \langle 0, 1)$. Let x be a fixed element of $\bigcup_{t\in\mathcal{N}} A_t$. By the definition of the generalized union there exists a t_0 such that $\frac{t_0}{t_0+1} \leqslant x < \frac{t_0+1}{t_0+2}$, and since $t_0 \geqslant 0$, we immediately get

$$0 \leqslant \frac{t_0}{t_0+1} \leqslant x < \frac{t_0+1}{t_0+2} < 1,$$

thus $x \in \langle 0, 1)$.

Now let x be a fixed element of the segment $\langle 0, 1)$. Consider the set $\left\{n \in \mathcal{N} : n > \frac{1-x}{x} - 1\right\}$. Obviously it is nonempty and therefore it has a least element. Call it t_0. Then $t_0 > \frac{x}{1-x} - 1 \geqslant t_0 - 1$, and so $t_0 + 1 > \frac{x}{1-x}$. Further, $(t_0+1)(1-x) > x$ $(1-x > 0)$. Finally, $t_0 + 1 > (t_0+2) \cdot x$ and since $t_0 + 2 > 0$ we get $x < \frac{t_0+1}{t_0+2}$.

On the other hand, $\frac{x}{1-x} \geqslant t_0$, and so $x \geqslant \frac{t_0}{t_0+1}$ $(t_0+1>0)$. Thus $x \in A_{t_0}$, which completes the proof. $\bigcup_{t\in\mathcal{R}^+} A_t = \langle 0, 1)$, whereas $\bigcap_{t\in\mathcal{R}^+} A_t = \bigcap_{t\in\mathcal{N}} A_t = O$.

6.9. $\bigcup_{t\in\mathcal{N}} A_t = (0, 3)$, $\quad \bigcap_{t\in\mathcal{N}} A_t = \langle 1, 2 \rangle$,

$\bigcup_{t\in\mathcal{R}^+} A_t = (0, 3)$, $\quad \bigcap_{t\in\mathcal{R}^+} A_t = \langle 1, 2 \rangle$.

6.10. $\bigcup_{t\in\mathcal{N}} A_t = (-1, +1)$, $\quad \bigcap_{t\in\mathcal{N}} A_t = \{0\}$,

$\bigcup_{t\in\mathcal{R}^+} A_t = (-1, +1)$, $\quad \bigcap_{t\in\mathcal{R}^+} A_t = \{0\}$.

6.11. $\bigcup_{t\in\mathcal{N}} A_t = (1, +\infty)$. $\quad \bigcap_{t\in\mathcal{N}} A_t = O$,

$\bigcup_{t\in\mathcal{R}^+} A_t = (1, +\infty)$, $\quad \bigcap_{t\in\mathcal{R}^+} A_t = O$.

Hint: Investigate the graph of the function $t^{1/t}$.

6.12. It is easy to show that $\bigcup_{t\in\mathcal{N}} A_t=(-11, +\infty)$. We prove that $\bigcap_{t\in\mathcal{N}} A_t=\langle-10, -3)$. Indeed, consider the function $f(t)=2t^2-6t+1$. Since it is a quadratic polynomial with a positive coefficient with t^2, our function possesses a global minimum. By well-known facts, that minimum is reached at $t=\frac{3}{2}$. Thus the least value for the integer t can be reached either for $t=1$ or for $t=2$. But $f(1)=f(2)=-3$. Thus, for every t, $f(t)\geqslant -3$.

Now let $x\in\bigcap_{t\in\mathcal{N}} A_t$. Then $x\in A_t$ for every t, in particular $x<-3$. On the other hand, if $x<-10$ then, considering the least t such that $t>\left|\frac{1}{x+10}\right|-1$ we get consecutively

$$t_0>-\frac{1}{x+10}-1 \quad (t_0+1)(x+10)<-1,$$

$$x+10<-\frac{1}{t_0+1}, \quad x<-10-\frac{1}{t_0+1},$$

and so $x\notin A_{t_0}$. Thus $-10\leqslant x<-3$.

However, if x is such that $-10\leqslant x<-3$, then, for an arbitrary natural t, $x>-10-\frac{1}{t+1}$, and so for an arbitrary $t\in\mathcal{N}$, $f(t)\geqslant-3$, and thus, for an arbitrary $t\in\mathcal{N}$, $x<2t^2-6t+1$, which completes the proof. On the other hand, $\bigcap_{t\in\mathcal{R}^+} A_t=\langle-10, -\frac{7}{2})$, which follows from the fact that the minimum value of $2t^2-6t+1$ is $-\frac{7}{2}$. On the other hand, $\bigcup_{t\in\mathcal{R}^+} A_t=(-11, +\infty)$, which is simple to prove.

6.13. $\bigcup_{t\in\mathcal{N}} A_t=\mathcal{R}^+-\{x: \bigvee_{n\in\mathcal{N}} x=n^2\}=\mathcal{R}^+-\{1, 4, 9, 16, 25, 36, 49, \ldots\}$, $\bigcup_{t\in\mathcal{R}^+} A_t=\mathcal{R}^+$, $\bigcap_{t\in\mathcal{N}} A_t=O=\bigcap_{t\in\mathcal{R}^+} A_t$.

6.14. $\bigcup_{t\in\mathcal{N}} A_t=\mathcal{R}=\bigcup_{t\in\mathcal{R}^+} A_t$, $\bigcap_{t\in\mathcal{N}} A_t=(-1, 2)$, $\bigcap_{t\in\mathcal{R}^+} A_t=\langle-1, \frac{15}{8})$.

6.15. $\bigcup_{t\in\mathcal{N}} A_t=\{x: \bigvee_{n\in\mathcal{Z}} (x=2n\pi+\frac{1}{2}\pi \vee x=n\pi)\}$, $\bigcup_{t\in\mathcal{R}^+} A_t=\{x: \sin x>0\}$, $\bigcap_{t\in\mathcal{N}} A_t=\bigcap_{t\in\mathcal{R}^+} A_t=O$.

6.16. $\bigcap_{t \in \mathscr{R}} A_t = O$, $\bigcup_{t \in \mathscr{R}^+} A_t = \{x : \bigvee_{n \in \mathscr{Z}} -\frac{1}{2}\pi + n\pi < x < \frac{1}{2}\pi + n\pi\}$.

6.17. $\bigcup_{t \in \mathscr{R}} A_t = \mathscr{R}^2$, $\bigcap_{t \in \mathscr{R}} A_t = \{\langle 0, 0 \rangle\}$.

6.18. $\bigcup_{t \in \mathscr{R}} A_t = \mathscr{R}^2$, $\bigcap_{t \in \mathscr{R}} A_t = O$.

6.19. $\bigcup_{t \in \mathscr{R}} A_t = \mathscr{R} \times \mathscr{R}^+$, $\bigcap_{t \in \mathscr{R}} A_t = \{\langle 0, 0 \rangle\}$.

6.20. $\bigcup_{t \in \mathscr{R}} A_t = \mathscr{R}^+ \times \mathscr{R}$, $\bigcap_{t \in \mathscr{R}} A_t = \{\langle 0, 0 \rangle\}$.

6.21. $\bigcup_{t \in \mathscr{R}} A_t = \mathscr{R}^2$, $\bigcap_{t \in \mathscr{R}} A_t = O$.

6.22. $\bigcup_{t \in \mathscr{R}} A_t = \mathscr{R}^2 - \{\langle 0, 0 \rangle\}$, $\bigcap_{t \in \mathscr{R}} A_t = O$.

6.23. $\bigcup_{t \in \mathscr{R}} A_t = \mathscr{R} \times \mathscr{R}^+$, $\bigcap_{t \in \mathscr{R}} A_t = \mathscr{R}^+$.

6.24. $\bigcup_{n,m} A_{n,m} = \mathscr{R}^+$, $\bigcap_{n,m} A_{n,m} = O$, $\bigcup_n \bigcap_m A_{n,m} = \bigcap_n \bigcup_m A_{n,m} = O$.

6.25. $\bigcup_{n,m} A_{n,m} = \mathscr{R}^+$, $\bigcap_{n,m} A_{n,m} = \bigcup_n \bigcap_m A_{n,m} = \bigcap_n \bigcup_m A_{n,m} = O$.

6.26. $\bigcup_n \bigcap_m A_{n,m} = \mathscr{R}^+$, $\bigcup_m \bigcap_n A_{n,m} = \bigcap_n \bigcup_m A_{n,m} = O$,

$\bigcap_m \bigcup_n A_{n,m} = \mathscr{R}^+ - \{k^2 : k \in \mathscr{N}\}$.

6.27. $\bigcup_n \bigcap_m A_{n,m} = \langle 1, \infty)$, $\bigcap_n \bigcup_m A_{n,m} = O$,

$\mathscr{R} - \bigcup_n \bigcap_m A_{n,m} = (-\infty, 1)$, $\mathscr{R} - \bigcap_n \bigcup_m A_{n,m} = \mathscr{R}$,

$\bigcap_m \bigcup_n A_{n,m} = \langle 1, \infty)$, $\bigcup_m \bigcap_n A_{n,m} = O$,

$\bigcap_m \bigcup_n -A_{n,m} = \mathscr{R}$, $\bigcup_m \bigcap_n -A_{n,m} = (-\infty, 1)$.

6.28. *Hint*: Use the tautology $\bigwedge_x \Phi(x) \Rightarrow \Phi(y)$.

6.29. *Hint*: Use the tautology $\Phi(y) \Rightarrow \bigvee_x \Phi(x)$.

6.42. No; consider $A_t = \{x \in \mathscr{R} : x \leqslant t\}$, $B_t = \{x \in \mathscr{R} : t < x\}$, $T = \mathscr{R}$.

6.43. No.

6.44. No.

6.45. No.

6.47. The inclusion holds but it cannot be replaced by equality.

6.57. $\mathfrak{P} A_t$ consists of one element, namely the sequence constantly equal to one.

6.58. $\mathfrak{P}A_t$ consists of all zero-one sequences.

6.59. $\mathfrak{P}A_t$ consists of all sequences such that, for all n, the nth term of the sequence is a natural number not bigger than n.

6.60. $\mathfrak{P}A_t$ is the set of all real functions with values in the segment $\langle 0, 1\rangle$.

6.61. $\mathfrak{P}A_t$ is the set of all real functions such that, for all x, $f(x) \leqslant x$.

6.62. $\mathfrak{P}A_t$ is the set of al functions defined for positive reals and such that, for all x, $|f(x)| \leqslant x$.

6.63. $\mathfrak{P}A_t$ consists of just one real function, the identity.

6.64. $\mathfrak{P}A_t$ consists of just one real function, namely that defined by $f(x) = -x$.

6.65. $\mathfrak{P}A_t$ consists of just one real function, namely that defined by $f(x) = x^2 + 1$.

6.67. *Note*: Use of the axiom of choice is necessary.

CHAPTER 7

CARDINAL NUMBERS

7.1. Consider the set $f=\{\langle 1,5\rangle, \langle 2,7\rangle\}$. As it is easy to observe, it is a function which is a bijection of A and B. Thus $A \sim B$.

7.2. *Hint*: Note that the two sets have 7 elements each and construct f as in 7.1.

7.4. *Hint*: Consider the following functions: Id_A, converse to f where f is a bijection of A and B, and $g \circ f$ where g is a bijection of B and C.

7.6. *Hint*: If f is a bijection of A_1 and B_1 and g is a bijection of of A_2 and B_2, then h defined by $h(\langle a, b\rangle)=\langle f(a), g(b)\rangle$ is a bijection of $A_1 \times A_2$ and $B_1 \times B_2$.

7.7. *Hint*: If f is a bijection of A_1 and B_1, and g is a bijection of A_2 and B_2, then (under the assumptions of our problem) $f \cup g$ is a bijection of $A_1 \cup A_2$ and $B_1 \cup B_2$.

7.8. *Hint*: If f is a bijection of A and B, then (under the assumptions of our problem) $f \cup \mathrm{Id}_c$ is a bijection of $A \cup C$ and $B \cup C$.

7.9. *Hint*: If f is a bijection of A and B, then $\tilde{f}$ defined by $\tilde{f}(A_1) = f * A_1$ is a bijection of $\mathscr{P}(A)$ and $\mathscr{P}(B)$.

7.10. We show that the cardinality of the set $A=\{x \in \mathscr{N} : 10|x\}$ is $\aleph_0$. To see this consider the function f defined by $f(x)=10 \cdot x$ for $x \in \mathscr{N}$. This is a bijection of $\mathscr{N}$ and A. Indeed, the domain of f is $\mathscr{N}$, the range of f is A and f is one-to-one, since $f(x_1)=f(x_2)$ means $10 \cdot x_1 = 10 \cdot x_2$, i.e. $x_1 = x_2$.

7.11. *Hint*: Consider the function $f(m)=\ln(m+1)$ and use the fact that log functions are monotone. The cardinality of our set is $\aleph_0$.

7.12. *Hint*: Note that our set consists of 0 and 1.

7.13. The cardinality of the set is $\aleph_0$.

7.16. Let f be a bijection of A and $\mathscr{N}$. Consider $f * B$. If B is empty then, obviously, B is countable. Assume that B is nonempty. Then $f * B$ is also nonempty. Inductively, we define a bijection g of the set $f * B$ and a certain initial segment of the set of natural numbers. This segment is either finite or the whole of $\mathscr{N}$. The composition $g * f$ injects B onto an initial segment of $\mathscr{N}$.

7.17. It is enough to show that $A \cup B$ is countable, since the other sets are its subsets (cf. 7.16). We can represent $A \cup B$ as $A \cup (B-A)$. $B-A$ is denumerable. Now, A is equipollent to a subset of the set of even numbers whereas $B-A$ is equipollent to a subset of the set of odd numbers. Finally, $A \cup B$ is equipollent to a subset of $\mathcal{N}$.

7.18. *Hint*: Using 7.14, represent the elements of the set $A \times B$ as pairs $\langle a_i, b_j \rangle$ where $a_i \in A$, $b_j \in B$, and then order those pairs into relation R as follows:

$$\langle a_i, b_j \rangle R \langle a_l, b_k \rangle \Leftrightarrow (i+j<l+k) \vee [(i+j=l+k) \wedge (i<l)].$$

The set $A \times B$ has cardinality $\aleph_0$ if A has cardinality $\aleph_0$ and B is nonempty, or conversely.

7.19. By the assumption there exists an $X_0 \subset X$ such that $\overline{\overline{X}}_0 = \aleph_0$. Let f be a bijection of X_0 and $\mathcal{N}$. Define $X_1 = f^{-1} * N$ par. It is easy to see that $X - X_1 = (X - X_0) \cup f^{-1} * \text{Par}$, $(X - X_0) \cap f^{-1} * \text{Par} = O$. We define for $x \in X - X_1$

$$g(x) = \begin{cases} x, & \text{if} \quad x \in X - X_0, \\ f^{-1}(f(x)/2), & \text{if} \quad x \in f^{-1} * \text{Par}. \end{cases}$$

It is clear that g is a bijection of $X - X_1$ and X.

7.21. Our set has cardinality $\aleph_0$ iff, for all n, $A_n \neq O$.

7.22. We define f in $\mathcal{Z}$ as follows

$$f(x) = \begin{cases} 2x, & \text{if} \quad x \in \mathcal{N}, \\ -2x-1. & \text{if} \quad x \in \mathcal{Z} - \mathcal{N}. \end{cases}$$

The reader will check that f is the required bijection.

7.23. *Hint*: Note that $\mathcal{Q}$ is equipollent to the subset of $\mathcal{Z} \times \mathcal{Z}$, and then use 7.18 and 7.22. The alternative way of solving the problem is to show that $\mathcal{Q}^+$ is denumerable and that $(\mathcal{Q} - \mathcal{Q}^+) \sim (\mathcal{Q}^+ - \{0\})$, and then to use 7.18.

7.26. *Hint*: Use 7.25 and 7.21. ([1])

7.27. *Hint*: Note that every non-degenerate segment of reals contains a rational.

7.28. *Hint*: Note that if f is continuous and has a (local) extremum at the point x_0, then there is a segment containing x_0 such that $f(x_0)$ is the global extremum of f inside that segment, and then use 7.27.

7.29. *Hint*: Note that the discontinuity points of a monotone function define on the axis $0x$ a family of parwise disjoint segments, and then use 7.27.

7.37. Our set has cardinality $\aleph_0$, if $A \neq O$.

7.38. *Hint*: Every such sequence is eventually constant and equal to zero. Then use 7.20.

7.40. Assume that f is a surjection of $\mathcal{N}$ onto A. Define the function h on the set A as follows:

$$h(a)=n \Leftrightarrow f(n)=a \wedge \bigwedge_m (m<n \rightarrow f(m) \neq a).$$

In this way we have defined an injection of A into $\mathcal{N}$. Using the reasoning of 7.16, we get a mapping of A onto an initial segment of $\mathcal{N}$. If g: $A \rightarrow \mathcal{N}$ and $g * A=\{0, \ldots, k-1\}$, then

$$f(n)=\begin{cases} g(n) & \text{if} \quad n<k, \\ g^{-1}(0) & \text{if} \quad n \geqslant k \end{cases}$$

maps $\mathcal{N}$ onto A. Finally, if $\bar{\bar{A}}=\aleph_0$ and g is a bijection of A and $\mathcal{N}$, then $f=g^{-1}$ is the required function.

7.43. *Hint*: Obtain the proof by using the following geometrical observation: If A and B are segments, then placing them as in the Fig. 44 we can map the point of A to the point of B which lies on the line passing through 0 and A. This is a bijection of A and B.

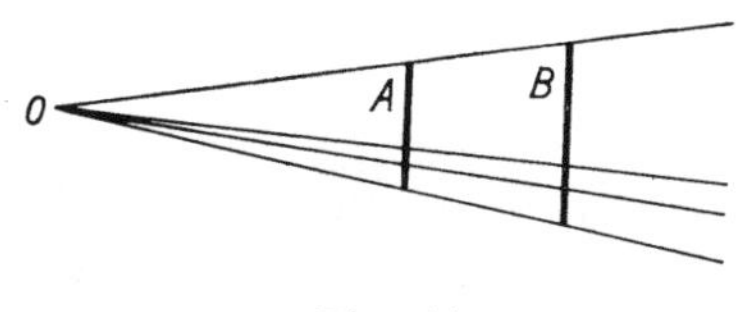

Fig. 44

7.45. Since every two open segments are equipollent (cf. 7.43), we can take as A the open segment $(0, 1)$. Similarly, we can take as B the half-open segment $(0, 1\rangle$. A bijection f of $(0, 1\rangle$ and $(0, 1)$ is defined as follows

$$f(x)=\begin{cases} \dfrac{1}{n+1} & \text{for } x=\dfrac{1}{n}, \text{ where } n \in \mathcal{N}-\{0\}, \\ x & \text{otherwise}. \end{cases}$$

This is the required bijection.

7.51. *Hint*: First show that the interior of the unit square has cardinality $\mathfrak{c}$.

To see this, represent its points as pairs of reals. The pair $\langle a, b\rangle$ is then mapped to $0, a_1 b_1 a_2 b_2 a_3 b_3 \dots a_n b_n \dots$, where $0, a_1 a_2 a_3 \dots$ is the infinite decimal fraction for a, and $0, b_1 b_2 b_3 \dots$ is such a fraction for b.

7.52. *Hint*: While showing that $A \times B$ has cardinality $\mathfrak{c}$, use the method of 7.51.

7.54. $A \times B$ has cardinality $\mathfrak{c}$ unless $B = O$; then $A \times B = O$.

7.57. Yes.

7.58. No; its cardinality is $\aleph_0$.

7.59.–7.67. Yes.

7.68. No; its cardinality is $\aleph_0$.

7.69–7.74. Yes.

7.75. *Hint*: Use induction and the reasoning of 7.51.

7.82. *Hint*: Use the axiom of choice.

7.83. *Note*: In the proof some version of the axiom of choice must be used. The set X satisfying the conditions of our problem is called *Dedekind-infinite*. Without some form of the axiom of choice one cannot prove that a set which is not Dedekind-infinite is finite.

7.85. $\overline{\overline{\mathfrak{P}_{n \in N} A_n}} = \mathfrak{c}$, iff the cardinality of the set $\{n \in \mathcal{N} : \bar{\bar{A}}_n \leqslant \aleph_0 \wedge \bar{\bar{A}}_n \geqslant 2\}$ is $\aleph_0$.

7.93. No.

7.95. Let f be a surjection of X onto $\mathscr{P}(X)$. Consider the set $Z = \{x \in X : x \notin f(x)\}$. Clearly $Z \subset X$ and so $Z \in \mathscr{P}(X)$. Since f is a surjection there must be an $x_0 \in X$ such that $Z = f(x_0)$. If $x_0 \in Z$ then $x_0 \notin f(x_0) = Z$, which is a contradiction. If $x_0 \notin Z$ then $x_0 \in f(x_0) = Z$, which is again a contradiction.

7.96. *Hint*: Use 7.95.

Note: The theorem in 7.96 is due to Cantor.

NOTE

[1] The set considered in this problem is called the *set of algebraic numbers.*

CHAPTER 8

ORDERINGS

8.1. (b) The number 1 is the least element and so the unique minimal one.

(c) If $x \in \mathcal{N}^+$ then $x|2x$.

(d) They are sets of the form: $\{k_1, k_1 \cdot k_2, k_1 \cdot k_2 \cdot k_3, \ldots\}$.

(e) They are subsets $X \subset \mathcal{N}^+$ such that, for every $x, y \in X$, $\mathrm{LCM}(x, y) \notin X$.

(f) The inclusion relation is an ordering in any family of sets. The minimal chains are the one-element sets. The following is the form of maximal chains: If such a set is ordered according to magnitude then consecutive number arises from the preceding one by multiplying the latter by a prime number.

8.2. The least element is O, the largest one is X.

8.3. (a) No. (b) No.

(c) The minimal elements are of the form $\{x\}$ for $x \in X$. The maximal elements are sets of the form $X - \{x\}$ for $x \in X$.

(d) The cardinality of each of those sets is equal to $\overline{\overline{X}}$. If $X = \{x, y\}$ then $U = \{\{x\}, \{y\}\}$. In this case the maximal and the minimal elements coincide.

8.4. Yes. If $a \in Y$ is maximal, minimal, etc. in $\langle X, R \rangle$ then it is maximal, minimal, etc. in $\langle Y, Y^2 \cap R \rangle$.

8.5. (a) To prove our theorem note that:

(1) The converse of a reflexive relation is reflexive.

(2) If $\langle x, y \rangle \in R^{-1}$ and $\langle y, x \rangle \in R^{-1}$ then $\langle y, x \rangle \in R$ and $\langle x, y \rangle \in R$, i.e. $x = y$.

(3) If $\langle x, y \rangle \in R^{-1}$ and $\langle y, z \rangle \in R^{-1}$ then $\langle z, y \rangle \in R$ and $\langle y, x \rangle \in R$; thus $\langle z, x \rangle \in R$ and $\langle x, z \rangle \in R^{-1}$, which shows the transitivity of R^{-1}.

(b) Yes. It is enough (in view of (a)) to show the connectedness of relation R. Now, $\bigwedge_{x, y} [\langle x, y \rangle \in R \vee \langle y, x \rangle \in R \vee x = y]$ is equivalent by the definition of R^{-1} to $\bigwedge_{x, y} [\langle y, x \rangle \in R^{-1} \vee \langle x, y \rangle \in R^{-1} \vee x = y]$, which means that R^{-1} is connected.

(c) No. It is enough to provide a counterexample. The set $\langle \mathcal{N}, \leqslant \rangle$ is a well-ordered set, but $\langle \mathcal{N}, \geqslant \rangle$ is not a well-ordered set.

(d) A minimal (the least) element of $\langle X, R \rangle$ becomes a maximal (the largest) element of $\langle X, R^{-1} \rangle$ and conversly.

8.6. (a) Assume that x_0 is the largest element of $\langle X, R \rangle$, i.e. $\bigwedge_x xRx_0$. If $x_0\ Rx$ holds for a certain x, then – by antisymmetry – $x_0 = x$, which was to be proved.

(b) This follows immediately from (a) and Problem 8.5 (d).

8.7. Yes. I_X is such a relation.

8.8. No. Here is the proof: According to our assumption $\overline{\overline{X}} \geqslant 2$ and so there are in X elements x and y such that $x \neq y$. Since R is connected xRy or yRx. We can assume xRy, and thus, as R is symmetric, also yRx. But R is antisymmetric, and so $x = y$, a contradiction.

8.9. The set X must be finite. If X were infinite, then it would have a subset Z ordered by $R \upharpoonright Z$ in the type of natural numbers. But then, in R^{-1}, Z would have no least element.

Note: The above reasoning uses some form of the axiom of choice, namely of what is known as "Tarski's principle of dependent choices": For every set X and relation $R \subset X^2$ such that $\bigwedge_x \bigvee_y xRy$ there exists a sequence $x \in {}^{\mathcal{N}}X$ such that $x_0\ Rx_1, x_1\ Rx_2, \ldots, x_n\ Rx_{n+1}, \ldots$.

8.10. *Hint*: Use induction.

8.11. Cf. Problem 8.22.

8.12. Cf. Problems 8.11 and 8.5 (d)).

8.13. (a) Proof by induction on the number of elements of the set X.

(1) If $\overline{\overline{X}} = 1$ then the theorem is obvious.

(2) Assume that our statement is true for every set X of cardinality n and every ordering relation R on X.

Consider now a set Y such that $\overline{\overline{Y}} = n+1$ and $R \subset Y^2$ such that $\langle Y, R \rangle$ is an ordered set. As Y is nonempty, pick $y_0 \in Y$. Consider the set

$$\langle Y - \{y_0\}, R \cap (Y - \{y_0\})^2 \rangle .$$

This set is ordered and nonempty (since we may assume that $n \geqslant 1$). By the inductive assumption there exists a maximal element y_1 in it. Consider now the set $\langle Y, R \rangle$. Either $y_1\ Ry_0$ or $\sim y_1\ Ry_0$ holds. In the former case y_0 is maximal in $\langle Y, R \rangle$, in the latter y_1 is maximal in $\langle Y, R \rangle$.

(b) Use (a) and Problem 8.5 (d)).

8.14. *Hint*: Modifying the reasoning of 8.13 (a)), we show that for a given $x \in X$ there exists a maximal element y such that xRy. We derive from this that if in $\langle X, R \rangle$ there is no largest element then there must exist at least two maximal elements in $\langle X, R \rangle$. The assumption that X is finite is sound – as shown in Problem 8.11.

8.15. *Hint*: The elements x and y are *incomparable* in R iff the set $\{x, y\}$ is an antichain. A subset Z of an antichain is again an antichain.

8.16. *Hint*: In a linearly ordered set any two elements are comparable.

8.17. This means that the set $\langle A, R \rangle$ is a linearly ordered set.

8.18. *Hint*: Use 8.4 and 8.10.

8.19. *Hint*: Use 8.4, 8.13 (a) and 8.13 (b)).

8.20. *Example*: $\langle X, I_X \rangle$ where $X = \{0, 1, \ldots, n-1\}$.

8.21. (a) (1) The relation R is reflexive since every sequence is its own initial segment.

(2) If x is an initial segment of y and conversely, then they must have the same length (since xRy implies $\mathrm{lh}x \leqslant \mathrm{lh}y$) and so they have the same terms.

(3) Transitivity is obvious.

(b) No. For every sequence x there exists a y such that xRy and $x \neq y$, for instance the y which arises from x by putting 0 after all terms of x.

(c) Yes. In fact there exists a least element, namely the empty sequence.

(d) Continuum.

8.22. (a) Yes, there is a unique maximal element: number 3. There are two minimal elements: numbers 2 and 3.

(b) No.

8.23. (b) The maximal elements are: 8, 9, 10, 11, 12, 13, 14, 15, the minimal elements are 2, 3, 5, 7.

(c) $\{2, 4, 8\}$, $\{2, 6, 12\}$, $\{3, 6, 12\}$, $\{2, 4, 12\}$.

8.24. (b) The least element of our set is the constant function O, there is no maximal element.

8.25. (b) $\langle F, R \rangle$ has maximal (minimal) elements iff $\langle X, S \rangle$ has maximal (minimal) elements. The form of the maximal (minimal) elements of $\langle F, R \rangle$ is the following: Let $I(J)$ be the set of maximal (minimal) elements of $\langle X, S \rangle$. Then every element of the set $T_I(T_J)$

is maximal (minimal) in $\langle F, R\rangle$. The converse is also true: the maximal (minimal) elements of $\langle F, R\rangle$ belong to $T_I(T_J)$.

(c) $\langle F, R\rangle$ possesses a largest element iff $\langle X, S\rangle$ possesses a largest element. An analogous proposition holds for the least elements. The largest element of $\langle F, R\rangle$ is a constant function equal to the largest element of $\langle X, S\rangle$.

8.26. (a) A_5 is maximal, A_0, A_1 and A_4 are minimal.

(b) A_5 is the largest element in our family, there is no least element.

8.27. (a) A_6 is maximal, A_1 and A_2 are minimal.

(b) A_6 is the largest element in our family, there is no least element.

8.28. (a) A_5 and A_6 are maximal. A_0 is minimal.

(b) There is no largest element. A_0 is the least element of our family.

8.29. Note that $A_0 = A_1 = O$ and so our family has a least element. A_2, A_3 and A_4 are maximal elements.

8.30. (b) Yes.

(c) 2 is the least element, there is no maximal element.

8.31. (a) 1 is the least element, there is no maximal element.

(b) Yes.

8.32. (b) Yes. It has the ordinal $\omega+\omega$.

8.33. (a) The constant function 0 is the least element. There is no maximal element; for every sequence $\boldsymbol{a}$ the sequence $\boldsymbol{b}$ defined by: $b_n = a_n + 1$ has the property $\boldsymbol{aRb}$ and $\boldsymbol{a} \neq \boldsymbol{b}$.

(b) No. The sequences $\boldsymbol{a} = \{1, 0, 0, 0, ...\}$ and $\boldsymbol{b} = \{0, 1, 0, 0, 0, ...\}$ are incomparable.

8.34. (a) Reflexivity and antisymmetry are obvious, connectedness follows from the fact that for the least k such that $a_k \neq b_k$ either $a_k < b_k$ or $b_k > a_k$.

Transitivity. Let k_1 be the least k such that $a_k \neq b_k$ and $a_{k_1} < b_{k_1}$ and k_2 the least k such that $b_k \neq c_k$ and $b_{k_2} < c_{k_2}$. If $k_1 \leqslant k_2$ then $a_{k_1} < b_{k_1} = c_{k_1}$. However, if $k_2 < k_1$ then $a_{k_2} = b_{k_2} < c_{k_2}$.

8.35. (b) No. Numbers 1 and i are incomparable.

8.36. (b) No.

8.37. $xRy \wedge \bigwedge_z [(xRz \wedge zRy) \Rightarrow (z = x \vee z = y)]$.

8.38. The successor of the element x is the least element of the set $\{y : xRy \wedge x \neq y\}$ provided this set is nonempty. Not every element (which is not a least one) has a predecessor. For instance the set $\{1 - 1|n :$

$n \in \mathcal{N} - \{0,1\}$ with the relation $\leqslant$ restricted to its elements isa well--ordered set. Yet number 1 has no predecessor

8.39. This condition, though necessary by 8.38, is not sufficient. Consider the set of integers $\langle \mathcal{Z}, \leqslant \rangle$. In this set every element has a successor but this is not a well-ordered set.

8.40. (a) Consider the set $\bigcup_{n \in \mathcal{N}} \{0, 1\}^n$ i.e. the set of all finite zero-one sequences. Define a relation R as follows: $\boldsymbol{a}R\boldsymbol{b}$ iff $\boldsymbol{a}=\boldsymbol{b}$ or the sequence $\boldsymbol{b}$ is an extension of the sequence $\boldsymbol{a}$.

(b) The set of chains in such a set has the cardinality of at least the continuum.

8.42. (a) We have to show that the connectedness property is preserved by similarity maps. Let z, $t \in Y$. By assumption $z=f(x)$ and $t=f(y)$ for some $x, y \in X$. Since R is connected, we have $xRy \vee yRz$, i.e. $f(x)Sf(y) \vee f(y)\,Sf(x)$, thus $zSt \vee tSz$.

(b) Let U be a nonempty subset of Y. Since f is a surjection $f^{-1} \times U \neq O$. Pick its least element x. It is easy to check that $f(x)$ is the least element of U.

8.43. The abstraction principle cannot be applied since it can only be applied to a relation, i.e. a set of pairs. In the system of set theory introduced by Kuratowski and Mostowski "Set Theory" it can be proved that the family of sets $\langle Y, S \rangle$ with the property: $\langle X, R \rangle \approx \langle Y, S \rangle$ does not form a set.

8.44. (a) Assume that a is a maximal element of $\langle X, R \rangle$. We claim that $f(a)$ is a maximal element of $\langle Y, S \rangle$. Indeed, otherwise there would be an element b such that $f(a)\ Sb \wedge f(a) \neq b$. Since f is a surjection, there exists a $c \in X$ such that $b=f(c)$. According to our assumption, $f(a)Sf(c)$ implies aRc. But $a \neq c$ since f is a function. Thus we get a contradiction of the fact that a is a maximal element.

(b) Assume that A is an antichain in $\langle X, R \rangle$, and consider $f*A$. If $f*A$ is not an antichain in $\langle Y, S \rangle$ then there exist elements $u, v \in f * A$ such that $uSv \vee vSu$. Now $u=f(x)$ and $v=f(y)$ for some $x, y \in A$. Thus $xRy \vee yRx$, a contradiction.

8.45. *Hint*: $(R^{-1})^{-1}=R$.

8.46. (a) Note that if $x<y$ then $x<\dfrac{x+y}{2}<y$.

8.47. Assume the converse. Then there exist sets $\langle X, R \rangle$ and $\langle Y, S \rangle$ such that $\langle X, R \rangle \approx \langle Y, S \rangle$, the set $\langle X, R \rangle$ is a dense set but $\langle Y, S \rangle$

is not dense. Let f be a similarity map. Since $\langle Y, S\rangle$ is not dense there are elements $z, t \in Y$ such that t is a successor of x. Pick x and y such that $z=f(x)$ and $t=f(y)$. By 8.44 (c), y is a successor of x and so $\langle X, R\rangle$ is not dense.

8.48. Consider x_0, the least element of the set $\langle X, R\rangle$. It is not the largest element, since $\overline{\overline{X}} \geqslant 2$. Let x_1 be the successor of x_0 in $\langle X, R\rangle$. Clearly $x_0 R x_2$ and $x_2 R x_1$ do not hold for any x_2 different from both x_0 and x_1.

8.49. (a) Our set is not dense. For instance, the sequence $\{0, 1, 0, 0, \ldots\}$ is one of the successors of the sequence $\{0, 0, 0, \ldots\}$.

(b) This set is dense.

Note: It follows that there exist a set X and relations R and S such that $\langle X, R\rangle$ is not dense, $\langle X, S\rangle$ is dense and $R \subset S$.

8.50. (a) Yes. If xRy then $xR\dfrac{x+y}{2}$ and $\dfrac{x+y}{2}Ry$.

(b) Yes.

8.51. No.

8.52. *Hint*: Well-order both sets in type ω and then construct a suitable mapping by the "back and forth" technique: Consider the elements from the set X in the even steps and the elements from $\mathcal{Q}$ in the odd steps, securing the choice of appropriate image (appropriate counterimage).

8.53. $f(x) = -(x+1)$ is a suitable map.

8.54. $f(x) = -x$ is a suitable map.

8.55. Cf. Problem 8.54.

8.56. Here is the idea of the proof: Pick any $x_0 \in X$ and define $f(x_0) = 0$. Consider x_1, which is the predecessor of x_0, and define $f(x_1) = -1$. Similarly, for x_2, the successor of x_0, define $f(x_2) = 1$ etc. Prove that the construction exhausts the whole set X.

8.57. No. Indeed, our set possesses a largest element whereas there is no largest natural number (cf. 8.44) a)).

8.58. (b) *Hint*: Consider Cantor's construction of reals (or Dedekind's construction of reals).

8.59. *Hint*: Use the following property of the set of rationals: For every linearly ordered set $\langle X, R\rangle$ such that $\overline{\overline{X}} = \aleph_0$ there exists an injection f: $X \to \mathcal{Q}$ such that

$$xRy \Leftrightarrow f(x) \leqslant f(y).$$

Then use Dedekind's construction of reals, appropriately modified.

8.60. It is enough to prove that

$$\bigwedge_{x,y} [f(x) S f(y) \Rightarrow xRy].$$

So assume that $f(x) S f(y) \wedge \sim xRy$. Since R is connected, yRx and $x \neq y$. By our assumption $f(y) S f(x)$ and $f(y) \neq f(x)$. But, by the antisymmetry property of S, $f(x) = f(y)$.

8.61. The set $\langle \mathcal{N}, \geqslant \rangle$ is not a well-ordered set. Indeed, there exists a nonempty subset of it without the least element; for instance $\mathcal{N}$ has no least element since there is no largest natural number.

8.62. *Hint*: Proof by contraposition. Assume that X is infinite, find a subset of it ordered in type ω and then use 8.61.

8.63. (1) Reflexivity. Let $x \in X$. In the set $\{x\}$, which is nonempty, there is a least element. It must be x, and thus xRx.

(2) Connectedness. Consider the set $\{x, y\}$. Its least element, say x, has the property xRy.

(3) Transitivity. Consider the set $\{x, y, z\}$.

Condition (b) cannot be omitted, since for every set X, $\bar{\bar{X}} \geqslant 2$, the relation X^2 satisfies condition e) but is not a well-ordering.

8.64. (a) They are the finite sets of the form $\{0, 1, \ldots, n\}$ and the empty set.

(6) They are the sets of the form $(-\infty, x) \cap \mathcal{Q}$, $(-\infty, x\rangle \cap \mathcal{Q}$ and the empty set.

8.65. *Hint*: Use 8.52.

8.67. Let Z be the set of those $x \in X$ for which $\sim \Phi(x)$. If Z is the empty set then all the elements of X have the property Φ. We show that the assumption $Z \neq O$ leads to a contradiction. Indeed, if $Z \neq O$ there is a least element z_0 in Z. Thus, by our construction, whenever tRz_0 and $t \neq z_0$ then $\Phi(t)$. This means that $t \in O_R(z_0) \Rightarrow \Phi(t)$. Thus $\Phi(z_0)$.

8.68. *Hint*: Use 8.67 choosing appropriate Φ.

8.69. *Hint*: Use 8.67 choosing appropriate Φ.

8.70. *Hint*: Construct an appropriate chain by transfinite induction as follows: Pick an arbitrary x_0; $x_{\alpha+1}$ is an arbitrary element of the set

$\{y\colon x_\alpha Ry \wedge x_\alpha \neq y\}$, provided this set is nonempty and x_λ is an arbitrary upper bound of the set $\{x_\mu\colon \mu<\lambda\}$. Prove that this procedure must terminate.

8.71. *Hint*: Prove that the set of relations S such that $R\subset S$ and $\langle X, S\rangle$ is an ordered set and, satisfies the assumptions of the Kuratowski–Zorn Lemma. The maximal element in this family is the required linear ordering.

8.72. *Hint*: Modify the method of 8.71, consider only those S in which a is a maximal element.

8.73. Use 8.5.

8.74. *Hint*: Prove that the family of all functions f such that:

1. The domain Df of f is a subset of T.
2. $\bigwedge\limits_{t\in Df} f(t)\in X_t$

with the relation R defined as follows: $fRg \Leftrightarrow g$ is an extension of f satisfies the assumptions of the Kuratowski–Zorn Lemma.

8.75. *Hint*: Prove that the family of all sets W such that

(a) $\quad W\subset \bigcup\limits_{t\in T} X_t,$

(b) $\quad \overline{\overline{W\cap X_t}}\leqslant 1$

for all $t\in T$ satisfies the assumptions of the Kuratowski–Zorn Lemma.

8.76. *Hint*: Prove that the family of all proper ideals in an arbitrary ring with unity, ordered by the inclusion relation, satisfies the assumption of the Kuratowski–Zorn Lemma.

8.77. *Hint*: Prove that the family of all linearly independent subsets of a fixed linear space satisfies the assumptions of the Kuratowski–Zorn Lemma.

8.78. *Hint*: Prove that the family of all antichains in a fixed ordered set $\langle X, R\rangle$, ordered by inclusion, satisfies the assumptions of the Kuratowski–Zorn lemma.

8.79. *Hint*: Given a set $T\subset X$, we consider its characteristics function χ_T, i.e. the function defined by

$$\chi_T(x)=\begin{cases}0 & \text{if} \quad x\in T,\\ 1 & \text{if} \quad x\in T.\end{cases}$$

8.80. (a) Proof as in 8.33.

(b) No. In particular, if $\overline{\overline{T}}\geqslant 2$ and, for every $t\in T$, $\overline{\overline{X}}_t\geqslant 2$ then the set $\langle \prod\limits_{t\in T} X_t, R\rangle$ is not linearly ordered.

CHAPTER 10

CARDINAL AND ORDINAL ARITHMETIC

10.1. We have to prove that if

$$\langle A, R\rangle \approx \langle A_1, R_1\rangle, \quad \langle B, S\rangle \approx \langle B_1, S_1\rangle,$$

$$A \cap B = A_1 \cap B_1 = O,$$

then

$$\langle A \cup B, R \cup S \cup (A \times B)\rangle \approx \langle A_1 \cup B_1, R_1 \cup S_1 \cup (A_1 \times B_1)\rangle.$$

Let f be a similarity map of $\langle A, R\rangle$ and $\langle A_1, R_1\rangle$ and g a similarity of $\langle B, S\rangle$ and $\langle B_1, S_1\rangle$. Since $A \cap B = O$, the union $f \cup g$ is a function.

Since $Rf \cap Rg = 0$, the fact that both f and g are injections implies that also $f \cup g$ is an injection.

By a similar reasoning we check that:

$$x\,[R \cup S \cup (A \times B)]\,y \Leftrightarrow f(x)\,[R_1 \cup S_1 \cup (A_1 \times B_1)]\,f(y).$$

10.2. $\alpha = \omega$, $\beta = 1$, $\gamma = 2$.

10.3. *Hint*: Given a set of type $\alpha + \beta$, we can see that it possesses exactly one initial segment of type α.

10.5. (a) $\langle \{1 - 1/n : n \in \mathcal{N} - \{0\}\} \cup \{1\}, \leqslant\rangle$,

(b) $\langle \{1 - 1/n : n \in \mathcal{N} - \{0\}\} \cup \{2 - 1/n : n \in \mathcal{N} - \{0\}\}, \leqslant\rangle$,

(c) $\langle \{1 - 1/n : n \in \mathcal{N} - \{0\}\} \cup \{2 - 1/n : n \in \mathcal{N} - \{0\}\} \cup \{2\}, \leqslant\rangle$.

10.6. *Hint*: $n + \omega$ is the order type of a subset of the set Z consisting of number greater than or equal to $-n$.

10.7. *Hint*: In an ordering of type ω every element possesses a finite number of predecessors. This is not true in the case of an ordering of type $\omega + \omega$.

10.8. $\alpha = 1$, $\beta = \omega$. *Note*: At least one of the numbers α and β must be infinite.

10.9. *Hint*: If f is a similarity of $\langle A, R\rangle$ and $\langle A_1, R_1\rangle$ and g is a similarity of sets $\langle B, S\rangle$ and $\langle B_1, S_1\rangle$ then h defined by $h(\langle x, y\rangle) = \langle f(x), g(y)\rangle$ is the required mapping.

10.10. *Hint*: Let $\overline{\langle A, R\rangle}=\alpha$, $\overline{\langle B, S\rangle}=\beta$. Let x_0 be the first element of the set B (in the ordering S). Then the set $A\times\{x_0\}$ is an initial segment of the set $\langle A\times B, T\rangle$ of type α.

10.11. $\beta=2$, $\gamma=3$, $\alpha=\omega$.

10.12. (a) In the set $N_1[x]$ introduce a relation T as follows:

$$(ax+b)\,T\,(a_1x+b_1)\Leftrightarrow(b<b_1)\vee(b=b_1\wedge a\leqslant a_1).$$

(b) Consider the set from a) and add at the end a set similar to the set $\mathcal{N}$ – for instance the set $\langle\{1-1/n\colon n\in\mathcal{N}\}, \leqslant\rangle$.

(c) Modify the example from (b) adding at the end a three-element set.

(d) *Hint*: Order the set $N_3[x]$ by the following relation T:

$$(ax^2+bx+c)\,T\,(a_1x^2+b_1x+c_1)\Leftrightarrow$$

$$\Leftrightarrow(c<c_1)\vee(c=c_1\wedge b<b_1)\vee(c=c_1\wedge b=b_1\wedge a\leqslant a_1).$$

This set is ordered in type $\omega\cdot\omega\cdot\omega$.

10.14. $2\cdot\omega$ is the type of an ordering in which there are ω pairs, and so it is type ω.

10.15. $2\cdot\omega=\omega$, $\omega\cdot 2=\omega+\omega$.

10.16. $\alpha=\omega\cdot\omega$, $\beta=\omega$.

10.18. $\alpha=\beta=1$, $\gamma=\omega$.

10.19. We have to prove that if a certain set of type α is similar to an initial segment of a set of type β, then every set of type α is similar to an initial segment of a set of type β.

Let $\langle A, R\rangle$ be a set of type α and assume that $\langle A, R\rangle$ is similar to an initial segment of a set $\langle B, S\rangle$ (of type β). Let $\overline{\langle A_1, R_1\rangle}=\alpha$ and $\overline{\langle B_1, S_1\rangle}=\beta$. Thus there exist similarity mappings f and g of $\langle A, R\rangle$ and $\langle A_1, R_1\rangle$ and of $\langle B, S\rangle$, and $\langle B_1, S_1\rangle$ respectively. Let t be such that $\langle A, R\rangle\approx\langle O_S(t), S\restriction O_s(t)\rangle$ and finally let h be the similarity map of those ordered sets. It is easy to check that $f^{-1}\circ h\circ g$ is a similarity map of $\langle A_1, R_1\rangle$ and $\langle O_{S_1}(g(t)), S_1\restriction O_{S_1}(g(t))\rangle$.

10.20. *Hint*: Use the reasoning of 10.19.

10.21. Here are the consecutive steps of the proof: we show first that $\alpha\leqslant\beta\Rightarrow\alpha\subset\beta$, then we prove that the intersection of two ordinals is an ordinal and finally that if $\beta-\alpha\neq O$ then α is a minimal element (see condition (3)) of $\beta-\alpha$.

10.22. *Hint*: Inclusion is a transitive relation.

10.24. *Hint*: Use 10.3.

10.25. *Hint*: Use 10.20.

10.26. $\alpha=2$, $\beta=1$, $\gamma=\omega$.

10.27. *Hint*: Use 10.24.

10.29. *Hint*: (b) In order to prove $\Leftarrow$ one uses 10.21 and the fact that condition (3) of the definition of an ordinal implies $\bigwedge_{\alpha} \sim \alpha \in \alpha$.

10.30. *Hint*: Use 10.3 and 10.20.

10.31. *Hint*: Proof by transfinite induction on α.

10.32. Assume that $\beta\cdot\gamma+\rho=\beta\cdot\gamma_1+\rho_1$, but $\gamma\neq\gamma_1$. We can assume that $\gamma<\gamma_1$, i.e. $\gamma+1\leqslant\gamma_1$. But $\rho<\beta$ and so $\beta\cdot\gamma+\rho<\beta\cdot\gamma+\beta=\beta(\gamma+1)$ $\leqslant\beta\cdot\gamma_1\leqslant\beta\cdot\gamma_1+\rho_1$. Thus

$$\beta\cdot\gamma+\rho\neq\beta\cdot\gamma_1+\rho_1.$$

But our assumption is that $\beta\cdot\gamma+\rho=\beta\cdot\gamma_1+\rho_1$, and so the assumption $\gamma\neq\gamma_1$ leads to a contradiction. So assume now that $\gamma=\gamma_1$ but $\rho\neq\rho_1$. We can assume $\rho<\rho_1$. The contradiction is now obtained by 10.27.

10.33. *Hint*: The proof by transfinite induction on γ.

10.34. *Hint*: The proof by transfinite induction on $\max(\alpha, \beta)$.

10.35. *Hint*: The proof by transfinite induction on α.

10.36. The symbol $<$ in the predecessor of the implication denotes the "less than" relation among ordinal numbers whereas in the successor it is used to denote the "less than" relation among cardinal numbers.

10.37. If $\beta+1=\alpha$ then α possesses a predecessor.

10.38. *Hint*: $\beta<\alpha\Rightarrow\bar{\beta}<\bar{\alpha}$. Simultaneously we have $\bar{\beta}\cdot\bar{\beta}=\bar{\beta}$.

10.40. Introduce in the set $\omega_\alpha\times\omega_\alpha$ the following relation R:

$$\begin{aligned}\langle\mu, \nu\rangle R\langle\mu_1, \nu_1\rangle\Leftrightarrow&[\max(\mu, \nu)<\max(\mu_1, \nu_1)]\vee\\ &\vee[(\max(\mu, \nu)=\max(\mu_1, \nu_1))\wedge\\ &\wedge(\mu=\max(\mu, \nu))\wedge(\nu_1=\max(\mu_1, \nu_1))]\vee\\ &\vee[\max(\mu, \nu)=\max(\mu_1, \nu_1)=\mu=\mu_1\wedge\\ &\qquad\wedge\nu<\nu_1]\vee\\ &\vee[\max(\mu, \nu)=\max(\mu_1, \nu_1)=\nu=\nu_1\wedge\\ &\qquad\wedge\mu\leqslant\mu_1].\end{aligned}$$

It is easy to check that this is a well-ordering relation. On the other hand, the segment determined by the pair $\langle\alpha, \beta\rangle$ has cardinality at most $\overline{\max(\alpha, \beta)}^2$.

10.41. *Hint*: We use 10.40.

10.42. *Hint*: Use 10.41 and the Cantor–Bernstein theorem.

10.44. (a) *Hint*: Use transfinite induction on γ.

(b) As in (a).

10.45. *Hint*: Use the transfinite induction on ξ.

10.46. Note that the union of ordinals is an ordinal itself.

10.47. *Hint*: What properties of the generalized union could you use?

10.48. Let ξ_0 be an arbitrary ordinal. Define by induction the sequence $\{\xi_n\}_{n\in\omega}$ as follows: $\xi_{n+1}=\varphi(\xi_n)$. Let $\xi=\lim_{n\in\mathcal{N}}\xi_n$. Then:

$$\varphi(\xi)=\varphi(\lim_{n\in\mathcal{N}}\xi_n)=\lim_{n\in\mathcal{N}}\varphi(\xi_n)=\lim_{n\in\mathcal{N}}\xi_{n+1}=\xi\,.$$

10.49. (a) $\overline{\alpha+\beta}=\overline{\alpha}+\overline{\beta}$.

(b) $\overline{\alpha\cdot\beta}=\overline{\alpha}\cdot\overline{\beta}$.

(c) If β is an infinite ordinal then the cardinal numbers under consideration need not be equal. In particular, $\overline{\alpha^\beta}$ is the cardinality of the set of those functions from β into α which take non-zero values only in a finite number of places. However if, β is finite then our equality holds.

10.50. Let X be an arbitrary set of ordinals. Then the union of X, $\bigcup X$, is again an ordinal. Its successor, $(\bigcup X)+1$, does not belong to X.

10.51. Since alephs are cardinals and they are numbered by means of ordinals, alephs do not form a set. Thus cardinals do not form set.

10.52. If an infinite set can be well-ordered then, in particular, it is equipollent to an ordinal, and so it is equipollent to an initial ordinal. Thus its cardinal is an aleph. Conversely, if the cardinality of X is an aleph, then there exists an α such that X is equipollent to ω_α. Thus X can be well-ordered.

10.53. The Zermelo theorem implies the axiom of choice as follows: Let X be a nonempty family of pairwise disjoint nonempty sets. Well-order the set $\bigcup X$. Since every element $U\in X$ is included in $\bigcup X$, it possesses a least element. The set consisting of the least elements of the sets from X is the required selector.

The proof of the converse implication runs as follows: First construct (using the axiom of choice) a choice function for the family $\mathscr{P}(X)-\{O\}$. Using such a function f, we construct, by transfinite induction, a well-ordering of X as follows:

$$x_0=f(X), \quad x_\alpha=f(X-\{x_\beta:\beta<\alpha\}).$$

10.54. *Hint*: It is enough to prove that $\alpha\in Z(\mathfrak{m})\wedge\beta<\alpha\Rightarrow\beta\in Z(\mathfrak{m})$.

10.55. Since $Z(\mathfrak{m})$ is an initial ordinal, it follows that if $\aleph(\mathfrak{m})\leqslant\mathfrak{m}$ then $Z(\mathfrak{m})$ belongs to $Z(\mathfrak{m})$, which is absurd.

10.56. *Hint*: Use 10.55.

10.57. *Hint*: Given $\alpha\in Z(\mathfrak{m})$ (where $\mathfrak{m}=\overline{\overline{M}}$), T_α is defined as the set of those subsets of M which possess a well-ordering of type α. Then, given an element of T_α we consider the family of all its initial segments (under an ordering of type α).

10.58. *Hint*: Given $\alpha\in Z(\mathfrak{m})$, consider the set of all well-orderings with the field included in the set M (of cardinality M) and of type α.

10.59. Let $\overline{\overline{M}}=\mathfrak{m}$, $\mathfrak{m}$ being infinite. Using the axiom of choice we get $\mathfrak{m}=\mathfrak{m}\cdot\mathfrak{m}$. Thus $2^{\mathfrak{m}}$ is the cardinal of the set of all relations with the field included in M. In particular, the set W of all well-orderings in M has cardinality $\leqslant 2^{\mathfrak{m}}$. Define in W the relation $\sim$ as follows:

$$R\sim S\Leftrightarrow\overline{R}=\overline{S}.$$

It is clear that $\sim$ is an equivalence and that: $\overline{\overline{W/_\approx}}=\aleph(\mathfrak{m})$ Thus $\aleph(\mathfrak{m})\leqslant\overline{\overline{W}}$ (again the axiom of choice is used) and so $\aleph(\mathfrak{m})\leqslant 2^{\mathfrak{m}}$.

10.60. *Hint*: Let $\overline{\overline{A}}=\mathfrak{m}$, $\overline{\overline{B}}=\aleph(\mathfrak{m})$. Decompose the set $A\times B$ into the union of sets of cardinality $\mathfrak{m}$ and $\aleph(\mathfrak{m})$.

10.61. $\mathfrak{n}=2^{\mathfrak{m}\aleph_0}$.

10.62. Consider an arbitrary infinite cardinal $\mathfrak{m}$. We show that it is an aleph (which is enough by 10.52 and 10.53). Let $\mathfrak{p}=\mathfrak{m}^{\aleph_0}$. We have

$$\mathfrak{p}+1=\mathfrak{m}^{\aleph_0}+1\leqslant\mathfrak{m}^{\aleph_0}\cdot\mathfrak{m}\leqslant\mathfrak{m}^{\aleph_0+1}=\mathfrak{m}^{\aleph_0}=\mathfrak{p},$$

Thus $\mathfrak{p}+1=\mathfrak{p}$. Since $\mathfrak{m}\leqslant\mathfrak{p}$, it is enough to show that $\mathfrak{p}$ is an aleph. Consider the number $[\mathfrak{p}+\aleph(\mathfrak{p})]^2$. It is equal to

$$\mathfrak{p}^2+2\mathfrak{p}\cdot\aleph(\mathfrak{p})+\aleph^2(\mathfrak{p}),$$

i.e.

$$\mathfrak{p}+\mathfrak{p}\cdot\aleph(\mathfrak{p})+\aleph(\mathfrak{p})=\mathfrak{p}\cdot\aleph(\mathfrak{p})+\aleph(\mathfrak{p})=\aleph(\mathfrak{p})[\mathfrak{p}+1]=\aleph(\mathfrak{p})\cdot\mathfrak{p}.$$

thus

$$\mathfrak{p}+\aleph(\mathfrak{p})=\aleph(\mathfrak{p})\cdot\mathfrak{p},$$

Hence $\mathfrak{p}$ is an aleph.

10.63. Let $\overline{\overline{X}}=\overline{\overline{Y}}=\mathfrak{m}$, $X\cap Y=O$. Decompose the set $\mathscr{P}(X\cup Y)$ into the union of disjoint sets Z_1 and Z_2, $\overline{\overline{Z}}_1=\mathfrak{n}$, $\overline{\overline{Z}}_2=\mathfrak{m}$. Then we have:

$$\mathscr{P}(X\cup Y)\sim\mathscr{P}(X)\times\mathscr{P}(Y).$$

Clearly Z_1 cannot contain a pair of the form $\langle t, u\rangle$ for every $t\in X$ since in that case we would be able to find a mapping of X onto $\mathscr{P}(X)$). Thus for some $t\in\mathscr{P}(X)$, all sets w with the property $w\cap X=t$ belong to Z_2. Thus $\overline{\overline{Z}}_2\geqslant\mathscr{P}(X)$.

10.64. *Hint*: Use 10.63.

10.65. $2^{\aleph_\alpha}\leqslant\aleph_\alpha^{\aleph_\alpha}\leqslant(2^{\aleph_\alpha})^{\aleph_\alpha}=2^{\aleph_\alpha^2}=2^{\aleph_\alpha}$.

10.66. $2^{\aleph_\omega}=2^{\sum_n \aleph_n}=\prod_n 2^{\aleph_n}=\prod_n 2^{\aleph_0}=(2^{\aleph_0})^{\aleph_0}=2^{\aleph_0}=\aleph_{\omega+1}$.

CHAPTER 11

FORMAL SYSTEMS AND THEIR PROPERTIES

11.1. *Hint*: Use 7.21.

11.6. *Hint*: Use the tautology $\alpha\Rightarrow[\beta\Rightarrow(\alpha\wedge\beta)]$.

11.7. *Hint*: Use the tautology:

$$[(\alpha\Rightarrow\beta)\wedge(\gamma\Rightarrow\beta)]\Rightarrow(\alpha\vee\gamma\Rightarrow\beta).$$

11.8. Let $\{\Phi_0, \ldots, \Phi_n\}$ be a proof of the formula $\Phi\Rightarrow\Psi$ from the set of formulas X. Then the sequence $\{\Phi_0, \ldots, \Phi_n, \Phi, \Psi\}$ is a proof of the formula Ψ from the set $X\cup\{\Phi\}$.

Conversely, let $\{\Phi_0, \ldots, \Phi_n\}$ be a proof of Ψ from $X\cup\{\Phi\}$. Then, by induction on i, one shows that $\Phi\Rightarrow\Phi_i\in \mathrm{Cn}\, X$ for all $i\leqslant n$. Since $\Phi_n=\Psi$, our proof is complete.

The above statement is called the "*Deduction theorem*".

11.14. *Hint*: Use the fact that a proof involves only a finite number of statements from the set X.

11.17. *Hint*: Use 11.14. The set $\mathscr{F}_{\mathbf{J}}$ is a theory, since $\mathrm{Cn}\,(X)\subset\mathscr{F}_{\mathbf{J}}$ for all $X\subset\mathscr{F}_{\mathbf{J}}$.

11.18. (a) No.

(b) No.

(c) Yes.

11.19. No. Consider $X=\boldsymbol{L}\cap\mathscr{F}_{\mathbf{J}}$ and Y consisting of X and a sentence Φ which is the negation of a sentence provable in logic (for instance the sequence $\bigvee_x x\neq x$). Thus $X\subset Y$ and Y is not a theory, since $Y=\mathscr{F}_{\mathbf{J}}$ though $\mathrm{Cn}\,(Y)=\mathscr{F}_{\mathbf{J}}$.

If X is a theory and $Y\subset X$, Y is not necessarily a theory, since for instance the empty set is not a theory (see Problem 11.86 (b)).

11.20. *Hint*: Use the tautologies

$$\bigwedge_x[\Phi(x)\Rightarrow\bigwedge_x\Phi(x)] \quad\text{and}\quad \bigwedge_x\Psi(x)\Rightarrow\Psi(x).$$

11.23. *Hint*: For the implication from the R.H.S. to the L.H.S. assume that there are sentences Φ and Ψ such that $\Phi\in X-Y$, $\Psi\in Y-X$, and consider the formula $\Phi\wedge\Psi$.

11.29. The system S consists of sentences provable in logic.

11.35. *Hint*: Use 11.14.

11.37. *Hint*: Let $\Psi \in S' \cap \mathscr{F}_J$ and let $\{\Psi_0, \ldots, \Psi_m\}$ be a proof of Ψ from S. Given a formula $\Psi_i(\ldots, f(x_1, \ldots, x_n), \ldots)$, consider the formula $\Psi_i' = \bigvee_y \Psi_i(\ldots, y, \ldots) \wedge \Phi(y, x_1, \ldots, x_n)$, where y is a variable not occurring in Ψ_i and Φ is the formula defining in S' the function f. Prove that $\Psi_m' = \Psi_m$ and that, for every $i \leqslant m$, $\Psi_i' \in \mathrm{Cn}(S \cup \{\Psi_0, \ldots, \ldots, \Psi_{i-1}\})$; then apply 11.2.

Note: The theorem proved in Problem 12.37 is called the "*Theorem on the elimination of function symbols*".

11.38. *Hint*: Use a method similar to that employed in the proof of 11.37. In this case, however, the formula Ψ_i' is obtained from Ψ_i by substituting the defining formula Φ for every occurrence of the predicate symbol R.

Note: This theorem is called the "*Theorem on the elimination of predicate symbols*".

11.39. *Hint*: Use 11.37 and 11.38.

11.40. Yes.

11.43. The relation $\leqslant$ has minimal elements. They are systems in a language which contains $=$ as the only extralogical symbol. There is no least element. There are no maximal elements.

11.44. *Hint*: Use 11.37 and 11.38.

11.45. The formula is not a definition of a function symbol. It follows from the existence of models of S in which the formula

$$\bigwedge_{x_3} \bigwedge_{x_1} [\bigvee_{x_2} x_1 \leqslant x_2 \wedge \bigvee_{x_2} x_3 \leqslant x_2 \Rightarrow x_1 = x_3]$$

fails.

Note: We use here the completeness theorem of Gödel (see the note to Problem 12.59).

11.48. *Hint*: Prove by induction on the number of quantifiers in the formula that for every formula Φ there exists a quantifier-free formula Φ' such that $\Phi \Leftrightarrow \Phi' \in S$.

Put the proof for the formulas in the quantifier normal form.

11.49–51. *Hint*: Use the reasoning of 11.48.

11.52.–54. *Hint*: See Problem 11.45.

11.55. The relation Int is neither antisymmetric nor symmetric.

To prove the former, pick any two systems which differ only in symbols. The latter is proved as follows: We need S_1 and S_2 such that S_1 Int S_2 but $\sim S_2$ Int S_1. Take any two systems S_1, S_2 in a language having as the only extralogical symbol and such that $S_1 \subset S_2$ and $S_1 \neq S_2$.

11.56. The relation $\mathrm{Int}/_{\sim}$ has a least element. It is the family of the systems which form an equivalence class of the system $S_{=}$ with the following axioms:

$$\bigwedge_x x=x,$$

$$\bigwedge_x \bigwedge_y x=y \Rightarrow y=x$$

$$\bigwedge_x \bigwedge_y \bigwedge_z (x=y \wedge y=z) \Rightarrow x=z.$$

11.57. *Hint*: Note that if $\Phi \in \mathrm{Cn}\, X$ and $\{\Phi_0, \ldots, \Phi_n\}$ is a proof of Φ then $\{\varphi\Phi_0, \ldots, \varphi\Phi_n\}$ is a proof of $\varphi\Phi$ in S.

11.58. *Hint*: Use 11.57 and the following: $\varphi(\sim\Phi) = \sim\varphi\Phi$.

11.59. Take $\Theta(x)=(x=x)$ and $\varphi(x \leqslant y) = [(xRy) \vee (x=y)]$ and, for the other interpretation $\Theta(x)=(x=x)$ and $\varphi(xRy) = [(x \leqslant y) \wedge (\sim x=y)]$.

11.60. *Hint*: Consider the improper extension S_1 arising by adding to S_1 the following definitions:

$$x<y \Leftrightarrow x \leqslant y \wedge \sim(x=y),$$

$$x \underset{n}{<} y \Leftrightarrow \bigvee_{x_1} \ldots \bigvee_{x_n} x<x_1 \wedge x_1<x_2 \wedge \ldots \wedge x_{n-1}<x_n \wedge x_n<y.$$

$$x \underset{n}{\leqslant} y \Leftrightarrow \bigvee_{x_1} \ldots \bigvee_{x_n} [x \leqslant x_1 \wedge \ldots \wedge x_n \leqslant y \wedge \bigwedge_z (x \leqslant z \wedge z \leqslant y$$

$$\Rightarrow x=z \vee x_1=z \vee \ldots \vee x_n=z \vee y=z)],$$

$$x \underset{n}{=} y \Leftrightarrow x \underset{n}{<} y \wedge x \underset{n}{\leqslant} y,$$

Then use the method employed in 11.48 and the result of 11.39.

11.63. *Hint*: Take $\Theta(x)=x \geqslant 0$ and $\varphi(x \leqslant y)=(x \leqslant y)$.

11.64. *Hint*: Take $\Theta(x)=(x=x)$ and $\varphi(x \leqslant y) = \bigvee_z x+z=y$.

11.65. *Hint*: Let Pn be the nth prime number. Take as $\Theta(x)$ the following formula: x is not divisible by the square of a prime number and for every sequence $m_1, \ldots, m_n$

if $Pm_1 | m_2 \wedge Pm_2 | m_3 \wedge \ldots \wedge Pm_{n-1} | m_n$ and $Pm_n | x$ then m_1 is not divisible by the square of a prime number.

The set $\{x \in \mathcal{N} : \Theta(x)\}$ is called the *set of hereditarily square-free numbers*. The function φ is defined as follows: $\varphi(x \in y) = Px|y$.

11.66. We outline the construction: Let A be the least set with the property: $O \in A$ and if $X \in A$, $Y \in A$ then $X \cup \{Y\} \in A$. (The set A is the family of hereditarily finite sets).

Define in A a relation $\sim$ as follows: $A \sim B \Leftrightarrow \bar{\bar{A}} = \bar{\bar{B}}$. Let N be the set of all equivalence classes of the relation $\sim$. $\Theta(x)$ is now: $x \in N$.

Define the operations $*$, $\circ$, $\times$ as follows:

If $a = [X]$ then $a* = [X \cup \{X\}]$.

If $a = [X]$ and $b = [Y]$ then $a \circ b = [X_1 \cup Y_1]$ where X_1 and Y_1 are, respectively, from a and b and $X_1 \cap Y_1 = O$.

If $a = [X]$ and $b = [Y]$ then $a \times b = [X \times Y]$.

Finally, φ is defined as follows:

$$\varphi(0) = [O],$$

$$\varphi(Sx) = X^*,$$

$$\varphi(x + y) = x \circ y,$$

$$\varphi(x \circ y) = x \times y.$$

CHAPTER 12

MODEL THEORY

12.1. The reason is that all the other connectives are definable by means of negation, conjunction, and the existential quantifier.

12.2. $\mathfrak{a} \models (\Phi \vee \Psi)[X]$ iff $\mathfrak{a} \models \Phi[X]$ or $\mathfrak{a} \models \Psi[X]$.

12.3. The satisfaction relation is a relationship bteween the formulas of a formal system (thus sequences of signs – symbols) and the sequences of elements of the universe of a relational system. Thus, for instance, $\Leftrightarrow$ is a sign of a formal language whereas $\equiv$ is an abbreviation of a metalanguage expression: if and only if.

The distinction between expressions of a language and those of a metalanguage – very important in metamathematical investigations – will not be very clear in the sequel. The meaning of a symbol should be obvious from the context.

12.4. $\mathfrak{a} \models (\Phi \Rightarrow \Psi)[X]$ iff $\mathfrak{a} \models \sim\Phi[X]$ or $\mathfrak{a} \models \Psi[X]$ (use the tautology $p \Rightarrow q \Leftrightarrow (\sim p \vee q)$), $\mathfrak{a} \models \bigwedge_{x_i} \Phi[X]$ iff for all $a \in A$, $\mathfrak{a} \models \Phi[X(^i_a)]$ (use the tautology: $\bigwedge_{x_i} \Phi \Leftrightarrow \sim \bigvee_{x_i} \sim \Phi$).

12.5. (a) No.

(b) No.

12.6. (a) No.

(b) Yes.

12.7. (a) No, no, yes.

(b) No, no, no.

(c) No, yes, yes.

12.8. *Hint*: Proof by induction on the degree of complexity of the formula.

12.10. The implication from left to right is obvious; Since $A \neq O$, pick $a \in A$, and then the constant sequence $\boldsymbol{a}$ defined by

$$\bigwedge_n a_n = a$$

satisfies Φ.

Conversely, if there exists a sequence $\boldsymbol{X}$ such that $\mathfrak{a} \models \Phi[\boldsymbol{X}]$ then every sequence $\boldsymbol{Y}$ satisfies Φ in $\mathfrak{a}$ (by 12.9).

12.11. *Hint*: Use 12.10.

12.12. *Hint*: Proof by induction on the number of free variables in Φ.

12.13. Assume $\mathfrak{a} \models \bigwedge_{v_0} \bigvee_{v_1} \Phi(v_0, v_1)$. Given $a \in A$ let $U_a = \{b : \mathfrak{a} \models \Phi[a, b, \ldots]\}$ (this definition is sound by 12.8).

Let f be a choice function for the family $\{U_a : a \in A\}$. Thus $f(a) \in U_a$ and f is the required function.

The converse is obvious.

12.14. Assume that it is not true that $\mathfrak{a} \models \Phi[X]$. Then, according to the definition, $\mathfrak{a} \models \sim \Phi[X]$.

12.15. Completeness follows from 12.14. Consistency follows from the fact that, according to the definition, $\sim\Phi \in \mathrm{Th}\ \mathfrak{a}$ iff $\Phi \notin \mathrm{Th}\ \mathfrak{a}$.

12.16. A is a two-element set; $A = \{0, 1\}$, the operation $+$ is given by

+	0	1
0	0	1
1	1	0

$B = \mathcal{N} - \{0\}$, $0' = 1$ and $+'$ is the usual addition.

2.17. *Hint*: Proof by induction on the length of the formal proof of formula Φ. Check that $\mathfrak{a} \models \Phi$ and $\mathfrak{a} \models \Phi \Rightarrow \Psi$ implies $\mathfrak{a} \models \Psi$ and that $\mathfrak{a} \models \Phi(x)$ implies $\mathfrak{a} \models \bigwedge_{y} \Phi(y)$ (cf. 12.12).

12.18. Assume that the proposition $\bigwedge_{x} (x + x = 0)$ may be proved from our set of sentences. Then, by 12.17 the sentence $\bigwedge_{x} (x + x = 0)$ is true in every system which contains at least two elements and in which the operation denoted by $+$ is commutative and associative. But in the system $\mathscr{Z}$ of integers our sentence is false. Thus it cannot be derived from our of sentences (as they are true in $\mathscr{Z}$). An appropriate example (12.16) works for the negation of our sentence.

12.19. According to the definition, Φ is independent of the set S of sentences if there exist systems $\mathfrak{a}$ and $\mathfrak{b}$ such that: $S \cup \{\Phi\} \subset \mathrm{Th}(\mathfrak{a})$ and $S \cup \{\sim\Phi\} \subset \mathrm{Th}(\mathfrak{b})$. As $\mathfrak{a} \models \Phi$ so $\Phi \in \mathrm{Th}(\mathfrak{a})$ thus $S \cup \{\Phi\} \subset \mathrm{Th}(\mathfrak{a})$. Similarly $S \cup \{\sim\Phi\} \subset \mathrm{Th}(\mathfrak{b})$.

12.20. The sentence A_3 says that the universe of the system consists exactly of three elements. This statement is true in the field $\mathscr{Z}_3$ (of the remainders modulo 3) and is false in the field $\mathscr{Q}$.

12.21. The sentence A_6 says that the universe of the system consists exactly of 6 elements. Yet one can prove that:

1° If the characteristics of a field is a nonzero integer then it is prime.

2° If a field P is finite then the cardinality of P is a power of its characteristics. Thus there is no 6-element field and so A_6 is not consistent with the axioms of fields. In particular, A_6 is not independent of it.

12.22. The sentence A_4 says that the universe of the system consists of four elements. As there exists a four-element field and there exist fields with a different number of elements, A_4 is independent of axioms of fields.

12.23. For instance: the group of permutations of a 3-element set is not a commutative group whereas the group $\langle \mathscr{Z}, + \rangle$ is a commutative group.

12.24. Our sentence says that the relation is dense. It is true in the system $\langle \mathscr{Q}, \leqslant \rangle$ but it is false in the system $\langle \mathscr{N}, \leqslant \rangle$.

12.25. No.

12.26. (a) First note that it is enough to consider the open formulas since by 12.12, if Φ is an open formula then $\mathfrak{a} \models \Phi$ iff $\mathfrak{a} \models \bigwedge_{v_{i_0}, \ldots, v_{i_n}} \Phi$. Open formulas arise from atomic formulas by using propositional connectives. By the definition of the satisfaction relation it is enough to prove our theorem for atomic formulas.

(b) If Φ is an existential formula then $\sim \Phi$ is equivalent to a universal formula. Thus if it is not true that Φ is satisfied in $\mathfrak{b}$ then $\sim \Phi$ is satisfied in $\mathfrak{b}$ and so (by (a)) $\sim \Phi$ is satisfied in $\mathfrak{a}$, which is absurd.

12.27. Note that there is an axiomatics of rings consisting of universal formulas only.

12.28. See 12.27.

12.29. If the sentence A_4 were equivalent to a universal sentence then it would be true in every subsystem of the four-element field, and thus in a two-element field.

12.30. Use the equality axioms.

12.31. *Hint*: Use 11.49. One can prove a stronger result, namely $\langle \mathscr{Q}, \leqslant \rangle \prec \langle \mathscr{R}, \leqslant \rangle$ (cf. 12.23), using the following property of reals:

For every rational a and b and every real c: if $a<c<b$ then there exists an isomorphism φ of $\langle \mathscr{R}, \leqslant \rangle$ such that $\varphi(a)=a$, $\varphi(b)=b$ and $\varphi(c) \in \mathscr{Q}$.

12.33. *Hint*: Use 12.11.

12.34. Let $\mathfrak{b}=\langle \mathscr{N}, \leqslant \rangle$, $\mathfrak{a}=\langle \mathscr{N}-\{0\}, \leqslant \rangle$. Clearly $\mathfrak{a} \subset \mathfrak{b}$. $\mathfrak{a} \equiv \mathfrak{b}$ holds since $\mathfrak{a}$ is isomorphic to $\mathfrak{b}$. Yet the constant sequence $a_n=1$ satisfies in $\mathfrak{a}$ the formula $\bigwedge_{v_0} (v_1 \leqslant v_0)$ but does not satisfy this formula in $\mathfrak{b}$.

12.35. (b) Since $\mathfrak{a} \prec \mathfrak{b}$ implies $\mathfrak{a} \subset \mathfrak{b}$, we have $\mathfrak{a} \subset \mathfrak{b}$ and $\mathfrak{b} \subset \mathfrak{a}$.

(c) Assume $\mathfrak{a} \models \Phi[X]$. Then $\mathfrak{b} \models \Phi[X]$ (as $\mathfrak{a} \prec \mathfrak{b}$). Since $\mathfrak{b} \prec \mathfrak{c}$, we have $\mathfrak{c} \models \Phi[X]$.

12.36. Since $A \subset B$, we have ${}^{\mathscr{N}}A \subset {}^{\mathscr{N}}B$. Assume $\mathfrak{a} \models \Phi[X]$. We have $\mathfrak{c} \models \Phi[X]$. But $X \in {}^{\mathscr{N}}B$ and $\mathfrak{c} \models \Phi[X]$ so $\mathfrak{b} \models \Phi[X]$.

12.37. The implication from left to right is obvious. Assume now that, for every formula Φ and every sequence $X \in {}^{\mathscr{N}}A$, $\mathfrak{b} \models \bigvee_{x_k} \Phi[X]$ implies the existence of an element $a \in A$ such that $\mathfrak{b} \models \Phi\left[X\binom{k}{a}\right]$. We show that

(*) $\qquad \mathfrak{b} \models \Phi[X] \Rightarrow \mathfrak{a} \models \Phi[X]$

for all $X \in {}^{\mathscr{N}}A$ (by induction on the degree of complexity of the formula Φ). We can assume that Φ is built by using only the connectives $\sim$, $\wedge$ and the quantifiers $\bigvee_{x_j}$. If Φ is atomic then (*) holds (by 12.26). If (*) holds for Φ then it holds for $\sim \Phi$ as well. The case of the connective $\wedge$ is again obvious. The last case – that of the existential quantifier – is just the assumption.

12.38. *Hint:* Use 12.8, i.e. the fact that the satisfaction depends on the finite part of the sequence X.

12.41. We need to show that if $f_1 \sim g_1, \ldots, f_{n_i} g_{n_i}$ and $U=\{t : R_i(f_1(t), \ldots, f_{n_i}(t))\} \in F$ then

$$S=\{t : R_i(g_1(t), \ldots, g_{n_i}(t))\} \in F.$$

Let $U_k=\{t : f_k(t)=g_k(t)\}$ $(k=1, 2, \ldots, n)$ and $V=U_1 \cap \ldots \cap U_{n_i} \cap U$. Since $U_1, \ldots, U_{n_i}$, U belong to the ultrafilter F, therefore V, being its intersection, also belongs to F. We show that $V \subset S$, and so S also belongs to F. If $t \in V$ then $f_1(t)=g_1(t), \ldots, f_{n_i}(t)=g_{n_i}(t)$ and $R_{t_i}[f_1(t), \ldots \ldots, f_{n_i}(t)]$. But then $R_{t_i}[g_1(t), \ldots, g_{n_i}(t)]$, and so $t \in S$. A similar reasoning shows that if $S \in F$ then also $U \in F$. The reasoning for the case of operations is similar.

12.42. Note that $f \sim_F g$ iff $f(i_0)=g(i_0)$. Thus every equivalence class is determined by an element of A_{i_0}. Then we see that $S_i([f_1], \ldots, [f_{n_i}])$ is equivalent to $R_{i, i_0}[f_1(i_0), \ldots, f_{n_i}(i_0)]$.

12.43. For atomic formulas our theorem is obvious. Given a sequence X and a formula Φ, consider the set $U_{\Phi, X}=\{t : \mathfrak{a}_t \models \Phi[x_0(t), x_1(t), \ldots]\}$. Note that $U_{\Psi \wedge \Phi, X}=U_{\Phi,X} \cup U_{\Psi, X}$, and so if both sets on the right hand side of our equality belong to F then also the set on the left hand side belongs to F. Now assume that it is not true that $\prod_{i \in I} \mathfrak{a}_i/F \models \Phi[X]$. Then, by inductive assumption, the set $U_{\Phi, X}$ does not belong to F. Thus the complement $I-U_{\Phi, X}$ belongs to the ultrafilter F, But $I-U_{\Phi, X}=\{t : \mathfrak{a}_t \models \sim \Phi[x_0(t), x_1(t), \ldots]\}$ and so $U_{\sim \Phi, X} \in F$.

For the existential quantifier case one applies 12.13.

12.44. Note that the mapping $\varphi([f]_{\sim F})=[f \restriction X]_{\sim F \restriction X}$ is an injection and preserves both relations and functions.

12.45. Assume that Φ is satisfied in $\mathfrak{a}$ by $\{[f_i]\}_{i \in \mathcal{N}}$, and the functions f_i are constants, $f_i(t)=a_i$. Thus $\{t : \mathfrak{a} \models \Phi[a_0, a_1, \ldots]\} \in F$. But as that set F belongs to the ultrafilter, it is not empty, and so $\mathfrak{a} \models \Phi[a_0, a_1, \ldots]$, i.e. $\varphi * \mathfrak{a} \models \Phi[[f_0], [f_1], \ldots]$.

12.46. Note that if φ is an imbedding then $\mathfrak{a}$ is isomorphic to $\varphi * \mathfrak{a}$ and so $\mathfrak{a} \equiv \varphi * \mathfrak{a}$. By 12.33. $\varphi * \mathfrak{a} \equiv \mathfrak{b}$, and so, by 12.30 (c)), $\mathfrak{a} \equiv \mathfrak{b}$.

12.47. Let $\bar{\bar{A}}=n$. Consider the statement

$$A_n = \bigvee_{v_0 \ldots v_{n-1}} \bigwedge_{v_n} [(v_0 \neq v_1 \wedge \ldots \wedge v_0 \neq v_{n-1} \wedge \ldots \wedge v_{n-2} \neq v_{n-1}) \wedge$$

$$\wedge (v_n = v_0 \vee \ldots \vee v_n = v_{n-1})].$$

This statement says that $\bar{\bar{A}}=n$, i.e. $\mathfrak{a} \models A_n \Leftrightarrow \bar{\bar{A}}=n$. By 12.45 and 12.46, $\mathfrak{a}^I/F \models A_n$. Thus $\overline{\overline{A^I/F}}=n$. It follows that the imbedding φ of 12.45 is a surjection, since there is no proper n-element subset of a n-element set.

12.48. As noted in 12.47, the sentence A_n characterizes the property $\bar{\bar{A}}=n$. Thus, if $\{i : \bar{\bar{A}}_i=n\} \in F$, then $\{i : \mathfrak{a}_i \models A_n\} \in F$, and so, by Łoś's theorem, $\prod_{i \in I} \mathfrak{a}_i/F \models A_n$, i.e. $\overline{\overline{\prod_{i \in I} A_i/F}}=n$, which implies that the ultraproduct is finite.

12.49. Let $\mathfrak{a}_i=\langle A_i, A_i^2 \rangle$ where $\bar{\bar{A}}_i=i+1$ for $i \in \mathcal{N}$. Let F be an arbitrary, nonprincipal ultrafilter on $\mathcal{N}$. Let B_i be the statement saying that $\bar{\bar{A}} \geqslant i$, for instance:

$$B_i: \bigvee_{v_0 \dots v_{i-1}} [v_0 \neq v_1 \wedge v_0 \neq v_2 \wedge v_0 \neq v_{i-1} \wedge v_1 \neq v_2 \wedge \dots \wedge v_1 \neq$$
$$\neq v_{i-1} \wedge \dots \wedge v_{i-2} \neq v_{i-1}].$$

It is easy to see that $\{l : \mathfrak{a}_l \models B_n\} \in F$ (indeed, its complement is finite). Thus $\prod_{i \in I} \mathfrak{a}_i/F \models B_l$ for all $l \in \mathcal{N}$. Thus $\overline{\overline{\prod_{i \in I} A_i/F}} \geqslant \aleph_0$, which completes the proof.

12.50. Let T be a subfamily of $\mathcal{P}(I)$ constructed as follows: First consider the family W consisting of sets of the form $a_1 \cap \dots \cap a_n$ ($a_i \in S$, $n \in \mathcal{N}$) and then define $X \in T \Leftrightarrow \bigvee_{Y \in W} (Y \subset X)$. It is easy to see that T is a filter. Indeed, if $X_1 \in T$, $X_2 \in T$ then $a_1 \cap \dots \cap \dots a_n \subset X_1$, $b_1 \cap \dots \cap b_k \subset X_2$, and so $a_1 \cap \dots \cap a_n \cap b_1 \cap \dots \cap b_k \subset X_1 \cap X_2$. Thus $X_1 \cap X_2 \in T$. If $X_1 \in T$ and $X_1 \subset X_2$ then $X_2 \in T$.

Now consider the family of all proper filters F such that $T \subset F$. This family satisfies the assumptions of the Kuratowski–Zorn lemma. Its maximal element is an ultrafilter extending T and, since $S \subset T$, it extends S as well.

12.51. Assume that $A_F = \bigcap F$ contains at least two different elements, say $a_1, a_2, a_1 \neq a_2$. Since F is an ultrafilter, either $\{a_1\} \in F$ or $I - \{a_1\} \in F$. If $\{a_1\} \in F$ then $A_F \subset \{a_1\}$. i.e. $A_F = \{a_1\}$ and $a_2 \notin A_F$. If $I - \{a_1\} \in F$ then $A_F \subset I - \{a_1\}$, and so $a_1 \notin A_F$.

12.52. It is enough (by 12.50) to show that our family possesses the finite intersection property. Let $X_1, \dots, X_n \in \mathcal{N}_\infty$. Consider $X = X_1 \cap \cap \dots \cap X_n$, we claim that it is nonempty. We simply show that $\mathcal{N} - X$ is finite. Indeed, $\mathcal{N} - X = (\mathcal{N} - X_1) \cup \dots \cup (\mathcal{N} - X_n)$ and the union of a finite number of finite sets is finite itself. Now let F be an ultrafilter containing $\mathcal{N}_\infty$. Since the sets $U_0 = \mathcal{N} - \{0\}$, $U_1 = \mathcal{N} - \{1\}, \dots, U_n = = \mathcal{N} - \{n\}$ all belong to the family $\mathcal{N}_\infty$ and thus to the ultrafilter F as well, $\bigcap F = O$.

12.53. We can assume that $A = \mathcal{N}$. We show that there exists a family $S \subset {}^{\mathcal{N}}\mathcal{N}$ such that $\overline{\overline{S}} = 2^{\aleph_0}$ and $f, g \in S, f \neq g$ implies $\{n : f(n) = g(n)\} < \aleph_0$. Indeed given a sequence $\varphi \in {}^{\mathcal{N}}2$, define $f_\varphi(n) = \sum_{m<n} f(m) 2^m$. It is easy to see that $\varphi \neq \psi$ implies that the set $\{n : f_\varphi(n) = f_\psi(n)\}$, is finite and it is equal to the set of numbers less than the least k for which $\varphi(k) \neq \psi(k)$. Since $\{n : f_\varphi(n) \neq f_\psi(n)\}$ belongs to $\mathcal{N}_\infty$, it belongs to F as well, and so

$[f_\varphi] \neq [f_\psi]$ whenever $\varphi \neq \psi$. Thus $\overline{\overline{{}^{\mathcal{N}}\mathcal{N}/F}} \geqslant \overline{\overline{\{[f_\varphi] : \varphi \in {}^{\mathcal{N}}2\}}} = 2^{\aleph_0}$. On the other hand, $\overline{\overline{{}^{\mathcal{N}}\mathcal{N}/F}} \leqslant \overline{\overline{{}^{\mathcal{N}}\mathcal{N}}} = 2^{\aleph_0}$. By the Cantor–Bernstein theorem $\overline{\overline{{}^{\mathcal{N}}\mathcal{N}/F}} = 2^{\aleph_0}$.

12.54. We show that for every X either $X \in \bigcup_{i \in I} F_i/G$, or $(\bigcup_{i \in I} A_i - X) \in \bigcup_{i \in I} F_i/G$. Indeed, let $U_X = \{i : X \cap A_i \in F_i\}$. If $U_X \notin G$ then $\{i : X \cap A_i \notin F_i\} \in G$, but $\{i : X \cap A_i \notin F_i\} = \{i : (\bigcup_{i \in I} A_i - X) \cap A_i \in F_i\}$.

12.55. Hint: If M is of cardinality $\mathfrak{m}$ then the set $N = \{X : X \subset M \wedge \wedge \overline{\overline{X}} < \aleph_0\}$ is also of cardinality $\mathfrak{m}$. Similarly the set $P = \{X : X \subset N \wedge \wedge \overline{\overline{X}} < \aleph_0\}$ is of cardinality $\mathfrak{m}$. Construct $2^{2^{\mathfrak{m}}}$ ultrafilters on the set P. Then show how one reconstructs an ultrafilter on M from an ultrafilter on P.

12.56. (a) *Hint*: If $\bigcap F_i = \{a\}$, $\bigcap F_2 = \{b\}$, then any permutation f of the set I for which $f(a) = b$ is the required isomorphism.

(b) Use 12.55 and the fact that $\mathfrak{m}^{\mathfrak{m}} = 2^{\mathfrak{m}} < 2^{2^{\mathfrak{m}}}$.

12.57. Recall that $\mathfrak{a}_i \equiv \mathfrak{b}_i \Leftrightarrow \mathrm{Th}(\mathfrak{a}_i) = \mathrm{Th}(\mathfrak{b}_i)$. Assume $\prod_{i \in I} \mathfrak{a}_i/F \models \Phi$. Then – by Łoś's theorem – $\{i : \mathfrak{a}_i \models \Phi\} \in F$, i.e. $\{i : \Phi \in \mathrm{Th}(\mathfrak{a}_i)\} \in F$. So $\{i : \Phi \in \mathrm{Th}(\mathfrak{b}_i)\} \in F$, which means $\{i : \mathfrak{b}_i \models \Phi\} \in F$, and so $\prod_{i \in I} \mathfrak{b}_i/F \models \Phi$.

It is enough to assume that $\{i : \mathfrak{a}_i \equiv \mathfrak{b}_i\} \in F$.

12.58. Let $\mathfrak{a} \in K$; by definition, $\mathrm{Th}\, K \subset \mathrm{Th}(\mathfrak{a})$ and so $\mathfrak{a} \in \mathrm{Mod\,Th}\, K$.

12.59. If $\mathrm{Mod}\, S \neq O$ then there exists a system $\mathfrak{a}$ such that $S \subset \mathrm{Th}\, \mathfrak{a}$. Since $\mathrm{Mod}\, S$ is the class of all $\mathfrak{a}$ such that $S \subset \mathrm{Th}(\mathfrak{a})$ we have $S \subset \bigcap_{\mathfrak{a} \in \mathrm{Mod}\, S} \mathrm{Th}(\mathfrak{a})$, i.e. $S \subset \mathrm{Th}\, (\mathrm{Mod}\, S)$.

12.60. The implication from left to right was proved in 12.15. To show the converse implication use the theorem of Gödel (completeness).

12.61. Assume that S has no model. Then, by the completeness theorem S is inconsistent. Thus there exists a proof D from S of the sentence $\Phi \wedge \sim \Phi$. The set of formulas from S used in the proof D is finite, let as call this set S_0. Then S_0 is inconsistent, and so by 12.15 has no model.

12.62. (a) $X \in B_{U_1} \cap B_{U_2} \Leftrightarrow X \in B_{U_1} \wedge X \in B_{U_2}$, which means $U_1 \subset X \wedge U_2 \subset X$ thus $U_1 \cup U_2 \subset X$.

The converse implication is obvious.

(b) *Hint*: Show that our family possesses the finite intersection property.

(c) Let S be a set of sentences such that, for every finite subset S_0, $\mathrm{Mod}\, S_0 \neq O$. Let $M_{S_0} \in \mathrm{Mod}\, S_0$.

Consider the family T of all finite subsets of the sets S. Consider an arbitrary ultrafilter F in T containing all B_{S_0} (cf. (b)). Now form the ultraproduct $\prod_{S_0 \in T} M_{S_0}/F$ since for all $\Phi \in S$, $B_{\{\Phi\}} \in F$ we have $\prod_{S_0 \in T} M_{S_0}/F \models$ $\models \Phi$, i.e. $S \subset \mathrm{Th}\,(\prod_{S_0 \in T} M_{S_0}/F)$.

12.63. *Hint*: The implication from right to left is obvious. To show the converse implication extend the language **J** by adding constants for the elements of A. Similarly, expand the model $\mathfrak{a}$ by adding constants for the elements of A. Let I be the set of all sentences (of the extended language) true in $\mathfrak{a}$. If the sentence $\Phi\ (c_{a_1}, \ldots, c_{a_n})$ is true in A, then there exist elements of the universe of the model $\mathfrak{b}$ satisfying the formula Φ. Now construct in the set I an ultrafilter F using the method described in the proof of 12.62, $\mathfrak{a}$ can be elementarily embedded in $\mathfrak{b}^I/F$.

12.64. *Hint*: Use the theorem of Frayne (12.63) and 12.47.

12.65. We need to show the following properties:

(a) $R_\xi \in F_x \wedge R_\xi \subset R_\zeta \Rightarrow R_\zeta \in F_x$,

(b) $R_\xi \in F_x \wedge R_\zeta \in F_x \Rightarrow R_\xi \cap R_\zeta \in F_x$,

(c) $R_\xi \in F_x \vee A - R_\xi \in F_x$.

(a) Under our assumptions $\mathfrak{a} \models \bigwedge_x [P_\xi(x) \Rightarrow P_\zeta(x)]$. Since $\mathfrak{a}'$ is an elementary extension of $\mathfrak{a}$, we have

$$\mathfrak{a}' \models \bigwedge_x [P_\xi(x) \Rightarrow P_\zeta(x)], \quad \text{i.e.} \quad R'_\xi \subset R'_\zeta,$$

so if $x \in R'_\xi$ then $x \in R'_\zeta$ i.e. $R'_\zeta \in F_x$. (Recall that P_ξ is the symbol used to denote R_ξ).

(b) Notice that $x \in R'_\xi \wedge x \in R'_\zeta \Rightarrow x \in R'_\xi \cap R'_\zeta$.

(c) Since $\mathfrak{a} \models \bigwedge_x [P_\xi(x) \vee \sim P_\xi(x)]$, we have $\mathfrak{a}' \models \bigwedge_x [P_\xi(x) \vee \sim P_\xi(x)]$. Finally $\mathfrak{a} \models \bigwedge_x \sim (P_\xi(x) \wedge \sim P_\xi(x))$, and F is a proper filter.

12.66. By definition $K \in EC$ iff, for a certain Φ, $K = \mathrm{Mod}\, \Phi$. Then $X - K = \mathrm{Mod} \sim \Phi$ (by 12.13).

12.67. *Hint*: Use 12.61.

12.68. *Hint*: Assume $K = \mathrm{Mod}\, S$ for a certain S. Show that $K = \bigcap_{\Phi \in S} \mathrm{Mod}\, \Phi$.

12.69. (a) By 12.45, $\mathrm{Th}(\mathfrak{a})=\mathrm{Th}(\mathfrak{a}^I/F)$. Thus if $S\subset\mathrm{Th}(\mathfrak{a})$ then $S\subset\mathrm{Th}(\mathfrak{a}^I/F)$.

(b) If $S\subset\mathrm{Th}(\mathfrak{a}_t)$ for all $t\in I$ then $\{t: S\subset\mathrm{Th}(\mathfrak{a}_t)\}=I$ and so I is a member of F.

(c) Note that $\mathfrak{a}\equiv\mathfrak{b}$ is equivalent to $\mathrm{Th}(\mathfrak{a})=\mathrm{Th}(\mathfrak{b})$.

12.70. *Hint*: Consider $\mathrm{Th}\,K$. We show that $K=\mathrm{Mod}\,\mathrm{Th}(K)$. $K\subset\mathrm{Mod}\,\mathrm{Th}(K)$ by 12.58. So let $\mathfrak{a}\in\mathrm{Mod}\,\mathrm{Th}(K)$. Then $\mathrm{Th}(K)\subset\mathrm{Th}(\mathfrak{a})$. Construct an ultrafilter on the set $\mathrm{Th}(\mathfrak{a})$ applying the method used in the proof of the theorem of Frayne (12.63), and picking appropriate models from K. We get an ultraproduct $\mathfrak{b}$ of a family of elements of K which is elementarily equivalent to the model $\mathfrak{a}$. Since K is closed with respect to ultraproducts, we have $\mathfrak{b}\in K$, and since $\mathfrak{a}\equiv\mathfrak{b}$, $\mathfrak{a}\in K$ as well.

12.71. *Hint*: If K is closed with respect to $\equiv$ then also $X-K$ is closed with respect to $\equiv$. Thus both K and $X-K$ are elementary. Using the compactness theorem, we now show that K (and so $X-K$ as well) are basic elementary.

12.72. *Hint*: Use 12.26 (a).

12.73. Implication from left to right is obvious. In order to show that a class satisfying the conditions (a), (b), (c) is a UC_Δ class consider the set T of all universal sentences true in all elements of K and show that $K=\mathrm{Mod}\,T$. We show that every element of $\mathrm{Mod}\,T$ can be imbedded in an ultraproduct of elements of K.

12.74. (a) Yes. There exists a universal axiomatization of the theory of groups.

(b) Yes.

(c) No. A subsystem of a field need not be a field, and so the class K of fields is not a UC_Δ class (cf. 12.72).

12.75. (a) No. K is not closed with respect to ultraproducts (cf. 12.49).

(b) Yes. $\mathrm{Th}\,K$ consists of the sentences B_n, $n\in\mathcal{N}$ (cf. 12.49).

(c) Yes. It is a UC_Δ class.

(d) No. The class of well-orderings is not closed with respect to ultraproducts. We show that, whenever F is a nonprincipal ultrafilter in $\mathcal{N}$, $\langle\mathcal{N},\leqslant\rangle^{\mathcal{N}}/F$ is not a well-ordered set. Indeed, consider the functions $f_0=[0,1,2,\ldots]$, $f_1=[0,0,1,2,3,\ldots]$, $f_2=[0,0,0,1,2,3,\ldots]$ etc. Thus $f_k(l)=\max(0,l-k)$. Now this sequence has the following properties:

1° $\bigwedge_{n,k} [n \neq k \Rightarrow \sim(f_n \sim_F f_k)]$,

2° $n < k \Rightarrow \{l : f_k(l) < f_n(l)\} \in F$.

Thus we have constructed an infinite descending sequence $\{[f_n]\}$ of elements of the ultrapower. Thus the ultrapower is not a well-ordered set.

Note: One can show that if the ultrapower $\langle N, \leqslant\rangle^I/F$ is a well-ordered set then F is a σ-complete ultrafilter, i.e. for every sequence $\{A_n\}_{n\in\omega}$ of elements of F, $\bigcap_{n\in\mathcal{N}} A_n \in F$. Ultrafilters on sets of "small" cardinality (for instance, smaller than the least inaccessible cardinal) with this property must be principal.

The sentence asserting the existence of a set I and of a non-principal, σ-complete ultrafilter F on I is called the *axiom of a measurable cardinal.* The existence of such I and F cannot be proved in the Zermelo–Fraenkel system of axioms of set theory. Negation of the axiom of measurable cardinal is consistent with ZF.

12.76. *Hint*: If $\overline{\overline{A}} \geqslant \aleph_0$, $\overline{\overline{I}} = \mathfrak{m}$, F is a nonprincipal ultrafilter on I and F contains all subsets of I with the complements of cardinality less than $\mathfrak{m}$, then $\overline{\overline{A^I/F}} > \mathfrak{m}$ Having constructed in this way an elementary extension of $\mathfrak{a}$ of cardinality greater than $\mathfrak{m}$, take an elementary substructure of $\mathfrak{a}^I/F$ of suitable power and use 12.36.

12.77. *Hint*: By a suitable choice of I_1, I_2 and ultrafilters F_1 and F_2 build nonisomorphic ultrapowers of $\mathfrak{a} \in \text{Mod}\, X$.

12.78. Assume that S is not complete. Thus there exists a sentence Φ and systems $\mathfrak{a}_1$ and $\mathfrak{a}_2$ such that:

$$S \cup \{\Phi\} \subset \text{Th}(\mathfrak{a}_1), \quad S \cup \{\sim\Phi\} \subset \text{Th}(\mathfrak{a}_2).$$

Now construct two models $\mathfrak{b}_1$, $\mathfrak{b}_2$ of cardinality $\mathfrak{m}$ such that $\mathfrak{b}_1 \equiv \mathfrak{a}_1$ and $\mathfrak{b}_2 \equiv \mathfrak{a}_2$ (use the Löwenheim–Skolem–Tarski theorem). By assumption, $\mathfrak{b}_1$ is isomorphic to $\mathfrak{b}_2$ and so $\text{Th}(\mathfrak{b}_1) = \text{Th}(\mathfrak{b}_2)$, a contradiction.

12.79. *Hint*: By the theorem of Cantor (8.52) every two denumerable, dense, linear orderings are isomorphic. Now use the Łoś–Vaught criterion.

12.80. Use the following theorem of Steinitz: For every uncountable cardinal $\mathfrak{m}$ and prime $p \in \mathcal{N}$ all algebraically closed fields of cardinality $\mathfrak{m}$ and characteristics p are isomorphic. Use the Łoś–Vaught criterion.

12.81. *Hint*: If dim $V=n$ then V is isomorphic to K^n. Then use 12.80.

12.82. If the set S is complete, then for every Φ, either $\Phi \in \mathrm{Cn}\, S$ or $\sim\Phi \in \mathrm{Cn}\, S$. By 12.17, whenever $S \subset \mathrm{Th}(\mathfrak{a})$, $\mathrm{Cn}\, S \subset \mathrm{Th}(\mathfrak{a})$ as well. Thus for all $\mathfrak{a}, \mathfrak{b} \in \mathrm{Mod}\, S$, $\mathfrak{a} \equiv \mathfrak{b}$. The other implication uses the Löwenheim–Skolem–Tarski theorem.

12.83. Assume that such a sentence exists. By compactness there must exist a finite subcollection of our family, say $\Phi_{i_0}, \ldots, \Phi_{i_l}$ such that $\{\Phi_{i_0}, \ldots, \Phi_{i_l}\} \vdash \Phi$. Since the sequence $\{\Phi_n\}_{n \in \mathcal{N}}$ is increasing, the statement $\Phi_k \Rightarrow \Psi$ must be true (where $k = \max(i_0, \ldots, i_l)$). But $\Psi \Rightarrow \Phi_{k+1}$ is also true, and thus $\Phi_k \Rightarrow \Phi_{k+1}$ is true, contradicting our assumptions.

12.84. *Hint*: Consider the sequence $\{\Phi_n\}_{n \in \mathcal{N}}$ as follows: The statement Φ_n says that the characteristic of a field is not smaller then p_n. Show that this sequence is increasing.

12.85. Note that if $\{\Phi_n\}_{n \in \mathcal{N}}$ is increasing then $\mathrm{Mod}\,\{\Phi_n : n \in \mathcal{N}\}$ is not *EC*.

12.86. The sentences $\{B_n : n \in \mathcal{N}\}$ (cf. 12.49) form an increasing sequence.

12.88. Theory of algebraically closed fields of fixed characteristics.

12.89. The set of sentences true in $\langle \mathcal{N}, \leqslant \rangle$ (i.e. $\mathrm{Th}(\langle \mathcal{N}, \leqslant \rangle)$).

12.90. *Hint* for $\Rightarrow$: Note that the expansion $\mathfrak{a}'$ of $\mathfrak{a}$ has the property

$$\mathfrak{a} \models \Phi(x_0 \ldots x_n)[a_0, a_1, \ldots] \Leftrightarrow \mathfrak{a}' \models \Phi(c_{a_0}, \ldots, c_{a_n}).$$

The above sentence is thus true in $\mathfrak{b}'$ and so $\mathfrak{b}' \models \Phi(c_{a_0}, \ldots, c_{a_n})$ thus $\mathfrak{b} \models \Phi(x_0, \ldots, x_n)\,[a_0, a_1, \ldots]$.

Hint for $\Leftarrow$: If X is not model–complete then consider a model $\mathfrak{a}$ and a formula Φ of $\mathbf{J}_{\mathfrak{a}'}$, independent of X. Given two models, $\mathfrak{b}_1'$ and $\mathfrak{b}_2'$ (in the language $\mathbf{J}_{\mathfrak{a}'}$) such that $\mathfrak{b}_1' \models \Phi$, $\mathfrak{b}_2' \models \sim\Phi$, we can assume that $\mathfrak{b}_1' \subset \mathfrak{b}_2'$. But then $\mathfrak{b}_1' \prec \mathfrak{b}_2'$, a contradiction.

12.91. *Hint*: Consider the formulas of $\mathbf{J}_{\mathcal{N}}$:

$$\Phi_n(x) : x = \underbrace{1 + \ldots + 1}_{n}.$$

Given a model M for arithmetic, there exists a unique x in M satisfying Φ_n.

12.92. *Hint*: First construct inside such a field an isomorphic copy of $\mathscr{Z}$ and then repeat the construction of rationals from integers.

12.94. *Hint*: An ordered ring has no divisors of zero.

12.95. It is enough to show that for every $\mathfrak{a}, \mathfrak{b} \in \operatorname{Mod} X$, $\mathfrak{a} \equiv \mathfrak{b}$. By assumption there exist $\mathfrak{a}'$, $\mathfrak{b}'$ such that $\mathfrak{a}' \subset \mathfrak{a}$, $\mathfrak{b}' \subset \mathfrak{b}$, $\mathfrak{a}'$ isomorphic to $\mathfrak{b}'$. But since X is model-complete, $\mathfrak{a}' \prec \mathfrak{a}$, $\mathfrak{b}' \prec \mathfrak{b}$. But $\mathfrak{a}' \equiv \mathfrak{b}'$, and so, by 12.33, $\mathfrak{a} \equiv \mathfrak{b}$.

12.96. Any theory model-complete but not complete (cf. 12.88).

12.97. *Hint*: If φ is an automorphism of $\mathfrak{b}$ then

$$\mathfrak{b} \models \Phi[X] \Leftrightarrow \mathfrak{b} \models \Phi[\varphi \circ X],$$

where $(\varphi \circ X)_n = \varphi(X_n)$. Now use 12.8.

12.98. This is in fact a generalization of 12.95. Indeed, if $\varphi * \mathfrak{a} \prec \mathfrak{c}$, $\varphi * \mathfrak{b} \prec \mathfrak{c}$ then $\varphi * \mathfrak{a} \equiv \varphi * \mathfrak{b}$ and so $\mathfrak{a} \equiv \mathfrak{b}$.

12.99. (a) We need to show that our space possesses a clopen basis. This, in fact, is contained in 12.66.

(b) Compactness follows from the compactness theorem. Our space is Hausdorff since the intersection of all complete, consistent sets of sentences which contain Φ is $\operatorname{Cn} \Phi$.

12.100. *Hint*: Given an infinite cardinal number $\mathfrak{m}$, there exist $\mathfrak{n} > \mathfrak{m}$ such that $\mathfrak{n}^2 = \mathfrak{n}$ (namely $\mathfrak{n} = 2^{\mathfrak{m} \cdot \aleph_0}$). Thus there exists a system $\mathfrak{a} = \langle A, P \rangle$ where A is a set of cardinality $\mathfrak{n}$ and P is a bijection of A^2 and A. By the downward Lowenheim–Skolem–Tarski theorem there exists a $\mathfrak{b} \prec \mathfrak{a}$, $\bar{\bar{\mathfrak{b}}} = \mathfrak{m}$. But then $\mathfrak{b} = \langle B, Q \rangle$ where $\bar{\bar{B}} = \mathfrak{m}$, Q is a bijection of B^2 and B. Thus $\mathfrak{m}^2 = \mathfrak{m}$ and so by 10.62 the axiom of choice holds.

12.101. If $K_1 = \operatorname{Mod} \Phi$, $K_2 = \operatorname{Mod} \Psi$ then $K_1 \cap K_2 = \operatorname{Mod}(\Phi \wedge \Psi)$, $K_1 \cup K_2 = \operatorname{Mod}(\Phi \vee \Psi)$.

12.102. If $K_1 = \operatorname{Mod} S$, $K_2 = \operatorname{Mod} T$ then $K_1 \cap K_2 = \operatorname{Mod}(S \cup T)$, $K_1 \cup K_2 = \operatorname{Mod} \{\Phi \vee \Psi : \Phi \in S \wedge \Psi \in T\}$.

12.103. 12.101 says that the union and the intersection of two sets from a basis is a basic set. 12.102 says that the union and the intersection of two closed sets in the space of 12.99 is closed.

12.104. Let $\mathfrak{a} = \langle A, \prec \rangle$ be an ordered set. Extend the language $\mathbf{J}_{\mathfrak{a}}$ by adding constants for the elements of A. Consider now the diagram of $\mathfrak{a}$, i.e. the set $T = \{c_a < c_b : a \prec b\}$. Given a finite set $T_0 \subset T$, there exists a linear ordering $\leqslant$ of a finite subset of A such that whenever $c_a < c_b \in T_0$ then $a \leqslant b$. Now use the completeness theorem.

CHAPTER 13

RECURSIVE FUNCTIONS

13.5. First note that every primitive recursive function arises from the initial functions by the application of a finite number of recursions and superpositions (in the case of partial recursive functions the minimum operation can be applied as well). Then use 7.21.

13.11. *Hint*: Define the predecessor function $P(x)$ by induction as follows:

$$P(0)=0, \quad P(x+1)=x.$$

13.12. *Hint*: $x \leqslant y \Leftrightarrow x \dot{-} y = 0$.

13.13. *Hint*: $x=y \Leftrightarrow (x \dot{-} y)+(y \dot{-} x)=0$.

13.14. *Hint*: $x<y \Leftrightarrow (x+1) \dot{-} y=0$.

13.18. *Hint*: Use 13.15.

13.22. *Hint*: Use 13.20.

13.23. *Hint*: Let $\bar{\bar{A}}=\aleph_0$ and $f * \mathcal{N}=A$ where f is recursive, consider the following function g:

$$g(0)=f(0),$$

$$g(n+1)=f(\mu m)[f(m)>g(n)]$$

and use 13.24.

13.24. *Hint*: Note that if f is an increasing function then $f * \mathcal{N}=A$ implies $n \in A \Leftrightarrow \bigvee_{m \leqslant n} n=f(m)$.

In order to get the other implication consider the following function g:

$$g(0)=(\mu n) f(n)=0,$$

$$g(n+1)=(\mu m)[m>g(n) \wedge f(m)=0].$$

13.33. *Hint*: Note that if

$$f(x, y)=(\mu z \leqslant x)[g(z, y)=0],$$

where g is a recursive function, then the function f can be defined as follows:

$$f(0, y)=0,$$

$$f(n+1, y)=\begin{cases} n+1 & \text{if } f(n, y)=0 \wedge g(0, y)\neq 0 \wedge \\ & \qquad \wedge g(n+1, y)=0, \\ f(n, y) & \text{otherwise}. \end{cases}$$

13.35. *Hint*: For $n=2$ use the functions defined in 5.88.

13.36. *Hint*: Use 13.35.

13.57. Let $f(x_1, \ldots, x_n, y)$ be defined as follows:

$$f(x_1, \ldots, 0)=g(x_1, \ldots, x_n),$$

$$f(x_1, \ldots, x_n, y+1)=h[x_1, \ldots, x_n, y, f(x_1, \ldots, x_n, y)].$$

Consider the following function Ψ:

$$\Psi(x, y)=J_{n+2}\{J_n^1(x), \ldots, J_n^h(x), y, f[J_n^1(x) \ldots J_n^h(x), y]\}.$$

Prove that the function $\Psi(x, y)$ can be defined by means of superposition and recursion of type

$$\Psi(x, 0)=\chi(x),$$

$$\Psi(x, y+1)=\Theta[\Psi(x, y)].$$

Finally, introduce a function $\varphi(x, y)$ such that

$$\varphi(x, 0)=x,$$

$$\varphi(x, y+1)=\Theta[\varphi(x, y)]$$

and prove that $\Psi(x, y)=\varphi[\chi(x), y]$.

13.39. *Hint*: Note that $0^2=0$ and $(x+1)^2=\varphi(x^2)+1$ where $\varphi(y)=$ $=y+2[\sqrt{y}]$. Then prove that φ can be obtained by iterating the following function Ψ:

$$\Psi(z)=z+1+2sq((x+4)-[\sqrt{x+4}]^2).$$

13.40. *Hint*: Consider the function $(x+y)^2+5x+3y+4$ and note that for $x\geqslant y$

$$(x+y+2)^2\leqslant(x+y)^2+5x+3y+4<(x+y+3)^2,$$

and so

$$q(x, y)=(x+y)^2+5x+3y+4-[\sqrt{(x+y)^2+5x+3y+4}]^2$$
$$=\begin{cases} x-y & \text{for } x \geqslant y \\ 3x+y+3 & \text{for } x<y. \end{cases}$$

On the other hand introducing the following function

$$sqy=\begin{cases} 0, & \text{if} \quad 2|x, \\ 1, & \text{if} \quad 2\nmid x \end{cases}$$

define the function

$$f(x)=x^2+E\left[\frac{x}{2}\right]$$

and finally $E\left[\frac{x}{2}\right]=q[f(x), x^2]$.

13.41. *Hint*: Use 13.39 and 13.40.

13.42. *Hint*: Note the following identity:

$$x \cdot y=\frac{q(q((x+y)^2 \cdot x^2) \cdot y^2)}{2}.$$

13.43. *Hint*: Use the functions q and sq.

13.44. *Hint*: In view of 13.37 it is enough to show that if f is defined as follows:

$$f(x, 0)=x,$$

$$f(x, y+1)=g[f(x, y)],$$

then f may be defined by means of the iteration operation. To see this define the functions $v(x)=Q(\sqrt{x})$ and $w(x, y)=((x+y)^2+y^2)+x$ and prove that

$$Q[w(x, y)]=x, \quad v[w(x+y)]=y$$

and that whenever $Q(x+1)\neq 0$ then $Q(x+1)=Q(x)+1$ and $v(x+1) =v(x)$.

Introduce the function $\Theta(x)=f[v(x), \Phi(x)]$ and prove that it can be defined from the initial functions by using the superposition and iteration operations. Finally see that $f(x, y)=\Theta[w(y, x)]$.

13.45. *Hint*: Use 13.35. Define the function $U(n, x)$ as follows:

$$U(0, x)=S(x), \quad U(1, x)=Q(x), \quad U(2, x)=J_2^1(x),$$
$$U(3, x)=J_2^2(x),$$

then for $n>3$

$$U(4n, x)=U(J_2^1(n), x)+U(J_2^2(n), x),$$
$$U(4n+1, x)=U\{J_2^1(n), U[J_2^2(n), x]\},$$
$$U(4n+2, 0)=U(n, 0),$$
$$U(4n+2, x+1)=U[n, U(n, x)].$$

Show that U is a recursive function which is universal for unary primitive recursive functions.

13.46. *Hint*: If $U(n, x)$ is a universal function for F then $U(n, n)+1$ cannot belong to F.

13.47. *Hint*: Use 13.45 and 13.46.

13.48. *Hint*: Proof by induction on the degree of complexity of the defining formula.

SUPPLEMENT 1

INDUCTION

S1.1. Let A be a nonempty set of natural numbers. Let Z be a set defined as follows:

$$Z=\{n\in\mathcal{N} : \bigwedge_{m\in A} n<m\}.$$

There are two possible cases:

(1) $0\in A$ and then A possesses a least element – it is the number 0.

(2) $0\notin A$ and then $0\in Z$ since 0, being the least number, is then smaller than all the elements of the set A. Now the set Z is disjoint from A since if $n\in Z\cap A$ then $n<n$, which is absurd. Moreover, it is easy to see that $n\in Z$ and $m<n$, then $m\in Z$.

We show now that Z must possess a largest element. Indeed, otherwise for every $k\in Z$ there exists a larger element k_1 in Z and, since $k+1\leqslant k_1$, $k+1$ also belongs to Z. Since $0\in Z$, by the induction principle Z contains all natural numbers and thus also the elements of A. But we have proved above that Z is disjoint from A.

Now, let k_0 be the largest element of Z. Thus, for all m in A, $k_0<m$ and there exists an m in A such that $k_0+1\geqslant m$. Thus $k_0+1=m_0$ for a certain $m_0\in A$. Since, for all m in A, $k_0<m$, it follows that for all m in A, $k_0+1\leqslant m$ and so m_0 is the least element of A.

S1.2. *Hint*: Consider the set $Z=\{n\in\mathcal{N} : \bigvee_{m\in A} n\leqslant m\}$ where A is a bounded, nonempty subset of $\mathcal{N}$. Using a reasoning similar to that of S1.1, show that if A has no largest element then $Z=\mathcal{N}$.

S1.3. *Hint*: Given a set Z satisfying our assumptions, form $Z'=Z\cup\{0, 1, ..., k-1\}$ and prove that Z' satisfies the assumptions of the induction principle.

S1.4. *Hint*: Given a set Z satisfying our assumptions, form the set $Z'=\{n\in\mathcal{N} : n<k\vee \bigwedge_{k\leqslant m<n} m\in Z\}$. Prove using the induction principle that $Z'\supset\mathcal{N}$ and that $Z'-\{0, ..., k-1\}\subset Z$.

S1.5. *Hint*: Given a set Z satisfying conditions (1) and (2), consider the set $A=\{n\in\mathcal{N} : n\notin Z\}$. Prove that if Z does not contain all the natural numbers then A has a least element, say n_0. Now $n_0 \neq 0$ and so $n_0 - 1 \in \mathcal{N}$ but $n_0 - 1 \in Z$ though $(n_0 - 1) + 1 \notin Z$ because A is disjoint from Z.

S1.6. *Hint*: Given a nonempty set A bounded from above, consider the set $Z = \{n \in \mathcal{N} : \bigwedge_{m \in A} m < n\}$.

S1.7. *Hint*: Given a set Z satisfying (1) and (2) of the generalized induction principle, consider the set $Z' = \{n \in \mathcal{N} : k \leqslant n \wedge n \notin Z\}$.

S1.8. *Hint*: Given a set A consisting of integers consider its decomposition $A = A_1 \cup A_2$ where A_1 consists of the negative elements of A. If $A_1 \neq 0$ then the set $A_1^* = \{-c : c \in A_1\}$ is a nonempty set of natural numbers, bounded from above.

S1.9. *Hint*: Use the reasoning of S1.8.

S1.10. *Hint*: As in S1.8 and S1.9 use the fact that the set Z of integers is the union of the set $\mathcal{N}$ and the set $\mathcal{N}^* = \{-n : n \in \mathcal{N}\}$.

S1.11. Let $Z = \left\{n \in \mathcal{N} : 0 + \ldots + n = \frac{n(n+1)}{2}\right\}$. It is easy to see that $0 \in Z$, and so the first assumption of the induction principle is satisfied.

Now let n be a natural number belonging to Z. This means that $0 + 1 + \ldots + n = \frac{n(n+1)}{2}$. Adding $n+1$ to both sides of the above equality, we get

$$\begin{aligned} 1 + 2 + \ldots + n + (n+1) &= \frac{n(n+1)}{2} + n + 1 \\ &= \frac{n(n+1) + 2(n+1)}{2} = \frac{(n+1)(n+2)}{2} \\ &= \frac{(n+1)((n+1)+1)}{2}. \end{aligned}$$

Thus $n+1$ also belongs to Z. In this way we have proved that the assumptions of the induction principle hold. But this means that $\mathcal{N} \subset Z$, i.e. the equality is true for all natural n.

S1.17. $k = 1$.

S1.36. *Hint*: Consider the set

$$Z = \left\{n \in \mathcal{N} : \bigwedge_{\theta} \sum_{i=0}^{n} \cos i\theta = \tfrac{1}{2} + \frac{\sin(n + \frac{1}{2})\theta}{\sin \frac{1}{2}\theta}\right\}.$$

S1.46. Consider the set $Z=\{n\in\mathcal{N}:2^n>n\}$. Since $2^0=1>0$, we have $0\in Z$. Now assume $n\in Z$ i.e. $2^n>n$. Multiplying both sides of our inequality by 2, we get $2^{n+1}>2n$, and so $2^{n+1}>n+n$. If $n=0$ then $2^{0+1}=2^1=2>0+1$, and if $n\geqslant 1$ then $2^{n+1}>n+n\geqslant n+1$. In both cases $2^{n+1}>n+1$, which finishes the proof of condition (2) of the induction principle, and so $Z=\mathcal{N}$.

S1.52. *Hint*: $6^{n+3}+7^{2n+3}=6^{n+2}6+7^{2n+1}(43+6)$.

S1.54. Let p be a fixed prime number and let $Z=\{n\in\mathcal{N}:p|n^p-n\}$. Clearly $0\in Z$, and so assume that $n\in Z$, i.e. $p|n^p-n$. We want to show that $p|(n+1)^p-(n+1)$. Using the binomial formula (S1.30), we have:

$$(n+1)^p-(n+1)=n^p+\binom{p}{i}n^{p-1}+\ldots+\binom{p}{p-1}n+1-n-1$$

$$=(n^p-n)+\sum_{j=1}^{p-1}\binom{p}{i}n^{p-i}.$$

Thus it is enough to show $p\left|\binom{p}{i}\right.$ for all $0<i<p$. Consider the expression $\binom{p}{i}=\frac{p\cdot(p-1)\cdot\ldots\cdot(p-i+1)}{1\cdot 2\cdot\ldots\cdot i}$. It is an integer (see S1.30). For simplicity let $k=\binom{p}{i}$. Then $\frac{p(p-1)\cdot\ldots\cdot(p-i+1)}{1\cdot 2\cdot\ldots\cdot i}=k$, and so $p(p-1)\cdot\ldots\cdot(p-$ $-i+1)=i!\cdot k$. It follows that $p|i!\cdot k$ and, since p is a prime number and $i<p$, p has no common divisors with $i!$. Thus $p|k$. Thus $p\left|\binom{p}{i}\right.$ which completes the proof.

S1.55. *Hint*: Use the following: For all n, $2|n^2+n$.

S1.58. *Hint*: $11^{n+3}+12^{2n+3}=11^{n+2}11+12^{2n+1}(133+11)$.

S1.61. *Hint*: Use the generalized induction principle with $k=3$.

S1.62. For 0 and $n>3$.

S1.63. For 0, 1 and $n>7$.

S1.64. For 0 and $n>5$.

S1.65. For $n>2$.

S1.66. Only for the numbers 3 and 4.

S1.67. The inequality never holds.

S1.68. For $n>6$.

S1.75. *Hint*: Use S1.18.

S1.77. *Hint*: Use S1.34, taking $f(n)=2^n$.

S1.78. *Hint*: Use S1.77.

S1.79. *Hint*: Use the following inequality:

$$\frac{a^2}{c}+\frac{b^2}{d}\geqslant\frac{(a+b)^2}{c+d} \qquad (a,b,c,d>0).$$

SUPPLEMENT 2

LATTICES AND BOOLEAN ALGEBRAS

S2.1. Assume $a=a\vee b$. Then $a\wedge b=(a\vee b)\wedge b$. By commutativity (L2) $a\wedge b=b\wedge(b\vee a)$. By L4 (a), $b\wedge(b\vee a)=a$, thus $a\wedge b=b$. The converse reasoning is analogous.

S2.2. We have to show that the relation $\leqslant$ is reflexive, antisymmetric, and transitive. To see the reflexivity note that $a\leqslant a$ is equivalent (by definition) to $a\vee a=a$. The latter formula is just our axiom L1 (b).

Assume now that $a\leqslant b$ and $b\leqslant a$, i.e.

$$a\vee b=b \quad \text{and} \quad b\vee a=a\,.$$

By the commutativity of the operation $\vee$ and the transitivity of the identity relation we get $a\vee b=b$ and $a\vee b=a$ and $a=b$.

We show the transitivity of the relation $\leqslant$. Assume $a\leqslant b$ and $b\leqslant c$, i.e. $a\vee b=b$ and $b\vee c=c$. Consider $a\vee c$. Since $b\vee c=c$, $a\vee c=a\vee(b\vee c)$. Since the operation $\vee$ is associative (L3 (b)) we get $a\vee c=(a\vee b)\vee c$, but since $a\vee b=b$, we have

$$(a\vee b)\vee c=b\vee c, \quad \text{i.e.} \quad a\vee c=b\vee c=c.$$

S2.3. Indeed, $a\vee b$ is a suitable c.

S2.4. (a) It is enough to show that if $d\leqslant a$ and $d\leqslant b$ then $d\leqslant a\wedge b$. By S2.1 we have $d\leqslant a\Leftrightarrow d\wedge a$, $d\leqslant b\Leftrightarrow d\wedge b=d$; thus $d\wedge(a\wedge b)=(d\wedge a)\wedge b=d\wedge b=d$ and so $d\leqslant a\wedge b$.

(b) *Hint*: Prove that $a\leqslant d$ and $b\leqslant d$ implies $a\vee b\leqslant d$.

S2.5. Consider the expression $(a\wedge b)\wedge a$. By L2a it is equal to $a\wedge(a\wedge b)$. Now apply L3 (a) and L1 (a), getting $(a\wedge a)\wedge b$ and finally $a\wedge b$. Thus $(a\wedge b)\wedge a=a\wedge b$, i.e. $a\wedge b=a$. The second and the third inequalities follow from S2.4.

S2.6. We need to check axioms (L1)–(L5) in our system. (L1) and (L2) are obvious. (L3) is translated in to the language of arithmetic as follows:

$$GCD[x, GCD(y, z)]=GCD[GCD(x, y), z],$$

$$LCM[x, LCM(y, z)]=LCM[LCM(x, y), z].$$

Both the above statements are true. Since

$$x \mid LCM(x, y)$$

and

$$x \mid y \Rightarrow (GCM(x, y) = x \wedge LCM(x, y) = y),$$

axiom (L4) is true as well.

Similarly check axiom (L5).

Note: One can show that our lattice is isomorphic to the so-called *weak Cartesian product* of a denumerable number of copies of the lattice $\langle \mathcal{N}, \wedge, \vee \rangle$ (where $\wedge$ and $\vee$ are generated by the natural ordering of $\mathcal{N}$). Indeed, the weak Cartesian product is the set of sequences eventually equal to 0. Introduce the operations $\wedge$ and $\vee$ on sequences as follows:

$$\{x_n\}_{n \in \mathcal{N}} \wedge \{y_n\}_{n \in \mathcal{N}} = \{\min(x_n, y_n)\}_{n \in \mathcal{N}},$$

$$\{x_n\}_{n \in \mathcal{N}} \vee \{y_n\}_{n \in \mathcal{N}} = \{\max(x_n, y_n)\}_{n \in \mathcal{N}}.$$

Since every natural number $x > 0$ is decomposed into the product $x = p_0^{x_0} p_1^{x_1} \dots p_n^{x_n}$ (where p_k is the kth prime number, i.e. $p_0 = 2$, $p_1 = 3$ etc.), the mapping: $x \to (x_0, x_1, \dots)$ is well defined. It is the required isomorphism.

S2.7. Introduce in the set X the operations $\min_{\leqslant}$ and $\max_{\leqslant}$ as follows:

$$\min_{\leqslant}(x, y) = \begin{cases} x & \text{if} \quad x \leqslant y, \\ y & \text{if} \quad y \leqslant x, \end{cases}$$

$$\max_{\leqslant}(x, y) = \begin{cases} x & \text{if} \quad y \leqslant x, \\ y & \text{if} \quad x \leqslant y. \end{cases}$$

The system $\langle X, \min_{\leqslant}, \max_{\leqslant} \rangle$ is a lattice and the derived ordering is identical with $\leqslant$.

S2.8. Proof by induction on the number of elements of the set X_0 (remember that X_0 is finite). If X_0 is a one-element set, then our theorem is obvious. So assume $\bar{\bar{X}}_0 = k + 1$. $X_0 = \{x_0, \dots, x_{k-1}, x_k\}$. The set $X_1 = \{x_0, \dots, x_{k-1}\}$ has cardinality k and so, by the inductive assumption, possesses an infimum, y. Consider the element $y \wedge x_k$. It is smaller than all elements of X_0 since it is smaller than all elements of X_1, and also it is smaller than x_k. So we just need to show that if t is smaller than all elements of X_0, then t is smaller than $y \wedge x_k$. But since t is smaller than any element of X_1 (indeed $X_1 \subset X_0$), we have $t \leqslant y$, and now just use S2.4a.

S2.9. Define the notion of the supremum as follows: If $X \subset L$ then t is a supremum of X if it satisfies the condition:

$$\bigwedge_x \bigwedge_y [(y \in X \Rightarrow y \leqslant x) \Rightarrow t \leqslant x].$$

Every finite subset $X \subset L$ possesses a supremum. The proof of this fact is analogous to that of S2.8.

S2.10. An example: Let L consist of subsets $X \subset \mathcal{N}$ such that $\mathcal{N} - X$ is a finite set. Define the operations $\wedge$ and $\vee$ as follows:

$$X \wedge Y = X \cap Y, \qquad X \vee Y = X \cup Y.$$

The family consisting of the sets $\mathcal{N} - \{0\}$, $\mathcal{N} - \{1\}, \ldots, \mathcal{N} - \{k\}, \ldots$ $\ldots, (k \in \mathcal{N})$ does not possess infimum since here the derived relation $\leqslant$ is the inclusion relation and the only set included in every element of the above family is the empty set, which is not an element of the class L.

Obviously the set L cannot be finite, since every subset of a finite set is finite and by S2.8 every finite subset of a lattice possesses an infimum.

Note 1: In our example every subclass of L possesses a supremum.

Note 2: The above construction should lead the reader to an example of a lattice L with the following properties: Every denumerable subclass of the lattice L possesses an infimum, but there exists a subclass of L without an infimum.

S2.11. *Hint*: Proof by induction on the number of elements of $X \cup Y$.

S2.12. The proof of the fact that the definition of the relation $\leqslant$ is sound: Assume $A \sim A_1$, $B \sim B_1$, $A - B$ finite. Then $A_1 - B_1 \subset (A_1 \dot{-} A) \cup$ $\cup (A - B) \cup (B_1 \dot{-} B)$. Since all the sets on the right-hand side are finite, their union is also finite, and so $A_1 - B_1$ is a finite set.

Define now $[A] \wedge [B] = [A \cap B]$, $[A] \vee [B] = [A \cup B]$. We need to show that the above definitions are sound (i.e. do not depend on the choice of representatives of the equivalence classes). The rest is obvious.

S2.13. This follows from the fact that our axiomatization has the duality property, i.e. substituting $\wedge$ for $\vee$ and conversely does not change the axiomatization. The result of our operation is called a dual lattice. Since the axioms (L5) (a) and (L5) (b) have the same duality property, the dual of a distributive lattice is again a distributive one. Finally, note that $(\mathscr{L}')=\mathscr{L}$.

S2.14. The common part of any family of closed sets is itself a closed set (thus an intersection of two closed sets is closed). A union of a finite number of sets is again closed. Distributivity follows from the usual distributivity law for the set operations $\cap$ and $\cup$.

S2.15. Assume that $\mathscr{L}$ is a distributive lattice and $a \leqslant c$. Thus $a \vee c = c$ and $a \vee (b \wedge c) = (a \vee b) \wedge (a \vee c) = (a \vee b) \wedge c$.

S2.16. It is easy to see that if the lattice $\mathscr{L}$ sublattice of the form described by Fig. 45 then $\mathscr{L}$ is not a modular lattice. Conversely, if $\mathscr{L}$ is not a modular lattice then there exst elements a, b, c such that $a \leqslant c$ but $a \vee (b \wedge c) \neq (a \vee b) \wedge c$. Take $d = b \vee c$ and $e = b \wedge c$. Then the elements a, b, c, d, e form a sublattice of the required form.

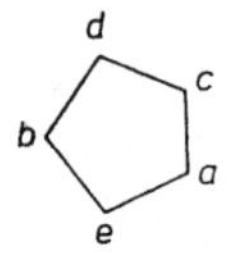

Fig. 45

S2.17. An example of a distributive lattice: $\langle \mathscr{P}(X), \cap, \cup \rangle$, where X is an arbitrary set.

An example of a non-distributive lattice: A non–modular lattice is clearly non-distributive. Consider the five–element lattice of S2.16.

We have proved in this way that both (L5) and its negation are consistent with the axioms (L1)–(L4). Thus (L5) is independent of (L1)–(L4).

S2.18. (a) Compare S2.16.

(b) Consider, the family of all the linear subspaces of a Hilbert space. Define the operations $\wedge$ and $\vee$ as follows: $A \wedge B = A \cap B$, $A \vee B$ is the least linear subspace containing both A and B. The lattice thus obtained is a modular lattice but not a distributive one. In this way we have proved the independence of (L5) from axioms (L1)–(L4), $(L4\frac{1}{2})$.

S2.19. The operations $\wedge$ and $\vee$ are just the set-theoretical operations $\cap$ and $\cup$.

S2.20. The operation $\wedge$ is the set-theoretical intersection $\cap$. Since we have proved (4.125) that intersection of two equivalences is an equivalence, this is a sound definition. The operation $\vee$ is defined in the different way; $R \vee S$ is the intersection of all the equivalences containing both R and S. Checking axioms (L1) and (L2) is obvious. The case of (L3) and (L4) is not immediate but fairly easy. Notice that $R = R \vee S$ implies $R \vee S \subset R$ and so $S \subset R$.

Conversely, if $S \subset R$ then $S \cup R = R$, and so R is the least equivalence containing both R and S, thus $S \leqslant R$.

S2.21. (a) Consider the expression $a \vee a$. By (L1) (a), $a \vee a = a \vee (a \wedge a)$. But now we use (L4) (b) and get $a \vee a = a$.

(b) It is enough to show that L1b follows from (L2), (L3) and (L4). Indeed, $a = a \vee (a \wedge b)$ (by (L4) (b) and $a \wedge a = a \wedge (a \vee (a \wedge b))$. But the right hand side of the last equation is the particular case of (L4) (a) (replace b by $a \wedge b$). Thus $a \wedge a = a$.

(c) In order to show (L5) (b), consider the expression $a \vee (b \wedge c)$. By (L4) (b), $a = a \vee (a \wedge c)$ and so

$$a \vee (b \wedge c) = [a \vee (a \wedge c)] \vee (b \wedge c) = a \vee [(a \wedge c) \vee (b \wedge c)]$$

(by (L3) (b)). Now use (L5) (a), getting

$$a \vee [c \wedge (a \vee b)] \quad \text{and} \quad [a \wedge (a \vee b)] \vee [c \wedge (a \vee b)],$$

and so (L5) (a) again $(a \vee c) \wedge (a \vee b)$, which by commutativity is $(a \vee b) \wedge$ $\wedge (a \vee c)$.

S2.22. (a) The infimum is the greatest element which is smaller than any element of x. Thus, in particular, it is smaller than all the elements $x \in X$.

(b) Reason as in (a).

(c) By the assumptions, $X \neq O$. Let $x \in X$. Then $\bigcap X \leqslant x$ (by (a)), $x \leqslant \bigcup X$ (by (b)). Thus $\bigcap X \leqslant \bigcup X$.

S2.23. *Hint*: If $x \in X$ then $x \leqslant a \vee x \leqslant \bigcup \{a \vee x : x \in X\}$. Since x is arbitrary, $\bigcup X \leqslant \bigcup \{a \vee x : x \in X\}$.

S2.25. (a) Notice that $\bigcap_{t \in T} a_t \leqslant a_t \vee b_t$ (for every $t \in T$); thus $\bigcap_{t \in T} a_t$ is smaller than any b_t and so $\bigcap_{t \in T} a_t \leqslant \bigcap_{t \in T} b_t$.

(b) Reason as in (a).

S2.26. We need to show that:

1° $\bigcup_{t\in T}(a_t\vee b_t)\leqslant \bigcup_{t\in T} a_t\vee \bigcup_{t\in T} b_t)$

and

2° $\bigcup_{t\in T} a_t\vee \bigcup_{t\in T} b_t\leqslant \bigcup_{t\in T}(a_t\vee b_t)$.

1° Since for every t, $a_t\leqslant \bigcup_{t\in T} a_t$ and $b_t\leqslant \bigcup_{t\in T} b_t$, so for every t

$$a_t\vee b_t\leqslant \bigcup_{t\in T} a_t\vee \bigcup_{t\in T} b_t.$$

Thus the supremum of $a_t\vee b_t$ is smaller than $\bigcup_{t\in T} a_t\vee \bigcup_{t\in T} b_t$.

2° Since for every t, $a_t\leqslant a_t\vee b_t$, we have

$$a_t\leqslant \bigcup_{t\in T}(a_t\vee b_t),$$

and so

$$\bigcup_{t\in T}(a_t\vee b_t)$$

s greater than a_t for all t, thus

$$\bigcup_{t\in T} a_t\leqslant \bigcup_{t\in T}(a_t\vee b_t).$$

Similarly $\bigcup_{t\in T} b_t\leqslant \bigcup_{t\in T}(a_t\vee b_t)$ and so the element $\bigcup_{t\in T} a_t\vee \bigcup_{t\in T} b_t$ is smaller than $\bigcup_{t\in T}(a_t\vee b_t)$.

(b) Reason as in (a).

(c) Since for every $t\in T$ we have $a_t\leqslant a_t\vee b_t$ and $b_t\leqslant a_t\vee b_t$, by S2.25 a we get:

$$\bigcap_{t\in T} a_t\leqslant \bigcap_{t\in T}(a_t\vee b_t), \qquad \bigcap_{t\in T} b_t\leqslant \bigcap_{t\in T}(a_t\vee b_t)$$

and finally

$$\bigcap_{t\in T} a_t\vee \bigcap_{t\in T} b_t\leqslant \bigcap_{t\in T}(a_t\vee b_t).$$

S2.27. The lattice $\langle \mathscr{P}(X), \cap, \cup\rangle$ is a complete lattice. For an example of incomplete lattice, see S2.10. The completeness of finite lattices was shown in S2.8.

S2.28. (b) Assume that a_0, a_1 have the properties

$$\bigwedge_x (a_0 \vee x = a_0), \quad \bigwedge_x (a_1 \vee x = a_1).$$

Since the operation $\vee$ is a commutative one, we have $\bigwedge_x (a_0 \vee x = a_0)$, $\bigwedge_x (x \vee a_1 = a_1)$. Substitute a_1 for x in the former formula, and a_0 for x in the latter formula. We get the equalities

$$a_0 \vee a_1 = a_0, \quad a_0 \vee a_1 = a_1,$$

and so $a_0 = a_1$.

(b) Cf. S2.6.

(c) Consider the dual lattice of the lattice of (b).

(d) Yes, if $\bar{\bar{L}} = 1$.

S2.29. (b) Assume that a is the unity in $\mathscr{L}'$, i.e. $\bigwedge_x (x \vee' a = a)$. Then, by the definition of $\vee'$, $\bigwedge_x (x \wedge a = a)$, i.e. a is the zero of $\mathscr{L}$.

(f) The dual of a modular lattice is itself a modular one. Indeed, if the lattice $\mathscr{L}'$ possesses a 5-element sublattice of the type described in S2.16, the lattice $\mathscr{L}$ possesses as a sublattice its dual, which happens to be the same 5-element lattice.

S2.31. Assume (I1) and (I2). Then one of the implications of $(\mathrm{I}\frac{1}{2})$ is simply statement (I1) and the other follows from the fact that

$$a \leqslant a \vee b \quad \text{and} \quad b \leqslant a \vee b.$$

Now assume $(\mathrm{I}1\frac{1}{2})$. Since $(p \Leftrightarrow q) \Rightarrow (p \Rightarrow q)$ is a tautology, $(\mathrm{I}1\frac{1}{2}$ implies (I1). Since $b \leqslant a$ is equivalent to $b \vee a = a$ and $a \in I$, we have $b \vee a \in I$ and so $b \in I$.

S2.32. (b) Since an ideal is by definition nonempty, if $x \in I$ then, since $0 \leqslant x$, $0 \in I$.

(c) Follows from (a) and (b).

S2.33. If $x \in (a)$ and $y \in (a)$ then $x \leqslant a$ and $y \leqslant a$ and so $x \vee y \leqslant a$, i.e. $x \vee y \in (a)$.

If $x \leqslant y$ and $y \in (a)$ then $x \leqslant y \leqslant a$, i.e. $x \leqslant a$ thus $x \in (a)$.

S2.34. If a lattice possesses the zero element then, by S2.32 (c), $\{0\}$ is the intersection of all the ideals in $\mathscr{L}$, and so this intersection is a nonempty set.

Conversely, assume that the intersection T of all the ideals in $\mathscr{L}$ is nonempty. Note that T must be a one-element set since $x \in T$, $y \in T$, implies, (x) being an ideal, $T \subset (x)$ and so $y \in (x)$ thus $y \leqslant x$. Similarly $x \leqslant y$ and so $x = y$. Thus T is a one-element set.

Now let u be the unique element of the set T. By the above reasoning, if $y \leqslant u$ then $y = u$. Consider an arbitrary $z \in L$, then $z \wedge u \leqslant u$, and thus $z \wedge u = u$. Since z is arbitrary, u must be the zero element of $\mathscr{L}$.

S2.35. Compare S2.30. Remember that $\wedge$ has two meanings: as a conjunction and as a lattice operation.

S2.36.–S2.38. Compare S2.31–S2.34.

S2.39. We show that every filter in a finite lattice is a principal filter.

Indeed, F is a finite set and so it possesses an infimum, t_F. We show that $F = (t_F)$. The inclusion $F \subset (t_F)$ is obvious.

Now, for the converse inclusion, notice that any filter is closed with respect to finite intersections the proof is analogous to that of S2.8.

Note: For every filter F, if $x \in F$ then $[x] \subset F$.

S2.40. An example: Consider the set of rationals $\mathscr{Q}$ with the lattice operations determined by the usual $\leqslant$ ordering. Consider the set $\{x : x < < \sqrt{2}\}$. It is an ideal in the lattice $\langle \mathscr{Q}, \wedge, \vee \rangle$ but since there is no largest rational number $< \sqrt{2}$ it is not a principal ideal.

S2.41. Notice that $0 = \bigcap L$, $1 = \bigcup L$, and then use S2.27.

S2.42. If x and y are $\leqslant$-incomparable then $\{x, y\}$ is not closed with respect to lattice operations and so it is not a sublattice of the lattice L.

S2.43. Assume $a \wedge b_0 = 0$, $a \vee b_0 = 1$, $a \wedge b_1 = 0$, $a \vee b_1 = 1$. Then $(a \wedge b_0) \vee b_1 = 0 \vee b_1$, i.e. $(a \vee b_1) \wedge (b_0 \vee b_1) = b_1$. Thus $1 \wedge (b_0 \vee b_1) = b_0 \vee b_1 = b_1$ and so $b_0 \leqslant b_1$. Similarly $b_1 \leqslant b_0$ and so $b_0 = b_1$.

S2.44. Use the reasoning of S2.43.

S2.45. $a \vee (\sim a) = 1$, $a \wedge (\sim a) = 0$, $(\sim a) \vee (\sim \sim a) = 1$, $(\sim a) \wedge (\sim \sim a) = 0$. Now use the commutativity of operations.

S2.46. We prove the first of our equalities. By S2.44 it is enough to prove that:

$$1^\circ \qquad (p \wedge q) \vee (\sim p \vee \sim q) = 1,$$

$$2^\circ \qquad (p \wedge q) \wedge (\sim p \vee \sim q) = 0.$$

$$1^\circ \quad (p\wedge q)\vee(\sim p\vee\sim q)=[p\vee(\sim p\vee\sim q)]\wedge[q\vee(\sim p\vee\sim q])$$
$$=[(p\vee\sim p)\vee\sim q]\wedge[(q\vee\sim q)\vee\sim p]$$
$$=[1\vee\sim q]\wedge[1\vee\sim p]=1\wedge 1=1\,,$$
$$2^\circ \quad (p\wedge q)\wedge(\sim p\vee\sim q)=[(p\wedge q)\wedge\sim p]\vee[(p\wedge q)\wedge\sim q]$$
$$=[(p\wedge\sim p)\wedge q]\vee[(q\wedge\sim q)\wedge p]$$
$$=[0\wedge q]\vee[0\wedge p]=0\vee 0=0\,.$$

S2.47. $\sim' x=\sim x$, $0'=1$, $1'=0$. A filter becomes an ideal; an ideal becomes a filter.

S2.48. *Note*: Though the proof is very easy, it should be closely examined, since our algebra has a large number of applications.

S2.49. This problem needs some experience. Below, you can see the main steps of the proof, i.e. the principal facts which have to be proved:

(a) If $P\subset Q$ then $X-\text{Cl}\,Q\subset X-\text{Cl}\,P$.

(b) If $P=\text{Int}\,P$ then $P\subset\text{Int Cl}\,P$.

(c) $X-\text{Cl}\,P=X-\text{Cl Int Cl}\,P$.

(d) $\text{Int Cl}(P\cap Q)=\text{Int Cl}\,P\cap\text{Int Cl}\,Q$.

S2.50. Use S2.49 (b).

S2.51. Assume $a=\bigcup_{t\in T} a_t$, thus, for all $t\in T$, $a_t\leqslant a$ and so $\sim a \leqslant \sim a_t$ (by S2.45, part 2).

Now, let $q\leqslant\sim a_t$ for all $t\in T$. Then $\sim q\geqslant\sim\sim a_t$ and so $\sim q\geqslant a_t$ (for all t). Thus $\sim q\geqslant\bigcup_{t\in T} a_t$ i.e. $\sim q\geqslant\sim a$, so $\sim\sim q\leqslant\sim a$, i.e. $q\leqslant\sim a$. Thus $\sim a$ is the infimum of $\{\sim a_t: t\in T\}$.

S2.52. Since we are proving an equivalence, we have to show two implications. .

(a) Assume $a\vee b\in F\Rightarrow a\in F\vee b\in F$. Since $1\in F$ (by S2.37 (b)), we have $a\vee\sim a\in F$ and thus $a\in F$ or $\sim a\in F$.

(b) Assume $a\in F\vee\sim a\in F$ (for all a). Let $a\vee b$ be an element of F. If both $a\notin F$ and $b\notin F$ then $\sim a\in F$ and $\sim b\in F$; thus $\sim a\wedge\sim b\in F$, and so by de Morgan's law $\sim(a\vee b)\in F$ thus $(a\vee b)\wedge\sim(a\vee b)\in F$, i.e. $0\in F$ which is a contradiction.

S2.53. We prove the converse theorem: Assume that F is a maximal filter, i.e. G extending F implies $F=G\vee G=A$. We have to show that F

is a prime filter, i.e. for all $a \in A$, $a \in F \vee \sim a \in F$. So assume that $a \notin F$ and $\sim a \notin F$. Consider the following set G: $G = \{x : \bigvee_{y \in F} y \wedge a \leqslant x\}$.

Clearly G is a filter. Also $F \subset G$ and $a \in G$. Since $a \notin F$, $F \neq G$ and so $G = A$, i.e. $\sim a \in G$. Thus there exists a $y \in F$ such that $y \wedge a \leqslant \sim a$. This immediately implies $(y \wedge a) \wedge a \leqslant \sim a \wedge a$, i.e. $y \wedge a \leqslant 0$. This, means, however, that $y \leqslant \sim a$ and, since $y \in F$, $\sim a \in F$, which is a contradiction.

S2.54. The proof uses the Kuratowski–Zorn lemma. It is enough to prove that the union of a chain of proper filters is itself a proper filter.

Indeed, let W be the union of a chain $\{F_t\}_{t \in T}$ of proper filters. If $a, b \in W$ then there exists a t such that $a \in F_t$ and $b \in F_t$ which is equivalent (by (F$1\frac{1}{2}$)) to $a \wedge b \in F_t$ and so $a \wedge b \in W$. Clearly W is a proper filter.

Thus we have proved that the assumptions of the Kuratowski–Zorn Lemma are satisfied; thus the maximal element exists. It is the required filter.

S2.55. *Hint*: Modify a little the proof of S2.54; instead of the family of all proper filters, consider only those filters which do not contain a; Prove that the set W does not contain a.

S2.56. We prove that $f^{-1} * \{1_1\}$ is a filter. Indeed, if $a, b \in f^{-1} * * \{1_1\}$ then $f(a) = f(b) = 1_1$ and, since $f(a \wedge b) = f(a) \wedge_1 f(b)$, we have $f(a \wedge b) = 1_1$.

Assume $a \wedge b \in f^{-1} * \{1_1\}$, then $f(a \wedge b) = 1_1$, i.e. $f(a) \wedge_1 f(b) = 1_1$. But $x \wedge y = 1$ implies $x = 1 \wedge y = 1$, thus $f(a) = 1_1, f(b) = 1_1$, i.e. $a \in f^{-1} * * \{1_1\}, b \in f^{-1} * \{1_1\}$.

S2.57. We prove the converse theorem. Assume that $h : \mathfrak{a} \to 2$ is a Boolean homeomorphism. Consider the set $h^{-1} * \{1\}$. By S2.56 it is a filter. Thus we have to show that, for arbitrary $x \in A$, $x \in h^{-1} * \{1\}$ or $\sim x \in h^{-1} * \{1\}$. Assume that this is not true, then $x \in h^{-1} * \{0\}$ and $\sim x \in h^{-1} * \{0\}$, but in that case

$$h(1) = h(x \vee \sim x) = h(x) \vee h(\sim x) = 0 \vee 0 = 0.$$

But $h(1) \neq 0$ as h is a Boolean homomorphism.

S2.58. Consider the set $\{0_{\mathfrak{a}}, 1_{\mathfrak{a}}\}$.

S2.59. It is clear that our condition is a sufficient one, since any valuation in the sense of chapter 1 is in fact a valuation in the Boolean algebra $\mathbf{2} = \{0, 1\}$.

Here is the proof of necessity. Otherwise there exists a Boolean algebra $\mathfrak{a}$ and a valuation v such that under the valuation v, the value of φ is equal to a certain $a \neq 1$. Using S2.55, we get a maximal filter F in $\mathfrak{a}$ such that $a \notin F$. Now, by S2.57, there exists a homomorphism h of the algebra $\mathfrak{a}$ into 2 such that $h(a)=0$. This in turn allows to construct a valuation $\bar{v}_1$ of φ into 2 such that $\bar{v}_1\varphi=0$ (take as $v_1(p)$ the element $h(v(p))$).

S2.60. Note that condition ($I1\frac{1}{2}$) is dual to ($F1\frac{1}{2}$).

S2.61. Reflexivity and symmetry are obvious. Transitivity follows by multiple use of the distributivity laws. In the set $\{[x] : x \in A\}$ define the operations as follows:

$$[a] \wedge_1 [b] = [a \wedge b], \quad [a] \vee_1 [b] = [a \vee b], \quad \sim_1 [a] = [\sim a].$$

It is easy to check that the above definitions does not depend on the choice of representatives.

S2.62. The proof follows the lines of the so-called "first isomoprhism theorem" from the algebra course.

S2.63. If you do not want to carry out the proof, notice that it is a particular case of S2.64, with $T=\{1, 2\}$ and, instead of $\mathfrak{a}_2$, its dual is considered.

S2.64. Note: Incidentally you have also proved the following statements:

1° A generalized product of lattices is a lattice.

2° If all the factors are modular then the product is a modular lattice (and similarly for distributivity). The converse also holds.

S2.65. In our algebra the atoms are sets of the form $\{t\}$ (for $t \in X$). Thus if $U \in \mathscr{P}(X)$ and $U \neq O$ then, for all $t \in U$, $\{t\} \subset U$, if the set X is empty then the theorem is obvious.

S2.66. A direct proof is possible, but the following non-effective proof works: Let $0 \neq x \in A$. Consider the principal filter $[x]$ determined by $[x]$. Now consider any maximal filter G containing $[x]$ (cf. S2.54). By S2.39 G is a principal filter. If $G=[t]$ then t is an atom and clearly $t \leqslant x$.

S2.67. Clearly h preserves the operations. In order to show that it is an isomorphism we have to prove that h is an injection.

Let $a \neq b$. We can assume that $a \wedge \sim b \neq 0$ and so there is an atom $t \leqslant a \wedge \sim b$. Thus $t \in h(a)$ and $t \notin h(b)$.

Notice that the completeness of the algebra $\mathfrak{a}$ intervenes while showing that h is a surjection, we infer that the following is true: Every atomic Boolean algebra is isomorphic to a subalgebra of the algebra $\langle \mathscr{P}(At), \cap, \cup, -, O, At \rangle$ where At is the set of atoms in the algebra $\mathfrak{a}$.

S2.68. The algebra $\mathfrak{a}$ is finite, and thus complete and atomic. Thus $\mathfrak{a}$ is iso norphic to the field of subsets of its set of atoms.

S2.69. Assume that a is an atom and $x \notin [a]$, i.e. $x \wedge a = 0$. Then $\sim x \wedge a$ cannot be zero since otherwise

$$(x \wedge a) \vee (\sim x \wedge a) = a = 0.$$

Thus $\sim x \wedge a \neq 0$, i.e. $a \leqslant \sim x$.

S2.71. (a) Assume $F \in X_{a \wedge b}$. Thus $a \wedge b \in F$ and by (F1$\frac{1}{2}$) $a \in F$ and $b \in F$, i.e. $F \in X_a$ and $F \in X_b$, so $X_{a \wedge b} \subset X_a \cap X_b$.

Assume $F \in X_a \cap X_b$. Then $a \in F$ and $b \in F$; so $a \wedge b \in F$ and thus $F \in X_{a \wedge b}$, i.e. $X_a \cap X_b \subset X_{a \wedge b}$.

(b) $F \in X_a \cup X_b$ is equivalent to $F \in X_a \vee F \in X_b$, i.e. $a \in F$ or $b \in F$; in any case $a \vee b \in F$, i.e. $F \in X_{a \vee b}$.

If $F \in X_{a \vee b}$ then $a \vee b \in F$ but, since F is a prime filter, $a \in F$ or $b \in F$; thus $F \in X_a \vee F \in X_b$, i.e. $F \in X_a \cup X_b$.

(c) If $F \in X - X_a$ then $F \notin X_a$, i.e. $a \notin F$. But F is a prime filter and so $\sim a \in F$, i.e. $F \in X_{\sim a}$. If $F \in X_{\sim a}$ then $\sim a \in F$. Since F is a proper filter, $a \notin F$, i.e. $F \notin X_a$.

S2.72. Prove that all the sets X_a are clopen. If $a \neq b$ then we can assume without loss of generality that $a \wedge \sim b \neq 0$, and so there exists a prime filter which contains a and does not contain b.

S2.73. This is a consequence of the construction of S2.71 and S2.72. This result is called the Theorem of Stone.

S2.74. Indeed, if a is an atom then there is no $x \neq 0$, $x \neq a$ such that $0 \leqslant x \leqslant a$.

S2.75. Clearly $a \dot{-} b = b \dot{-} a$. Associativity of the addition and distributivity are quite simple. The subtraction operation is identical with addition. Indeed,

$$a + a = a + a^2 = a^2 + a^2 + a^2 + a^2 = (a + a) + (a + a),$$

Thus

$$a + a = 0.$$

S2.76. We check the associativity law:

$$\begin{aligned} a\cdot'(b+'c) &= a\cdot'(b+c-2bc)\\ &= a\cdot b+a\cdot c-2abc\\ &= a\cdot b+a\cdot c-2a^2bc\\ &= a\cdot b+a\cdot c-2(ab)\cdot(ac)=(a\cdot'b)+'(a\cdot'c). \end{aligned}$$

S2.77. We check that: $a\wedge(a\vee b)=a$.

$$a\wedge(a\vee b)=a\cdot(a+b+ab)=a^2+ab+a^2b=a+ab+ab=a,$$

since $ab+ab=0$.

S2.78. Proof by induction on the degree of complexity of the polynomial.

(a) The variables 0 and 1 are normal polynomials.

(b) Assume that f_1 and g_1 are normal polynomials corresponding to f and g.

For the polynomial $f\vee g$, it is the polynomial $f_1\vee g_1$.

For the polynomial $f\wedge g$ apply the distributivity law to the polynomial $f_1\wedge g_1$. Similarly for $\sim f$: first apply the De Morgan laws and then the distributivity laws.

S2.79. The relation $\sim$ defined by: $\varphi\sim\psi$ iff ($\varphi\Leftrightarrow\psi$ is a tautology) is an equivalence. The equivalence classes form a Boolean algebra. Problem 1.80 is a special case of problem S2.78 for our Boolean algebra.

S2.80. Clearly 0 and 1 are Boolean elements. If a and b are Boolean then $(a\wedge b)\vee(\sim a\vee\sim b)=1$ and $(a\wedge b)\wedge(\sim a\vee\sim b)=0$ (cf. S2.43, S2.46).

We show similarly that $a\vee b$ is a Boolean element. If a is a Boolean element then so is $\sim a$ since $\sim\sim a=a$.

S2.81. If $t=\langle x_1,\ldots,x_n\rangle$ and $x_1\geqslant\ldots\geqslant x_n$ and u is the complement of t, $u=\langle y_1,\ldots,y_n\rangle$ then $x_1\cup y_1=\ldots=x_n\cup y_n=1$, $x_1\cap y_1=\ldots=x_n\cap y_n=0$. Thus $y_1=\sim x_1,\ldots,y_n=\sim x_n$ and so $y_1\leqslant y_2\leqslant\ldots\leqslant y_n$. As $y_1\geqslant\ldots\geqslant y_n$, $y_1=\ldots=y_n$ and $x_1=\ldots=x_n$. Now it is clear that the mapping $x\to\langle x,\ldots,x\rangle$ is an isomorphism of B and the algebra of Boolean elements of P.

S2.82. No, for instance 0 or 1 need not belong to a sublattice; For instance, let $\mathfrak{a}=\langle\mathscr{P}(\mathscr{N}),\cap,\cup,-,O,\mathscr{N}\rangle$ and let $\mathscr{L}=\langle\mathscr{P}(\mathrm{Par}),\cap,\cup\rangle$. Then $\mathscr{L}$ is a sublattice of the lattice $\langle\mathscr{P}(\mathscr{N}),\cap,\cup\rangle$, but it is clear that $\langle\mathscr{P}(\mathrm{Par}),\cap,\cup,-,O,\mathrm{Par}\rangle$ is not a subalgebra of the algebra $\mathfrak{a}$.

REFERENCES

Chang, C. C., and H. J. Keisler (1973) *Model Theory*, North-Holland, Amsterdam.

Drake, F. R. (1973) *Set Theory*, North-Holland, Amsterdam.

Fraenkel, A. A. (1953) *Abstract Set Theory*, North-Holland, Amsterdam.

Fraïsse, R. (1971) *Course de logique mathématique*, Gauthier-Villars, Paris.

Jech, T. T. (1978) *Set Theory*, Academic Press, New York.

Kuratowski, K., and A. Mostowski (1976) *Set Theory*, PWN–North-Holland, Warszawa-Amsterdam.

Levy, A. (1979) *Basic Set Theory*, Springer-Verlag, Berlin–Heidelberg–New York.

Shoenfield, J. R. (1967) *Mathematical Logic*, Addison–Wesley, Reading.

Sierpiński, W. (1965) *Cardinal and Ordinal Numbers*, PWN, Warszawa.

Sikorski, R. (1964) Boolean Algebras, Springer-Verlag, Heidelberg–Berlin.

INDEX

LIST OF SYMBOLS

$\mathcal{N}$ set of natural numbers
Par set of even natural numbers
NPae set of odd natural numbers
$\mathcal{Z}$ set of integers
$\mathcal{Q}$ set of rational numbers
$\mathcal{Q}^+$ set of nonnegative rational numbers
$\mathcal{R}$ set of real numbers
$\mathcal{R}^+$ set of nonnegative real numbers
$\mathcal{C}$ set of complex numbers
$\mathcal{N}\,[x]$ set of polynomials with natural coefficients
$\mathcal{Z}\,[x]$ set of polynomials with integer coefficients
$\mathcal{Q}\,[x]$ set of polynomials with rational coefficients
$\mathcal{R}\,[x]$ set of polynomials with real coefficients
$\mathcal{R}_n\,[x]$ set of polynomials with real coefficients and of degree at most ***n***
$\mathcal{C}_\infty\,[x]$ set of real functions with derivatives of all degrees
Z_k the ring of remainders modulo k
(a, b) open segment with the endpoints a and b
$\langle a, b)$ left-closed segment with the endpoints a and b
$(a, b\rangle$ right-closed segment with the endpoints a and b
$\langle a, b\rangle$ closed segment with the endpoints a and b
$a|b$ divisebelity relation (a divides b)
GCD greatest common divisor
LCM least common multiple
Im z imaginary part of z
Re z real part of z
$E[x]$ integer part of x